Current Developments in Mathematical Sciences

(Volume 1)

Frontiers in Fractional Calculus

Edited by

Sachin Bhalekar

Department of Mathematics, Shivaji University, Kolhapur, India

Current Developments in Mathematical Sciences

Volume 1

Frontiers in Fractional Calculus

Editor: Sachin Bhalekar

ISSN (Online): 2589-272X

ISSN (Print): 2589-2711

ISBN (Online): 978-1-68108-599-9

ISBN (Print): 978-1-68108-600-2

General:

1. Any dispute or claim arising out of or in connection with this License Agreement or the Work (including non-contractual disputes or claims) will be governed by and construed in accordance with the laws of the U.A.E. as applied in the Emirate of Dubai. Each party agrees that the courts of the Emirate of Dubai shall have exclusive jurisdiction to settle any dispute or claim arising out of or in connection with this License Agreement or the Work (including non-contractual disputes or claims).

2. Your rights under this License Agreement will automatically terminate without notice and without the need for a court order if at any point you breach any terms of this License Agreement. In no event will any delay or failure by Bentham Science Publishers in enforcing your compliance with this License Agreement constitute a waiver of any of its rights.

3. You acknowledge that you have read this License Agreement, and agree to be bound by its terms and conditions. To the extent that any other terms and conditions presented on any website of Bentham Science Publishers conflict with, or are inconsistent with, the terms and conditions set out in this License Agreement, you acknowledge that the terms and conditions set out in this License Agreement shall prevail.

Bentham Science Publishers Ltd.
Executive Suite Y - 2
PO Box 7917, Saif Zone
Sharjah, U.A.E.
Email: subscriptions@benthamscience.org

**BENTHAM
SCIENCE**

CONTENTS

FOREWORD...i

PREFACE..ii

LIST OF CONTRIBUTORS...iv

PART I Fractional Diffusion Equation

CHAPTER 1   SOLVING FRACTIONAL DIFFUSION EQUATION BY WAVELET METHODS.......1

Zhijiang Zhang and Weihua Deng

1.1. INTRODUCTION 1

1.2. PRELIMINARIES 3

1.3. UNIFORM SCHEMES...8

1.4. MULTISCALE ADAPTIVE SCHEME 15

1.5. NUMERICAL SIMULATIONS 17

1.6. CONCLUSION AND DISCUSSION 25

APPENDIX 27

CONFLICT OF INTEREST 30

ACKNOWLEDGEMENTS 30

REFERENCES 30

CHAPTER 2   A SURVEY ON THE RECENT RESULTS REGARDING MAXIMUM PRINCIPLES FOR THE TIME-FRACTIONAL DIFFUSION EQUATIONS 33

Yuri Luchko, Masahiro Yamamoto

2.1. INTRODUCTION 33

2.2. TIME-FRACTIONAL DIFFUSION EQUATIONS WITH THE CAPUTO FRACTIONAL DERIVATIVE 35

2.3. TIME-FRACTIONAL DIFFUSION EQUATIONS WITH THE RIEMANN-LIOUVILLE FRACTIONAL DERIVATIVE 43

2.4. TIME-FRACTIONAL DIFFUSION EQUATIONS WITH THE GENERAL FRACTIONAL DERIVATIVE 56

2.5. TIME-SPACE FRACTIONAL DIFFUSION EQUATIONS..................................63

2.6. DISCUSSIONS AND OPEN PROBLEMS 66

CONFLICT OF INTEREST 67

ACKNOWLEDGEMENTS 67

REFERENCES 67

CHAPTER 3   INTEGRAL-BALANCE SOLUTION TO NONLINEAR SUBDIFFUSION EQUATION 70

Jordan Hristov

3.1. INTRODUCTION 70

3.2. PROBLEM FORMULATION...72

3.3. INTEGRAL METHOD TO THE NON-LINEAR DIFFUSION EQUATION 73

    3.3.1. Preliminary Treatment of the Diffusion Term 73

        3.3.1.1. Single-integration Approach (HBIM) 74

        3.3.1.2. Double-Integration Approach (DIM) 74

3.4. SOLUTION OF THE PROBLEM BY DIM 75

    3.4.1. Approximation of the Fractional-time Double-balance Integral (FTDBI) 75

        3.4.1.1. FT-DBI with Riemann-Liouville Time-fractional Derivative 75

        3.4.1.2. FT-DBI with Caputo Time-fractional Derivative 76

    3.4.2. Penetration Depth 76

        3.4.2.1. DIM Solution 76

        3.4.2.2. Penetration Depth: HBIM Solution 78

        3.4.2.3. Alternative Frozen Front Approach (FFA) 80

        3.4.2.4. Matching the Penetration Depths of DIM and HBIM 83

3.5. APPROXIMATE PROFILE, SIMILARITY VARIABLE AND RELATED ISSUES 83

    3.5.1. Error Measure of Approximate Solution 85

    3.5.2. Approximate Fractional Derivative 85

    3.5.3. The Exponent n at the Boundaries of the Diffusion Layer 86

    3.5.4. Minimization of the Global Error of Approximation 88

3.6. NUMERICAL EXPERIMENTS AND DATA ANALYSIS 91

    3.6.1. Profiles Expression through the Similarity Variable $\eta_\mu = x/\sqrt{D_{0\mu}t^\mu}$ 92

    3.6.2. Competition Between the Subdiffusion and the Degenerate Diffusion 94

    3.6.3. Profile Expressions by the Effective Similarity Variable $\eta_{eff} = x/\sqrt{D_{eff(\mu,m)}t^\mu}$ 96

    3.6.4. Direct Expression of the Profiles by Physical Variables x and t 98

    3.6.5. Comparative analysis 101

3.7. CONCLUSION 102

CONFLICT OF INTEREST 103

ACKNOWLEDGEMENTS 103

REFERENCES 103

PART II Analysis

CHAPTER 4   **ANALYSIS OF SOLUTION AND SYSTEM IDENTIFICATION OF COUPLED FRACTIONAL DELAY DIFFERENTIAL EQUATION BY SHIFTED JACOBI POLYNOMIALS** 106

B. Ganesh Priya, P. Muthukumar and P. Balasubramaniam

4.1. INTRODUCTION 106

4.2. PRELIMINARIES 107

    4.2.1. Fractional Calculus 107

    4.2.2. Shifted Jacobi Polynomials 108

        4.2.2.1. Properties of Shifted Jacobi Polynomials 108

4.2.2.2. Function Approximation Using the Shifted Jacobi Polynomials 111

4.3. ESTIMATION OF UPPER BOUNDS FOR ERROR 113

4.4. COUPLED FDDES WITH TWO DIFFERENT ORDERS 116

    4.4.1. Numerical Solution of Coupled FDDEs 117

    4.4.2. System Identification of Coupled FDDEs 122

4.5. NUMERICAL EXAMPLES 125

4.6. CONCLUSION 129

CONFLICT OF INTEREST 130

ACKNOWLEDGEMENTS 130

REFERENCES 130

CHAPTER 5   MONOTONE ITERATION PRINCIPLE IN THE THEORY OF HADAMARD FRACTIONAL DELAY DIFFERENTIAL EQUATIONS

CHAPTER 5   **MONOTONE ITERATION PRINCIPLE IN THE THEORY OF HADAMARD FRACTIONAL DELAY DIFFERENTIAL EQUATIONS** 132

Kishor D. Kucche

5.1. INTRODUCTION 132

5.2. PRELIMINARIES 134

    5.2.1. Hadamard Fractional Derivative 134

    5.2.2. Partially Ordered Normed Linear Spaces (PONLS) 135

5.3. HADAMARD FRACTIONAL DELAY DIFFERENTIAL EQUATIONS 139

    5.3.1. An Existence Result for HFDDE 141

    5.3.2. An Uniqueness Result for HFDDE 146

5.4. LINEAR PERTURBATIONS OF FIRST TYPE 148

5.5. NEUTRAL HADAMARD FRACTIONAL DIFFERENTIAL EQUATIONS (NHFDE) 151

5.6. CONCLUSION 155

CONFLICT OF INTEREST 156

ACKNOWLEDGEMENTS 156

REFERENCES 156

CHAPTER 6   DYNAMICS OF FRACTIONAL ORDER MODIFIED BHALEKAR-GEJJI SYSTEM

CHAPTER 6   **DYNAMICS OF FRACTIONAL ORDER MODIFIED BHALEKAR-GEJJI SYSTEM** 159

Sachin Bhalekar

6.1. INTRODUCTION 159

6.2. RELATED WORK 161

6.3. PRELIMINARIES 164

6.4. NUMERICAL METHODS FOR FDDEs 164

    6.4.1. Fractional Adams Method (FAM) 164

    6.4.2. New Predictor-Corrector Method (NPCM) 165

6.5. STABILITY ANALYSIS 166

6.6. MODIFIED BHALEKAR-GEJJI SYSTEM 168

6.7. ANALYSIS OF BG SYSTEM 168

6.8. CHAOS IN MODIFIED BG SYSTEM OF FRACTIONAL ORDER 169

    6.8.1. Commensurate Order System 169

    6.8.2. Incommensurate Order System 172

6.9. CHAOS CONTROL 173

6.10. HYBRID SYNCHRONIZATION 176

6.11. CONCLUSION 180

CONFLICT OF INTEREST 180

ACKNOWLEDGEMENTS 180

REFERENCES 180

CHAPTER 7   GRUNWALD-LETNIKOV DERIVATIVE: ANALYSIS IN RANGE OF FIRST ORDER 183

Radosław Cioc

7.1. INTRODUCTION 183

7.2. GRÜNWALD-LETNIKOV DERIVATIVE FROM THE FIRST ORDER DERIVATIVE POINT OF VIEW 184

7.3. PROPERTIES of η ORDER G-L DERIVATIVE 188

    7.3.1. Estimation of $f(t\eta)$ 188

7.4. CONCLUSION 196

CONFLICT OF INTEREST 197

ACKNOWLEDGEMENTS 197

REFERENCES 197

PART III Computational Techniques

CHAPTER 8   GPU COMPUTING OF SPECIAL MATHEMATICAL FUNCTIONS USED IN FRACTIONAL CALCULUS 199

Parag Patil, Navin Singhaniya, Chaitanya Jage, Vishwesh A. Vyawahare, Mukesh D. Patil and P S V Nataraj

8.1. INTRODUCTION 199

8.2. SPECIAL MATHEMATICAL FUNCTIONS USED IN FRACTIONAL CALCULUS 201

    8.2.1. Gamma Function 202

    8.2.2. Mittag-Leffler Function 202

    8.2.3. Pochhammer Symbol 204

    8.2.4. Gauss Hypergeometric Function 205

    8.2.5. Confluent Hypergeometric Function 205

    8.2.6. Dawson's Function 206

    8.2.7. Bessel Wright Function 206

8.3. GRAPHICS PROCESSING UNIT (GPU) 207

    8.3.1. GPU Computing with MATLAB 208

8.4. RESULTS 211

8.5. DISCUSSION 221

8.6. CONCLUSION..222

CONFLICT OF INTEREST 223

ACKNOWLEDGEMENTS 223

APPENDIX 223

REFERENCES 230

PART IV Review

CHAPTER 9   NEW ITERATIVE METHOD: A REVIEW 233

Varsha Daftardar-Gejji, Manoj Kumar

9.1. INTRODUCTION 233

9.2. NEW ITERATIVE METHOD...234

9.3. TAYLOR SERIES AND NIM...235

   9.3.1. Convergence of New Iterative Method..237

9.4. APPLICATIONS OF NEW ITERATIVE METHOD 237

9.5. SOLVING NON-LINEAR ALGEBRAIC EQUATIONS 238

9.6. SOLVING INITIAL VALUE PROBLEMS 240

   9.6.1. Notations and Preliminaries..240

   9.6.2. Non-linear Partial Differential Equations 241

   9.6.3. Non-linear Fractional Differential Equations 246

9.7. SOLVING BOUNDARY VALUE PROBLEMS 253

9.8. SOLVING INTEGRO-DIFFERENTIAL EQUATIONS...255

9.9. METHODS BASED ON NIM..258

   9.9.1. Three-step Iterative Method 258

   9.9.2. Iterative Laplace Transform Method 259

   9.9.3. New Predictor-corrector Method..261

   9.9.4. New Numerical Methods..262

9.10. CONCLUSIONS 265

CONFLICT OF INTEREST 265

ACKNOWLEDGEMENTS 265

REFERENCES 265

CHAPTER 10   FRACTIONAL DERIVATIVE WITH NON-SINGULAR KERNELS: FROM THE SEMINAL DEFINITION OF CAPUTO AND FABRIZIO AND BEYOND WITH EMPHASIS ON DIFFUSION PROBLEMS 269

Jordan Hristov

10.1. INTRODUCTION 269

   10.1.1. Fractional Derivatives with Non-Singular Kernels..270

      10.1.1.1. Basic Definition of Caputo-Fabrizio Time-fractional Derivative.......................270

      10.1.1.2. Basic Properties of $_{cf}^{c}D_t^{\alpha}$ 271

10.1.1.3. Normalization Function M(α), Associated Fractional Integral and an Alternative Definition of $^c_{cf}D^\alpha_t$ 273

10.1.1.4. Relation to the Associated Ordinary Derivative 274

10.1.1.5. Laplace Transform 275

10.1.1.6. Fractional Derivative of Elementary and Transcendental Functions 275

10.1.2. Fractional Derivative with Non-singular Kernel of Riemann-Liouville Type 277

10.1.3. Some Applications-Briefly 280

10.1.4. Generalized Fractional Derivative with Non-singular Kernel *via* the Mittag-Leffler Function 281

10.1.5. Fractional Operator of Bi-Order 283

10.2. DIFFUSION MODELS WITH TIME-FRACTIONAL CAPUTO-FABRIZIO DERIVATIVE ..284

10.2.1. Diffusion Models with Constitutive Equations *Flux-Gradient* 284

10.2.1.1. The Diffusion Equation of Caputo-Fabrizio [3] 285

10.2.1.2. Diffusion Equation of with Jeffrey's Kernel in the Constitutive Equation 286

10.2.1.3. Diffusion Equation with Cattaneo Memory Kernel (Single-Memory Model) 291

10.2.1.4. Diffusion Model with Composite Memory Kernel (Two-Memories Model) 297

10.2.1.5. Brief Concluding Remarks 300

10.3. DIFFUSION EQUATION WITH A SPATIAL MEMORY ONLY 301

10.3.1. Fraction Gradient Operators of Caputo and Fabrizio with Non-Singular Kernels 301

10.3.2. From the Cattaneo Concept of Time-Memory Effects to a Spatial Jeffrey Kernel 302

10.3.3. Towards the Spatial Fractional Derivative of Caputo-Fabrizio Type 304

10.3.4. Steady-State Heat Conduction Equation with a Spatial Memory 307

10.3.4.1. Short-Range Memory Effects: Space Memory Only 307

10.3.4.2. Space Memory with Extended Relaxation Function (Jeffrey Memory Kernel) 310

10.3.4.3. The Ratio λ/L and the Fractional Order μ 312

10.3.4.4. The Redistribution Coefficient m and the Fractional Order μ 313

10.3.4.5. Temperature Profiles 313

10.4. FORMALISTIC FRACTIONALIZATION OF THE DIFFUSION EQUATION: SOME COMMENTS AND SUGGESTIONS BY ANALOGY 316

10.4.1. Examples Coming from Fractional Models with Singular Kernels and Consequent Formalisms 316

10.4.2. Examples Suggesting Flux-Gradient Relations in Caputo-Fabrizio Sense 318

10.4.3. Some Brief Comments on the Formalistic Fractionalization 320

10.5. WHAT TYPE OF RELAXATIONS ARE MODELLED BY THE NON-SINGULAR MEMORY KERNELS AND THE NEW DERIVATIVES? 320

10.5.1. The Stretched-Exponentially Kohlrausch Relaxation Function 321

10.5.2. Exponential Kernel of the Caputo-Fabrizio Derivative: What Type of Fading Memory is Modelled? 325

10.5.3. Real-World Examples with Exponentially Decaying Relaxation Functions 328

10.5.3.1. Example 1: The Caputo Set of Constitutive Equations for Plastic Media 328

10.5.3.2. Example 2: Anomalous Diffusion of Vapors Through Solid Polymers 330

10.5.3.3. Example 3: Diffusion of Solvents Through Swelling Solid Polymers 332

10.5.3.4. Example 4: Darcy Flow with Relaxation Effect 333

10.5.3.5. Some Briefs on the Real-World Examples 335

10.6. CONCLUDING REMARKS 335

CONFLICT OF INTEREST 336

ACKNOWLEDGEMENTS 336

REFERENCES 336

**CHAPTER 11 FRACTIONAL ORDER NONLINEAR SYSTEMS: SOME OPEN PROBLEMS
IN NUMERICAL COMPUTATIONS AND CHAOS THEORY** 342

Sachin Bhalekar

11.1. INTRODUCTION 342

11.2. SOLUTION METHODS 343

11.2.1. Exact Methods 343

11.2.2. Approximate Analytical Methods 343

11.2.3. Numerical Methods 343

11.3. CHAOS IN FONS 344

11.3.1. Detection of Chaos 345

11.4. CONCLUSION 346

CONFLICT OF INTEREST 346

ACKNOWLEDGEMENTS 346

REFERENCES 346

SUBJECT INDEX 348

FOREWORD

Fractional calculus, in allowing integrals and derivatives of any positive order (the term fractional is kept only for historical reasons), can be considered a branch of mathematical physics that deals with integro-differential equations, where integrals are of convolution type and exhibit mainly singular kernels of power law or logarithm type.

It is a subject that has gained considerably popularity and importance in the past few decades in diverse fields of science and engineering.

The purpose of this book is to establish a collection of articles that reflect some mathematical and conceptual developments in the field of fractional calculus and explore the scope for applications in applied sciences. The book is divided in 4 sections as follows where the authors are outlined.

I congratulate the Editor, PhD Sachin Bhalekar who was able to collect a variety of topics of relevance to the reader.

Francesco Mainardi
Professor of Mathematical Physics
University of Bologna
Italy

PREFACE

Fractional Calculus is a very popular subject among the Mathematicians and Applied Scientists. The nonlocal operators in fractional calculus provided challenging research topics for Mathematicians. On the other hand, the applied scientists found these operators more useful than classical integer order ones.

It is indeed a happy moment to present this book which contains eleven chapters on different aspects of Fractional Calculus.

The first part of this book contains three chapters on **Fractional Diffusion Equations**. In chapter 1, Zhang and Deng discussed the solutions of fractional diffusion equations using wavelet methods. The maximum principle for time fractional diffusion equations is proposed by Luchko and Yamamoto in chapter 2. Chapter 3, by Hristov is on the nonlinear sub-diffusion equations.

The second part **Analysis** contains four chapters. The shifted Jacobi polynomials are used by Ganesh Priya, Muthukumar and Balasubramaniam in chapter 4 to analyze the solution and system identification of coupled fractional delay differential equations. In chapter 5, Kucche presented the monotone iteration principle in the theory of Hadamard fractional delay differential equations. Bhalekar, in chapter 6 analyzed the dynamics of fractional order modified Bhalekar-Gejji System. Cioc analyzed the Grunwald-Letnikov derivatives in chapter 7.

Third part of this book is **Computational Techniques.** It contains chapter 8 by the research group: P. Patil, N. Singhaniya, C. Jage, V. Vyawahare, M. Patil and P.S.V. Nataraj. They presented the GPU computing of special mathematical functions used in fractional calculus.

The last part of the book is **Review** which contains three chapters. Daftardar-Gejji and Kumar presented a review on the popular iterative method NIM in chapter 9. A review on fractional derivative with non-singular kernels is taken by Hristov in chapter 10. Some open problems in fractional order nonlinear system are discussed by Bhalekar in chapter 11.

I am very thankful to all the contributors of this book for their valuable work. The foreword for this book is written by Prof. Fransesco Mainardi. I am indebted to him for his guidance and support. Mr. Shehzad Naqvi and Mr. Omer Shafi of Bentham Science Publishers helped me in the publication process. I am thankful to Bentham Science Publishers for publishing this book.

I wish that the book will provide basic knowledge to the readers, introduce higher topics and present applications of this subject.

Sachin Bhalekar
Department of Mathematics
Shivaji University, Kolhapur
India

List of Contributors

B. Ganesh Priya Department of Mathematics, The Gandhigram Rural Institute-Deemed University, Gandhigram 624 302, India

Chaitanya Jage Department of Electronics Engineering, Ramrao Adik Institute of Technology, D. Y. Patil Vidyanagar, Nerul, Navi Mumbai 400 706, India

Jordan Hristov Dept. Chem. Eng, UCTM, Sofia 1756, 8 Kl.Ohridsky Blvd, Bulgaria

Kishor D. Kucche Department of Mathematics, Shivaji University, Kolhapur 416004 India

Manoj Kumar Department of Mathematics, Savitribai Phule Pune University, Pune - 411007, India;
National Defense Academy, Khadakwasala Pune- 411023, India

Masahiro Yamamoto Graduate School of Mathematical Sciences, the University of Tokyo, Japan

Mukesh D. Patil Department of Electronics and Telecommunication Engineering, Ramrao Adik
Institute of Technology, D. Y. Patil Vidyanagar, Nerul, Navi Mumbai 400 706, India

Navin Singhaniya Department of Electronics Engineering, Ramrao Adik Institute of Technology, D. Y. Patil Vidyanagar, Nerul, Navi Mumbai 400 706, India

P. Balasubramaniam Department of Mathematics, The Gandhigram Rural Institute-Deemed University, Gandhigram 624 302, India

P. Muthukumar Department of Mathematics, The Gandhigram Rural Institute-Deemed University, Gandhigram 624 302, India

P. S. V. Nataraj IDP in Systems and Control Engineering, Indian Institute of Technology Bombay, Mumbai 400 076, India

Parag Patil Department of Electronics Engineering, Ramrao Adik Institute of Technology, D. Y. Patil Vidyanagar, Nerul, Navi Mumbai 400 706, India

Radosław Cioᶜ Faculty of Transport and Electrical Engineering, Kazimierz Pulaski University of Technology and Humanities in Radom Malczewskiego Str. 29, Radom 26-600, Poland

Sachin Bhalekar Department of Mathematics, Shivaji University, Kolhapur 416004 India

Varsha Daftardar-Gejji Department of Mathematics, Savitribai Phule Pune University, Pune - 411007, India

Vishwesh A. Vyawahare Department of Electronics Engineering, Ramrao Adik Institute of Technology, D. Y. Patil Vidyanagar, Nerul, Navi Mumbai 400 706, India

Weihua Deng School of Mathematics and Statistics, Gansu Key Laboratory of Applied Mathematics and Complex Systems, Lanzhou University, Lanzhou 730000, P.R. China

Yuri Luchko Department of Mathematics, TU of Applied Sciences Berlin, Germany

Zhijiang Zhang School of Mathematics and Statistics, Gansu Key Laboratory of Applied Mathematics and Complex Systems, Lanzhou University, Lanzhou 730000, P.R. China

CHAPTER 1

Solving Fractional Diffusion Equation by Wavelet Methods

Zhijiang Zhang and Weihua Deng[*]

School of Mathematics and Statistics, Gansu Key Laboratory of Applied Mathematics and Complex Systems, Lanzhou University, Lanzhou 730000, P.R. China

Abstract: The wavelet numerical methods for solving the classic differential equations have been well developed, but their application in solving fractional differential equations is still in its infancy. In this chapter we tentatively investigate the advantages of the spline wavelet basis functions in solving the fractional PDEs. Our contributions are as follows: 1. the techniques of efficiently generating stiffness matrix with computational cost $\mathcal{O}(N)$ (N denotes the degree of freedom) are provided for first, second, and any order bases; 2. the effectiveness of the wavelet preconditioning and the wavelet adaptivity for solving the fractional PDEs is discussed theoretically or numerically.

Keywords: Fractional PDEs, Wavelet preconditioning, Wavelet adaptivity, FFT and FWT.

AMS Subject Classification: 35R11, 65T60, 65F08.

1.1. INTRODUCTION

Fractional Calculus are now winning more and more interest. They appear in modeling diverse physical systems such as viscoelastic materials, bioengineering applications, porous or fractured media, turbulence *etc.* [16, 18, 31]. In addition, it has been found that the anomalous diffusion is one of the most ubiquitous phenomena in nature [18, 34], and many governing fractional PDEs has been derived. For example, the continuous time random walk (CTRW), a fundamental model in statistic physics, is a stochastic process with arbitrary distributions of jump lengths and waiting times. When the jump length and/or waiting time distribution(s) are/is power law and the second order moment of jump lengths and/or the first order moment of waiting times are/is divergent, the CTRW describes the anomalous diffusion, *i.e.*, the super and sub diffusive cases; and its Fokker-Planck equation has space and/or time fractional derivative(s) [2].

---

[*]**Corresponding author Weihua Deng:** School of Mathematics and Statistics, Gansu Key Laboratory of Applied Mathematics and Complex Systems, Lanzhou University, Lanzhou 730000, P.R. China; E-mail address: dengwh@lzu.edu.cn

Sachin Bhalekar (Ed.)

The obtained analytical solutions of fractional PDEs are usually in the form of transcendental functions or infinite series; and in much more cases, the analytical solutions are not available. Then the approximation and numerical techniques for solving the fractional PDEs become essential and have been developed very fast recently, such as, the finite difference method [17, 26, 36], the finite element method [7, 9, 32], and the spectral method [13, 35]. But unlike the integer-order PDEs, the computational expenses and nonuniform regularity are still the big challenges that one faces in numerically solving the fractional PDEs, owing to the nonlocality of the fractional derivatives and the possible singularity of the exact solution; and based on the preconditioning and fast transform techniques to develop high efficient methods seems to be a new trend. The preconditioning techniques are discussed in [19, 33], where the Krylov subspace projection is their common theme. Fast Fourier transform method and the multigrid method are provided in [30] and [20], respectively.

On the other hand, although the advantages of the wavelet numerical methods for solving the classical differential and integral equations have been well studied [1, 3, 6, 10, 22, 27], there seems to be very limited works [11, 24, 25, 29] to use them to solve fractional differential equations. In this chapter, we aim at numerically digging out the potential advantages of spline wavelets in treating the fractional operators, including wavelet preconditioning and adaptivity. For avoiding all non indispensable complications, we present the main ideas and techniques in their simplest form and restrict ourselves to the following fractional diffusion equation with absorbing boundaries (see [8, p. 046104-4]):

$$\begin{cases} \mathcal{B}u + q(x)u = f \ on \ \Omega, \\ \qquad\qquad u = 0 \ on \ \mathbb{R}\backslash\Omega, \end{cases} \tag{1.1}$$

where $\Omega = (0,1)$, and $\mathcal{B}$ is a $(2 - \beta)$-th ($0 \leq \beta < 1$) order differential operator

$$\mathcal{B}u := -\kappa_\beta D\left(p \ _{-\infty}D_x^{1-\beta}u - (1 - p) \ _xD_\infty^{1-\beta}u\right) \tag{1.2}$$

with $\kappa_\beta > 0$ being the generalized diffusivity, $q(x) \geq 0$, $0 \leq p \leq 1$, D represents a single spatial derivative; $_{-\infty}D_x^{1-\beta}$ and $_xD_\infty^{1-\beta}$ are the left and right Riemann-Liouville fractional derivative operators, being, respectively, defined as

$$_{-\infty}D_x^{1-\beta}u := \frac{1}{\Gamma(\beta)}\frac{d}{dx}\int_{-\infty}^x (x - s)^{\beta-1}u(s) \ ds, \tag{1.3}$$

$$_xD_\infty^{1-\beta}u := -\frac{1}{\Gamma(\beta)}\frac{d}{dx}\int_x^\infty (s - x)^{\beta-1}u(s) \ ds. \tag{1.4}$$

Because of the absorbing boundary conditions, in the following we will simply rewrite (1.1) as

$$\begin{cases} \mathcal{B}u + q(x)u = f, & x \in \Omega, \\ u(0) = u(1) = 0 \end{cases} \tag{1.5}$$

with

$$\mathcal{B}u := -\kappa_\beta D\left(p\ _0 D_x^{1-\beta} u - (1-p)\ _x D_1^{1-\beta} u \right) \tag{1.6}$$

and give the numerical schemes based on this latter form. Since the basic theoretical framework for the variational solution of (1.5) has been presented in [9, 28], it will enable us to put focus on the wavelet numerical methods themselves. We will test the following two results: 1) stiffness matrix of fractional operator is quasi-Toeplitz under the spline scaling bases because of the shift-invariant property, then one can implement the wavelet preconditioning effectively by the aid of the FFT and the fast wavelet transform (FWT). 2) the wavelet coefficients still indicate the local regularity of the solutions of the fractional PDEs well, especially for the boundaries weak singularities, and they can be used as the indicator in the adaptive mesh refinement for controlling the entire computational process and increasing the efficiency.

This chapter is organized as follows. In Section 1.2, we give a brief recall to the spline scaling and wavelet functions. They have the closed formulaes, which is of course attractive for the fractional operators. In Section 1.3, we study the computational formulation with respect to the uniform grids. We discuss the effective way to construct the algebraic system, and then provide the effective implementation of wavelet preconditioning . In Section 1.4, we give a heuristically adaptive algorithm and show how singularities can be easily detected by wavelet, and put our attention on its efficiency by proposing and testing the adaptive algorithms that concentrate the degrees of freedom in the neighborhood of boundary singularities. The numerical results are shown in Section 1.5 and we conclude the paper with some remarks in the last section.

1.2. PRELIMINARIES

As usual, let $\mathcal{H}^s(\mathbb{R})(s > 0)$ be the Sobolev space of order s on $\mathbb{R}$, and $\mathcal{H}^s(\Omega)$ the space of the restriction of the functions from $\mathcal{H}^s(\mathbb{R})$ [9]; and $H_0^s(\Omega)$ is defined as the closure of $C_0^\infty(\Omega)$ w.r.t. $\|\cdot\|_{\mathcal{H}^s(\Omega)}$. $(\cdot,\cdot)$ and $\|\cdot\|$ denote inner product and norm on $L_2(\Omega)$, respectively; and $\langle\cdot,\cdot\rangle$ denotes the duality pairing. By $A\lesssim B$ we mean that A can be bounded by a multiple of B, independent of the parameters they may depend on, and $A \sim B$ denotes $A \lesssim B$ and $B \lesssim A$. Assuming that $v > 0$, then the left and right Riemann-Liouville integrals are, respectively, given as $_0D_x^{-v}u(x) = \frac{1}{\Gamma(v)} \int_0^x (x - \xi)^{v-1} u(\xi) \, d\xi$ and $_xD_1^{-v}u(x) = \frac{1}{\Gamma(v)} \int_x^1 (\xi - x)^{v-1} u(\xi) \, d\xi$; and for any $u(x), v(x) \in L_2(\Omega)$, it follows that $\left(_0D_x^{-v}u, v\right) = \left(u, {_xD_1^{-v}}v\right)$ (see [7, Lemma 2.13]).

Definition (Riesz basis). (see [6, p. 463]) A family $\{e_n\}_{n\in\mathbb{Z}}$ is a Riesz basis of a Hilbert space H, if and only if it spans H and there exist $0 < C_1 \le C_2$ such that for all finite sequence $\{x_i\}$, one has

$$C_1 \sum_i |x_i|^2 \le \left\| \sum_i x_i e_i \right\|_H^2 \le C_1 \sum_i |x_i|^2. \tag{1.7}$$

In the following, we collect/present some essential properties of the basis functions used in this chapter. All of them can be obtained from [23, 21, 22], but for convenience, here we use a slightly different way to describe them. To get a deeper understanding, refer also to [27, 5, 6].

For $d, j \in \mathbb{N}$ and $d \ge 2, j \ge J_0$, let

$$T_d^j := \{t_k^j\}_{k=-d+1}^{2^j+d-1} = \{\underbrace{0, \dots, 0}_{d}, 1 \times 2^{-j}, 2 \times 2^{-j}, \dots, (2^j - 1) \times 2^{-j}, \underbrace{1, \dots, 1}_{d}\}$$

be the dyadic knot sequence on Ω with multiplicity d at the end points, and $[a_1, \dots, a_n; f(t)]_t$ be the d-th order *divided difference* of f at the knots $a_1, \dots, a_n$ concerning the variable t, $t_+^l := (\max\{0, t\})^l$. Then the d-th order interval B-spline function sets $\Phi_j = \left\{ \phi_{j,k}, k \in \Delta_j = \{1, \dots, 2^j + d - 3\} \right\}$ satisfying homogeneous boundary conditions can be generated by defining (see [23, Lemma 3.1 and subsection 5.3] and [22, section (B)])

$$\phi_{j,k}(x) := 2^{\frac{j}{2}} \left(t_{k+1}^j - t_{k-d+1}^j \right) \left[t_{k-d+1}^j, \dots, t_{k+1}^j; (t - x)_+^{d-1} \right]_t.$$

Proposition 1.1. *(see [21, Sections 3.2-3.4], [22, Section (B)], and [27, p. 37-38, Section 2.6]) The interval B-spline functions Φ_j satisfy the following properties:*

- *The system Φ_j is compact support and locally finite, i.e., supp $\phi_{j,k} = \left[t_{k-d+1}^j, t_{k+1}^j \right]$ and $\#\{\phi_{j,k}: \text{supp}\phi_{j,k} \cap \text{supp}\phi_{j,i}\} \lesssim 1$.*

- *There are $d - 2$ boundary functions for 0 ($k = 1, \cdots, d - 2$) and $d - 2$ boundary functions for 1 ($k = 2^j, \cdots, 2^j + d - 3$). Further, for $k =$*

$1, \cdots, d-2$, *it holds that* $\phi_{j,2^j-k+d-2}(x) = \phi_{j,k}(1-x)$ *and* $\phi_{j+1,k}(x) = \sqrt{2}\phi_{j,k}(2x)$, *in fact, these suggests that one only needs to determine* $d-2$ *functions* $b_1(x), b_2(x), \cdots, b_{d-2}(x)$, *and then let* $\phi_{j,k}(x) = 2^{j/2}b_k(2^j x)$ *and* $\phi_{j,2^j-k+d-2} = 2^{j/2}b_k(2^j(1-x))$.

- *The inner functions* $(k = d-1, d, \cdots, 2^j - 1)$ *are just dilations and translations of the d-th order cardinal B-splines* $N_d(x) := \int_0^1 N_{d-1}(x - \xi)d\xi$ *with* $N_1(x) := \chi_{[0,1]}(x)$, *i.e.* $\phi_{j,k} = 2^{\frac{j}{2}}N_d(2^j x - (k - d + 1))$. *Note that* $N_d(x)$ *is refinable with* $N_d(x) = 2^{1-d}\sum_{m=0}^{d}\binom{d}{m}N_d(2x - m)$ *and symmetric with* $N_d\left(\frac{d}{2} - x\right) = N_d\left(\frac{d}{2} + x\right)$; *and it also satisfies* $N_d(x) = \frac{1}{(d-1)!}\sum_{m=0}^{d}(-1)^m\binom{d}{m}(x - m)_+^{d-1}$.

Let $S_j = span\{\Phi_j\}$. Then sequence $S_j (j \geq J_0)$ forms a multiresolution analysis (MRA) of $L_2(I)$ (see [23, Definition 2.8 and Subsection 5.3] and [22, Remark 5.1]). In particular, one has $clos_{L_2(I)}\bigcup_{j \geq J_0} S_j = L_2(\Omega)$ and $S_j \subset S_{j+1}$. The latter means that there exist two-scale matrices $M_{j,0} \in \mathbb{R}^{(2^{j+1}+d-3)\times(2^j+d-3)}$, such that

$$\Phi_j^T = \Phi_{j+1}^T M_{j,0}.$$

Meanwhile, one can construct another MRA sequence (see [23, section 4] and [22, section V(C)]) $\tilde{S}_j = span\{\tilde{\Phi}_j\}(j \geq J_0)$ (of $L_2(\Omega)$) with $\tilde{\Phi}_j = \{\tilde{\phi}_{j,k}, k \in \Delta_j\}$ satisfying the duality relation $\langle\phi_{j,k_1}, \tilde{\phi}_{j,k_2}\rangle = \delta_{k_1,k_2}, k_1, k_2 \in \Delta_j$. We call S_j primal MRA and $\tilde{S}_j$ dual MRA, and correspondingly, Φ_j primal scaling functions and $\tilde{\Phi}_j$ dual scaling functions. Though one can not get the closed formulas of $\tilde{\phi}_{j,k}$, there still exists two-scale matrices $\tilde{M}_{j,0} \in \mathbb{R}^{(2^{j+1}+d-3)\times(2^j+d-3)}$ such that $\tilde{\Phi}_j^T = \tilde{\Phi}_{j+1}^T\tilde{M}_{j,0}$, and $\tilde{\Phi}_j$ has the polynomial exactness of order $\tilde{d}$ (see [22, Lemma 5.6]; and see also [23, Definition 2.3]). Further, starting from $M_{j,0}$ and $\tilde{M}_{j,0}$ and using the method of stable completion (see [23, Section 6], [22, Section (D)], and [27, Section 8.2]), one can construct the two-scale matrices $M_{j,1}, \tilde{M}_{j,1} \in \mathbb{R}^{(2^{j+1}+d-3)\times(2^j)}$, such that

$$\Psi_j^T := \Phi_{j+1}^T M_{j,1} \text{ and } \tilde{\Psi}_j^T := \tilde{\Phi}_{j+1}^T\tilde{M}_{j,1} \tag{1.8}$$

are, respectively, uniform Riesz bases of the 2^j dimensional wavelet spaces $W_j = span\{\Psi_j\}$ and $\widetilde{W}_j = span\{\widetilde{\Psi}_j\}$. One has

$$S_{j+1} = S_j + W_j, \quad W_j \perp \tilde{S}_j, \tag{1.9}$$

$$\tilde{S}_{j+1} = \tilde{S}_j + \widetilde{W}_j, \quad \widetilde{W}_j \perp S_j, \tag{1.10}$$

and the biorthogonal relation $\left(\psi_{j_1,k_1}, \tilde{\psi}_{j_2,k_2}\right) = \delta_{j_1,j_2}\delta_{k_1,k_2}$. Moreover, letting $\nabla_j = \{0,1,\cdots,2^j-1\}$ and denoting $\Psi_j = \{\psi_{j,k}, k \in \nabla_j\}$ and $\widetilde{\Psi}_j := \{\tilde{\psi}_{j,k}, k \in \nabla_j\}$, then Ψ_j and $\widetilde{\Psi}_j$ have the properties similar to Proposition 1.1. We call Ψ_j primal wavelets, and $\widetilde{\Psi}_j$ dual wavelets. From (1.9) and (1.10), it is easy to see that the multiscale function sets $\Psi^J := \Phi_{J_0} \cup \cup_{j=J_0}^{J-1} \Psi_j$ is another bases of S_J; similarly for $\tilde{S}_J$.

Proposition 1.2. *By the compact support of $\phi_{j,k}$ and $\psi_{j,k}$, the Jackson and Bernstein estimates of S_J (see [23, Subsection 5.2], [22, Section], and [6, p. 622-631]), and the polynomial exactness of Φ_J (see [22, Lemma 5.6] and [6, p. 623-624]), it holds that the following properties:*

- **Biorthogonal projection** *(Its proof is similar to [6, Chapter , Sections 26 and 27]): Define $P_j: L_2(\Omega) \to S_j, P_j v := \sum_{k\in\Delta_j} \left(v, \tilde{\phi}_{j,k}\right)\phi_{j,k}$. Then*

$$\left\| v - P_j v \right\|_{H^s(\Omega)} \lesssim 2^{j(s-\gamma)}\|v\|_{H^\gamma(\Omega)}, 0 \leq s \leq 1, s \leq \gamma \leq d. \tag{1.11}$$

- **Norm equivalences** *(See [23, Lemma 5.2 and Remark 5.3] and [22, P. 1392]; and see also [27, Theorem 5.12]): For $u \in H_0^s(\Omega), s \in [0,1]$, and*

$$u = \sum_{k\in\Delta_{J_0}} c_{J_0,k}\phi_{J_0,k} + \sum_{j\geq J_0} \sum_{k\in\nabla_j} d_{j,k}\psi_{j,k}, \tag{1.12}$$

one has

$$\|u\|_{H^s(\Omega)}^2 \sim \sum_{k\in\Delta_{J_0}} 2^{2J_0 s}|c_{J_0,k}|^2 + \sum_{j\geq J_0} \sum_{k\in\nabla_j} 2^{2js}|d_{j,k}|^2. \tag{1.13}$$

This suggests that $\left(2^{-J_0 s}\Phi_{J_0}\right) \cup \cup_{j=J_0}^{\infty} \left(2^{-js}\Psi_j\right)$ is a Riesz basis of $H_0^s(\Omega)$.

- **Compressibility** *(Its proof is similar to [27, p. 148, Proposition 5.9] and [6, p. 593, Remark 29.5]): For $u \in H_0^s(\Omega), 0 < s \leq d$, there exists*

$$|d_{j,k}| = |(u, \tilde{\psi}_{j,k})| \lesssim 2^{-js} \|u^{(s)}\|_{L_2(supp\tilde{\psi}_{j,k})}, \tag{1.14}$$

which means that the size of wavelet coefficients allows to characterize the local smoothness of u at point x by the size of the wavelet coefficients $d_{j,k}$ such that the support of $\psi_{j,k}$ or of $\tilde{\psi}_{j,k}$ contains x.

We point out that starting from the scaling function sets $\Phi_j (j \geq J_0)$, some other special wavelets can also be constructed. For example, for $d = 2$, if we define

$$\psi^{2,0}(x) = 9/10\, N_2(2x) - 3/5\, N_2(2x-1) + 1/10\, N_2(2x-2), \tag{1.15}$$

$$\psi^2(x) = N_2(2x) - 3/5(N_2(2x+1) + N_2(2x-1))$$

$$+ 1/10(N_2(2x+2) + N_2(2x-2)), \tag{1.16}$$

then $\Psi_j := \left\{ 2^{\frac{j}{2}} \psi_j^{2,0}(2^j x), 2^{\frac{j}{2}} \psi^2(2^j x - k)|_{k=1}^{2^j-2}, 2^{\frac{j}{2}} \psi^{2,0}(2^j(1-x)) \right\}$ is the semi-orthogonal wavelets given in [6, P. 530]; and if we let $\psi(x) = N_2(2x)$, then $\Psi_j := \left\{ \psi_{j,k} = 2^{\frac{j}{2}} \psi(2^j x - k), k \in \nabla_j \right\}$ is the spline interpolation wavelets given in [6, p. 433-437]. Because the semiorthogonal wavelets are special biorthogonal wavelets (*i.e.* $S_j = \tilde{S}_j, W_j = \widetilde{W}_j$; see [27, p. 161-163]), the norm equivalence in Proposition 1.2 still holds; and for the interpolation wavelets, they satisfy the norm equivalence for $s \in \left(\frac{1}{2}, \frac{3}{2}\right)$ (see [6, p. 605]).

Since Φ_J and Ψ^J are equivalent bases of S_J, there exists matrix M (the FWT matrix) such that $(\Psi^J)^T = \Phi_J^T M$, and which can be explicitly given as

$$M = M_{J-1}\begin{pmatrix} M_{J-2} & 0 \\ 0 & I_{J-2} \end{pmatrix}\begin{pmatrix} M_{J-3} & 0 \\ 0 & I_{J-3} \end{pmatrix} \cdots \begin{pmatrix} M_{J_0} & 0 \\ 0 & I_{J_0} \end{pmatrix}. \tag{1.17}$$

Here $M_j := (M_{j,0}, M_{j,1})$, and I_j denotes the $(2^J - 2^{j+1}) \times (2^J - 2^{j+1})$ unit matrix. For our numerical methods, the dual scaling functions and wavelets will not be used, but we need the the two-scale relationship M_j to perform the FWT and to obtain the expressions of wavelet basis functions. For the inner scaling functions, by the refinable relation given in Proposition 1.1, it follows that

$$\phi_{j,k}(x) = 1/\sqrt{2} \sum_{m=0}^{d} 2^{1-d} \binom{d}{m} \phi_{j+1,2k+m-d+1}. \tag{1.18}$$

Clearly, for $d = 2$, no boundary scaling functions are involved; and for $d = 3$, using the fact (see [21, Section 3.2])

$$b_1(x) = -3/2\, x_+^2 + 2x_+ + 2(x-1)_+^2 - 1/2\, (x-2)_+^2, \qquad (1.19)$$

$$b_1(x) = 1/2\, b_1(2x) + 3/4\, N_3(2x) + 1/4\, N_3(2x-1), \qquad (1.20)$$

and the symmetry $N_d(x) = N_d(d-x)$, one has

$$\phi_{j,1}(x) = 1/\sqrt{2}\, (1/2\, \phi_{j+1,1}(x) + 3/4\, \phi_{j+1,2}(x) + 1/4\, \phi_{j+1,3}), \qquad (1.21)$$

$$\phi_{j,2^j}(x) = 1/\sqrt{2}\, (1/4\, \phi_{j+1,2^{j+1}-2} + 3/4\, \phi_{j+1,2^{j+1}-1}(x) + 1/2\, \phi_{j+1,2^{j+1}}(x)). \qquad (1.22)$$

Similarly for the general d. For the semiorthogonal wavelets, by (1.15) and (1.16), one has

$$\psi_{j,0}(x) = 1/\sqrt{2}\, (9/10\, \phi_{j+1,1} - 3/5\, \phi_{j+1,2} + 1/10\, \phi_{j+1,3}), \qquad (1.23)$$

$$\psi_{j,k}(x) = 1/\sqrt{2}\, \big(\phi_{j+1,2k+1} - 3/5(\phi_{j+1,2k} + \phi_{j+1,2k+2}) + 1/10(\phi_{j+1,2k-1} + \phi_{j+1,2k+3})\big), \qquad (1.24)$$

$$\psi_{j,2^j-1}(x) = 1/\sqrt{2}\, \big(1/10\, \phi_{j+1,2^{j+1}-2} - 3/5\, \phi_{j+1,2^{j+1}-1} + 9/10\, \phi_{j+1,2^{j+1}}\big), \qquad (1.25)$$

and we will give the part of the expressions of the biorthogonal wavelets used here in Appendix A1. In practice, the FWT can be implemented like the reconstruction algorithms in [6, Chapter 1, Sections 2 and 3] with the complexity $\mathcal{O}(2^J)$.

1.3. UNIFORM SCHEMES

The nonlocal property of fractional operator makes the matrix of its discretizations inevitably dense. We will show that the chosen basis functions render the matrix to have a special structure, which greatly reduces the costs of computing and storing the entires. In this sense, these kind of bases are superior to the other possible bases, such as the usually used finite element or spectral polynomial bases. Meanwhile, based on the benefits of these bases, a simple diagonal preconditioner and the fast transforms (FFT and FWT) are used to enhance the effectiveness of the widely used iterative schemes.

Considering (1.5), it has the variational formulation: Find $u \in H_0^\alpha(\Omega)$ with $\alpha = 1 - \beta/2$ $(0 \leq \beta < 1)$, such that

$$a(u, v) = (f, v) \quad \forall v \in H_0^\alpha(\Omega), \tag{1.26}$$

where [9, 28]

$$a(u, v) = \kappa_\beta \left\langle p \, _0D_x^{-\beta/2} Du,_x D_1^{-\beta/2} Dv \right\rangle$$

$$+ \kappa_\beta \left\langle (1 - p)_x D_1^{-\beta/2} Du,_0 D_x^{-\beta/2} Dv \right\rangle + (qu, v). \tag{1.27}$$

The bilinear form $a(\cdot, \cdot) \colon H_0^\alpha(\Omega) \times H_0^\alpha(\Omega) \to \mathbb{R}$ is continuous and coercive, *i.e.*,

$$|a(u, v)| \overset{<}{\sim} \| u \|_\alpha \| v \|_\alpha , \quad a(u, u) \overset{>}{\sim} \| u \|_\alpha^2. \tag{1.28}$$

For $f \in L_2(\Omega)$, Eq. (1.26) admits a unique solution. Letting S_J be a subspace of $H_0^\alpha(\Omega)$ with order d, the multiresolution Galerkin method (MGM) is to find that $u_J \in S_J$ satisfies

$$a(u_J, v_J) = (f, v_J) \quad \forall \, v_J \in S_J. \tag{1.29}$$

If u is sufficiently smooth, by the biorthogonal projection (1.11) and the Cea''s lemma, one gets

$$\| u - u_J \|_\alpha \overset{<}{\sim} \inf_{v_J \in S_J} \| u - v_J \|_\alpha \overset{<}{\sim} 2^{J(\alpha-d)} \| u \|_{H^d(\Omega)}. \tag{1.30}$$

For space discretization, one can either use the scaling bases Φ_J or the multiscale bases Ψ^J, *i.e.*, let

$$u_J = \sum_{k \in \Delta_J} c_{J,k} \phi_{J,k} \quad \text{or} \quad u_J = \sum_{k \in \Delta_{J_0}} c_{J_0,k} \phi_{J_0,k} + \sum_{j=J_0}^{J-1} \sum_{k \in \nabla_j} d_{j,k} \psi_{j,k}, \tag{1.31}$$

generating the following linear systems, respectively,

$$A_J \mathbf{c}_J = F_J, \tag{1.32}$$

$$\hat{A}_J \mathbf{d}_J = \hat{F}_J, \tag{1.33}$$

where $A_j = a(\Phi_j, \Phi_j), F_j = (f, \Phi_j), \hat{A}_j = a(\Psi^j, \Psi^j), \hat{F}_j = (f, \Psi^j)$, and $\mathbf{d}_j = (\mathbf{c}_{J_0}, \mathbf{d}_{J_0}, \dots, \mathbf{d}_{J-1})$ with $\mathbf{c}_j = (c_{j,k})_{k \in \Delta_j}, \mathbf{d}_j = (d_{j,k})_{k \in \nabla_j}$. There are the following Lemmas.

Lemma 1.1. *Let* $\phi(x) \in H_0^1(\Omega)$, $\mathrm{supp}\phi(x) = [0, d]$ *and* $\phi_{J,k}(x) := 2^{J/2}\phi(2^J x - k), 0 \leq k \leq 2^J - d, k \in \mathbb{N}$. *Then* $a(\phi_{J,k_1}, \phi_{J,k_2}) = a(\phi_{J,k_1'}, \phi_{J,k_2'})$ *if and only if* $k_2 - k_1 = k_2' - k_1'$.

Proof. For $\phi(x) \in H_0^1(\Omega)$, there holds

$$\left({}_0D_x^{-\beta/2}D\phi_{J,k_1}, {}_xD_1^{-\beta/2}D\phi_{J,k_2}\right) = \left({}_0D_x^{-\beta}D\phi_{J,k_1}, D\phi_{J,k_2}\right)$$

$$= \frac{1}{\Gamma(\beta)}\int_0^1\int_0^x (x-s)^{\beta-1}\phi'_{J,k_1}(s)\,\mathrm{d}s\,\phi'_{J,k_2}(x)\,\mathrm{d}x$$

$$= \frac{2^{3J}}{\Gamma(\beta)}\int_{2^{-J}k_2}^{2^{-J}(d+k_2)}\int_0^x (x-s)^{\beta-1}\phi'(2^J s - k_1)\,\mathrm{d}s\,\phi'(2^J x - k_2)\,\mathrm{d}x$$

$$= \frac{2^{2J}}{\Gamma(\beta)}\int_0^d\int_0^{2^{-J}(x+k_2)} (2^{-J}(k_2+x)-s)^{\beta-1}\phi'(2^J s - k_1)\,\mathrm{d}s\,\phi'(x)\,\mathrm{d}x$$

$$= \frac{2^{2J\alpha}}{\Gamma(\beta)}\int_0^d\int_{-k_1}^{x+k_2-k_1} (k_2+x-s-k_1)^{\beta-1}\phi'(s)\,\mathrm{d}s\,\phi'(x)\,\mathrm{d}x$$

$$= \frac{2^{2J\alpha}}{\Gamma(\beta)}\int_0^d\int_0^{x+k_2-k_1} (x-s+k_2-k_1)^{\beta-1}\phi'(s)\,\mathrm{d}s\,\phi'(x)\,\mathrm{d}x,$$

which just depends on the value of $k_2 - k_1$. The second part of (1.27) can be expressed by its first part, *i.e.*,

$$\left({}_xD_1^{-\beta/2}D\phi_{j,k_2}, {}_0D_x^{-\beta/2}D\phi_{j,k_1}\right) = \left({}_0D_x^{-\beta/2}D\phi_{j,k_1}, {}_xD_1^{-\beta/2}D\phi_{j,k_2}\right). \quad \textbf{(1.34)}$$

Then the desired result is obtained.

Lemma 1.2. *Let* $\phi(x)$ *and* $\phi_{J,k}(x)$ *be given as above, and* $\phi(d/2 - x) = \phi(d/2 + x)$. *Define* $\theta_{J,i}(x) := 2^{J/2}\theta_i(2^J x)$ *and* $\tilde{\theta}_{J,i}(x) := 2^{J/2}\theta_i(2^J(1-x))$ *with* $\theta_i(x) \in H_0^1(\Omega)$ *and* $\mathrm{supp}\theta_i(x) = [0, d_i]$, *where* $0 < d_i < d$ *and* $i = 1,2$. *Then*

$$\left({}_0D_x^{-\beta/2}D\theta_{J,i}, {}_xD_1^{-\beta/2}D\phi_{J,k}\right) = \left({}_0D_x^{-\beta/2}D\phi_{J,2^J-d-k}, {}_xD_1^{-\beta/2}D\tilde{\theta}_{J,i}\right). \quad \textbf{(1.35)}$$

Proof. Similar to Lemma 1.1, it follows that

$$\left({}_0D_x^{-\beta/2}D\theta_{J,i}, \; {}_xD_1^{-\beta/2}D\phi_{J,k} \right)$$

$$= \frac{1}{\Gamma(\beta)} \int_0^1 \int_0^x (x-s)^{\beta-1}\theta'_{J,i}(s) \, ds \, \phi'_{J,k}(x) \, dx$$

$$= \frac{2^{2J\alpha}}{\Gamma(\beta)} \int_0^d \int_0^{x+k} (x+k-s)^{\beta-1}\theta'_i(s) \, ds \, \phi'(x) \, dx.$$

By the properties of symmetry and compact support, there exists

$$\left({}_0D_x^{-\beta/2}D\phi_{J,2^J-d-k}, \; {}_xD_1^{-\beta/2}D\tilde{\theta}_{J,i} \right)$$

$$= \frac{1}{\Gamma(\beta)} \int_0^1 \int_0^x (x-s)^{\beta-1}\phi'_{J,2^J-d-k}(s) \, ds \, \tilde{\theta}'_{J,i}(x) \, dx$$

$$= \frac{2^{2J}}{\Gamma(\beta)} \int_{d_i}^0 \int_0^{1-2^{-J}x} (1-s-2^{-J}x)^{\beta-1}\phi'(2^J s - 2^J + d + k) \, ds \, \theta'_i(x) \, dx$$

$$= \frac{2^{2J\alpha}}{\Gamma(\beta)} \int_0^{d_i} \int_{\max\{0,x-k\}}^{\min\{2^J-k,d\}} (s+k-x)^{\beta-1}\phi'(s) \, ds \, \theta'_i(x) \, dx$$

$$= \frac{2^{2J\alpha}}{\Gamma(\beta)} \int_0^d \int_0^{x+k} (x+k-s)^{\beta-1}\theta'_i(s) \, ds \, \phi'(x) \, dx,$$

where the Fubini-Tonelli theorem and $\min\{2^J - k, d\} = d$ are used.

Similarly, for any $i_1, i_2 \in \{1,2\}$, one has

$$\left({}_0D_x^{-\beta/2}D\theta_{J,i_1}, \; {}_xD_1^{-\beta/2}D\theta_{J,i_2} \right) = \left({}_0D_x^{-\beta/2}D\tilde{\theta}_{J,i_2}, \; {}_xD_1^{-\beta/2}D\tilde{\theta}_{J,i_1} \right), \qquad \textbf{(1.36)}$$

$$\left({}_0D_x^{-\beta/2}D\phi_{J,k}, \; {}_xD_1^{-\beta/2}D\theta_{J,i} \right) = \left({}_0D_x^{-\beta/2}D\tilde{\theta}_{J,i}, \; {}_xD_1^{-\beta/2}D\phi_{J,2^J-d-k} \right); \quad \textbf{(1.37)}$$

and by the compact support of Φ_J, it is easy to check that

$$\left({}_0D_x^{-\beta/2}D\phi_{J,k_1}, \; {}_xD_1^{-\beta/2}D\phi_{J,k_2} \right) = 0 \quad \forall k_2 - k_1 \leq -d. \qquad \textbf{(1.38)}$$

Now, from Proposition 1.1 and the above lemmas, one knows that the matrix $A_l := \left({}_0D_x^{-\beta/2}D\Phi_J, \; {}_xD_1^{-\beta/2}D\Phi_J \right)$ has a quasi-Toeplitz structure, that is, it is a

Toeplitz matrix after removing very few rows and columns near the boundaries. More precisely, for $d = 2$, it is a full Toeplitz matrix, but for $d = 3$ and $d = 4$, they have the following structures, respectively,

$$\begin{pmatrix} a_1 & r(\mathbf{a_2})^T & 0 \\ \mathbf{a_1} & H_{(2^J-2)\times(2^J-2)} & \mathbf{a_2} \\ a_2 & r(\mathbf{a_1})^T & a_1 \end{pmatrix}_{2^J\times2^J} , \tag{1.39}$$

$$\begin{pmatrix} a_1 & a_2 & r(\mathbf{a_1})^T & 0 & 0 \\ a_3 & a_4 & r(\mathbf{a_2})^T & 0 & 0 \\ \mathbf{a_3} & \mathbf{a_4} & H_{(2^J-3)\times(2^J-3)} & \mathbf{a_2} & \mathbf{a_1} \\ a_5 & a_6 & r(\mathbf{a_4})^T & a_4 & a_2 \\ a_7 & a_5 & r(\mathbf{a_3})^T & a_3 & a_1 \end{pmatrix}_{(2^J+1)\times(2^J+1)} , \tag{1.40}$$

where a_i are real numbers; $\mathbf{a_i}$ are vectors, $r(\mathbf{a_i})$ the reverse order of $\mathbf{a_i}$; and $H_{N\times N}$ is Toeplitz matrix.

It should be noted that the inner basis functions are obtained by dilating and translating of a single function and the boundary basis functions are symmetric. This fact is essential for obtaining the above results, recalling that they do not hold for the general finite element (except linear element) and spectral polynomial basis functions. Obviously, it follows that

$$A_r := \left({}_xD_1^{-\beta/2}D\Phi_J, \, {}_0D_x^{-\beta/2}D\Phi_J \right) = \left({}_0D_x^{-\beta/2}D\Phi_J, \, {}_xD_1^{-\beta/2}D\Phi_J \right)^T = A_l^T, \tag{1.41}$$

Remark The structure of Φ_J also allows one to compute its Riemann-Liouville fractional derivative easily. As an example, we present the techniques for $d = 4$ in Appendix A2, being similar for other values of d. Then combining with (40), the left differential matrix $A_l = \left({}_0D_x^{1-\beta}\Phi_J, D\Phi_J \right)$ can be calculated exactly or numerically with the cost $\mathcal{O}(N)$, being superior to the traditional finite element and spectral approximation with the cost $\mathcal{O}(N^2)$ [9, 13].

For an algebraic system with dense matrix, a well convergent iterative method generally has the computational cost $\mathcal{O}(N\log(N))$ or $\mathcal{O}(N^2)$, which is much less than the cost $\mathcal{O}(N^3)$ of the direct method. Moreover, a well conditional number and 'bunching of eigenvalues' usually bring good numerical stability and fast convergence speed. In general, for a linear system $Ax = b$, a satisfactory preconditioned system $Bx' = b'$ should have the property

$\| B \| \leq C, \| B^{-1} \| \leq C, C$ *is a moderate $-$ sized constant independent of N ;*

and the computational cost for the preconditioning step is cheap. Here, both the matrix A_J and $\hat{A}_J$ are dense, and their condition numbers are of order $\mathcal{O}(2^{2J\alpha})$; see Table **1.5** for Example 2. But with the aid of wavelet bases, by the norm equivalence, a simple diagonal scaling can lead to a good preconditioned system. In fact, define

$$\nabla^J = \Delta_{J_0} \cup \nabla_{J_0} \cup \cdots \cup \nabla_{J-1}, \tag{1.42}$$

$$K = diag(\underbrace{2^{-J_0\alpha}, \dots, 2^{-J_0\alpha}}_{\#\Delta_{J_0}}, \underbrace{2^{-J_0\alpha}, \dots, 2^{-J_0\alpha}}_{\#\nabla_{J_0}}, \dots, \underbrace{2^{-(J-1)\alpha}, \dots, 2^{-(J-1)\alpha}}_{\#\nabla_{J-1}}). \tag{1.43}$$

Combining the ellipticity (1.28), norm equivalence (1.13), and the Riesz representation theorem, one gets that for all $\mathbf{x} \in l_2(\nabla^J)$,

$$\| K\hat{A}_J K\mathbf{x} \|_{l_2(\nabla^J)} = \sup_{\mathbf{y}\in l_2(\nabla^J)} \frac{\langle K\hat{A}_J K\mathbf{x}, \mathbf{y}\rangle_{l_2(\nabla^J)}}{\|\mathbf{y}\|_{l_2(\nabla^J)}}$$

$$= \sup_{\mathbf{y}\in l_2(\nabla^J)} \frac{a(\mathbf{x}^T K\Phi^J, \mathbf{y}^T K\Phi^J)}{\|\mathbf{y}\|_{l_2(\nabla^J)}} \lesssim \frac{\|\mathbf{x}^T K\Phi^J\|_\alpha \|\mathbf{y}^T K\Phi^J\|_\alpha}{\|\mathbf{y}\|_{l_2(\nabla^J)}} \lesssim \| \mathbf{x} \|_{l_2(\nabla^J)},$$

$$\| K\hat{A}_J K\mathbf{x} \|_{l_2(\nabla^J)} \gtrsim \frac{\|\mathbf{x}^T K\Phi^J\|_\alpha \|\mathbf{x}^T K\Phi^J\|_\alpha}{\|\mathbf{x}\|_{l_2(\nabla^J)}} \gtrsim \| \mathbf{x} \|_{l_2(\nabla^J)}.$$

Therefore, there exist C_1, C_2 not depending on J such that

$$C_1 \| \mathbf{x} \|_{l_2(\nabla^J)} \leq \| K\hat{A}_J K\mathbf{x} \|_{l_2(\nabla^J)} \leq C_2 \| \mathbf{x} \|_{l_2(\nabla^J)}. \tag{1.44}$$

Now, one arrives at

$$\| K\hat{A}_J K \| \lesssim C_2, \| (K\hat{A}_J K)^{-1} \| \lesssim (1/C_1), \tag{1.45}$$

$$cond_2(K\hat{A}_J Ku_J) = \| K\hat{A}_J K \| \| (K\hat{A}_J K)^{-1} \| \lesssim (C_2/C_1). \tag{1.46}$$

By (1.13) and (1.28), it also holds that $a(\psi_{j,k}, \psi_{j,k}) \sim \|\psi_{j,k}\|_{H^s(\Omega)}^2 \sim 2^{2js}$. Therefore, one can also define matrix K by the inverse square root of the diagonal of $\hat{A}_J$, and (1.44) and (1.46) still hold. Usually the current K performs better since it uses the information directly from the stiffness matrix, and we will use it in

Section 1.5. Moreover, the cost of generating K is only $\mathcal{O}(J)$; this is because that by using the translation property of the inner wavelet on the same level, one just needs to calculate the entries $a(\psi_{j,k}, \psi_{j,k})$ near the boundaries and one in the inner part without the necessity to assemble $\hat{A}_J$.

Now, one can rewrite (1.33) as the two-sided preconditioned form

$$\underbrace{K\hat{A}_J K}\, K^{-1}\mathbf{d}_J = K\hat{F}_J. \tag{1.47}$$

Further using the fact that $\mathbf{c}_J = M\mathbf{d}_J$, one gets that

$$\underbrace{KM^T A_J MK}\, K^{-1}M^{-1}\mathbf{c}_J = KM^T F_J. \tag{1.48}$$

A straightforward product of A_J or $\hat{A}_J$ to a given vector needs a computational cost $\mathcal{O}(2^{2J})$. But if one uses the quasi-Toeplitz structure of the matrix, the computational cost can be reduced to $\mathcal{O}(J2^J)$. In fact, one can rewrite A_J as

$$A_J = diag(K_1)A_l + diag(K_2)A_r + A_q, \tag{1.49}$$

where K_1 and K_2 denote the coefficient vectors, formed by the coefficient of space fractional derivative taking values at the discretized intervals, A_q is a sparse matrix produced by the reaction term qu, and A_l and A_r are quasi-Toeplitz matrices. Using the FFT to the quasi-Toeplitz matrix-vector product makes the computational cost $\mathcal{O}(J2^J)$ [4]. Finally, because the FWT (having the matrix representation M or M^T, which denotes the primal reconstruction or the dual decomposition (see [6, Chapter, Sections 2 and 3] and [27, Section 5.4]) can be implemented with the cost $\mathcal{O}(2^J)$, if the CG scheme (symmetric) is applied to (48) or to the corresponding normal equation (non-symmetric), the well conditioned number of the matrix implies that the convergence rate is independent of the level J; then we can solve it with the total operations $\mathcal{O}(J2^J)$. For the general iterative schemes, such as GMRES or Bi-CGSTAB, usually one can show that the system with clustered spectrum and well conditioned number after preconditioning has an accelerated convergence. What's more, compared with the most existing preconditioners which require the solving of a linear system (see, *e.g.*, the ILU [15] and the Strang [14]), the wavelet preconditioning operation reduces to the matrix-vector product, where FWT can be used. The implementation process (for $d = 2$) are given in Algorithm 1.

Algorithm 1. Solving (1.5) with wavelet preconditioning

1:   Chosen J_0, J and $d = 2$

2:   Generate A_q, F_J, K, and the first row and first column of A_l (*i.e*, the first column and first row of A_r)

3:   Solve $\underbrace{KM^T A_J MK}\, v = KM^T F_J$ by the appropriate iterative solver, where we compute $KM^T A_J MKv$ by $K\left(M^T \left(A_J\left(M(Kv)\right)\right)\right)$, and the FFT is used to compute the matrix-vectors product involving A_l and A_r, the FWT are used to compute $M^T F_J$ and matrix-vector products involving M and M^T. Denote the final iteration solution still as v.

4:   Compute c_J by $c_J = M(Kv)$.

1.4. MULTISCALE ADAPTIVE SCHEME

Unlike the integer-order PDE, for sufficiently smooth (limited on Ω) source term f, the exact solution of the fractional PDE may show some kind of singularity. The authors in [12] have preliminarily investigated these for the simplest static fractional diffusion equation with one-sided Riemann-Liouville derivative (a complete mathematical justification of the regularity of the fractional equations with two-sided fractional derivatives or more complex forms is still missing). It seems that designing the adaptive algorithms for solving fractional equations is an urgent task. Unfortunately, there are very limited (or no) works focus on these. Here, as an attempt, we numerically show the effectiveness of wavelet in dealing with such problems, and we expect that this will inspire more enthusiasm in this field.

Note that the compressibility in Proposition 1.2, we simply take the wavelet coefficients of the (to be determined) solution as the local error indicator of the adaptive algorithm: small coefficient implies good local regularity while big one indicates the opposite, and increase or decrease the number of wavelet basis functions in corresponding regions. We give the concrete implementation process in Algorithm 2. One of the main features of Algorithm 2 is that the finest grid resolution can be automatically determined by the given tolerance $\varepsilon(j)$. In order to gain a more robust and faster algebra solver, $\psi_{j,k}$ has been scaled by the inverse square root of $a\left(\psi_{j,k}, \psi_{j,k}\right)$, and the multiscale approximation of the solution at the current scale has been as an initial guess for the iteration in the finer scale obtained after adding the wavelets. For constructing the refined index set, thanks to the tree structure of wavelet singularity detection (see [27, Section 7.1]), we first include a coarsening step by thresholding the latest available wavelet coefficients to get a significant index set, then add all their children. If $j = l + 1$ and $k \in \{2\lambda, 2\lambda + 1\}$, then the wavelet indexed by (j, k) is called a child of the wavelet indexed by (l, λ). One can further extend the index set by including the horizontal neighbors of the wavelet indices already included. Such an extended index set associated with the index (l, λ) is called an adjacent zone, which is denoted by $\mathcal{N}_{l,\lambda}$. In Algorithm 2, the index set is continuously updated to resolve the local structures that appear in the solution. One can dynamically adjust the number and locations of the wavelets used in the wavelet expansion, reducing significantly the cost of the scheme while providing enough resolution in the regions where the solution varies significantly. The m-th approximation of the solution is given by

$$\hat{u}_{J_0+m} = \sum_{\lambda \in \Delta_{J_0}} c^m_{J_0,\lambda}\, \phi_{J_0,\lambda} \; + \sum_{(l,\lambda)\in\Lambda^m} d^m_{l,\lambda}\hat{\psi}_{l,\lambda}, \tag{1.50}$$

where $\Delta_{J_0} \cup \Lambda^m$ is the irregular index set. Finally, after getting the sufficiently accurate approximation, the corresponding single scaling representation can be got by the FWT.

Algorithm 2. Solving (1.5) with wavelet Adaptivity

1:   Given It_{max}, J_0, and $\varepsilon(j)$ for $j = J_0, J_0 + 1, \dots, J_0 + It_{max}$

2:   $m = 0$

3:   Solve the equation (1.33) in space S_{J_0+1} to get the initial approximation coefficients $(c^0_{J_0}, d^0_{J_0})$ and the index $\Lambda^0 = \{(J_0, \lambda), \lambda \in \nabla_{J_0}\}$

4:   Determine the initial significant index set by
$$\Lambda = \left\{ (J_0, \lambda) : (J_0, \lambda) \in \Lambda^0, \left| d^0_{J_0,\lambda} \right| \geq \varepsilon(J_0) \right\}$$

5:   **repeat**

6:   Check the adjacent zone index set $\mathcal{N}_{l,\lambda}$ of each $(l, \lambda) \in \Lambda$; denote $\Lambda_{\mathcal{N}} = \cup_{(l,\lambda)\in\Lambda} \mathcal{N}_{l,\lambda}$ and establish $\Lambda^{m+1} = \Lambda \cup \Lambda_{\mathcal{N}}$

7:   **for** $(l, \lambda) \in \Lambda^{m+1}$ **do**

8:   $^*d^{m+1}_{l,\lambda} = \begin{cases} d^m_{l,\lambda}, & (l, \lambda) \in \Lambda^m \\ 0, & \text{otherwise} \end{cases}$

9:     **end for**

10:    $^*c_{J_0}^{m+1} = c_{J_0}^m$

11:    Solve the algebraic matrix equation, resulted from the discretization in the approximation space $\hat{V}_{J_0+m+1}(\Omega) = \mathrm{span}\left\{\Phi_{J_0} \cup \bigcup_{(l,\lambda)\in\Lambda^{m+1}} \psi_{l,\lambda}\right\}$, by appropriate iterative scheme with the initial guess $\left(^*c_{J_0}^{m+1}, ^*d_{l,\lambda}^{m+1}|_{(l,\lambda)\in\Lambda^{m+1}}\right)$, and denote the solution vector by $\left(^*c_{J_0}^{m+1}, ^*d_{l,\lambda}^{m+1}|_{(l,\lambda)\in\Lambda^{m+1}}\right)$

12:    Determine $\Lambda = \{(l,\lambda): (l,\lambda) \in \Lambda^{m+1}, |d_{l,\lambda}^m| \geq \varepsilon(l)\}$

13:    $m = m + 1$

14:    **until** $m > It_{max}$ or $\Lambda = \Phi$ (empty set)

1.5. NUMERICAL SIMULATIONS

In this section, we give some examples to illustrate the accuracy and efficiency of the proposed numerical schemes. Example 1 is used to discuss the convergence orders of (1.5) approximated with the scaling B-spline basis functions. We use Example 2 to show the powerfulness of the provided wavelet preconditioner. And Example 3 is used to illustrate the effectiveness of the presented wavelet adaptive schemes. Since the reaction term $qu(x)$ in (1.5) do not cause any calculation and theoretical difficulties, we always choose $q(x) = 0$ below.

Example 1.

Consider the MGM for (1.5) with $q = 0$ and $p = \kappa_\beta = 1$, and the source term

$$f(x) = \frac{2x^\beta}{\Gamma(\beta+1)} - \frac{\Gamma(v+1)}{\Gamma(v+\beta-1)} x^{v+\beta-2}, \quad x \in \Omega.$$

For $x \in \Omega$, the exact solution of the problem is $u(x) = x^v - x^2$. It is well known that if $v > 0$ and $v \notin \mathbb{N}$, then $u \in H^{v+1/2-\varepsilon}(\Omega)$. For $\beta = 4/5$, the numerical results are listed in Tables **1.1** and **1.2**, which confirm that if the analytical solution is smooth enough, the convergence order is d and $d - \alpha$ in the L_2 and H^α-norm, respectively. Otherwise the convergence order is limited by the regularity of the solution, but the approximation accuracy is improved when the high order bases are used.

Table 1.1. Numerical results of (1.5), solved by MGM, with $q = 0$, $p = \kappa_\beta = 1$, and $\beta = 4/5$.

d	J	$v = 4$				$v = 17/10$	
		L_2-Err	L_2-Rate	$a(u - u_J, u - u_J)^{1/2}$	H^α-Rate	L_2-Err	L_2-Rate
	6	17589e-04	—	8.5677e-04	—	1.4535e-05	—
$d = 2$	7	4.3968e-05	2.0001	3.1635e-04	1.4374	3.6314e-06	2.0009
	8	1.0993e-05	1.9999	1.1829e-04	1.4192	9.0287e-07	2.0079
	6	6.2317e-07	—	6.2807e-06	—	1.0342e-06	—
$d = 3$	7	7.7779e-08	3.0021	1.1896e-06	2.4004	2.2509e-07	2.2000
	8	9.7152e-09	3.0011	2.2531e-07	2.4005	4.8988e-08	2.2000

Table 1.2. Numerical results of (1.5), solved by MGM, with $q = 0$, $p = \kappa_\beta = 1$, and $\beta = 4/5$.

J	$v = 11/10$				$v = 21/10$			
	$d = 3$		$d = 4$		$d = 3$		$d = 4$	
	L2-Err	L_2-Rate	L_2-Err	L_2-Rate	L2-Err	L_2-Rate	L_2-Err	L_2-Rate
6	1.4385e-05	—	8.0390e-06	—	1.2656e-07	—	3.2703e-08	—
7	4.7453e-06	1.6000	2.6516e-06	1.6002	2.0865e-08	2.6007	5.3930e-09	2.6002
8	1.5654e-06	1.6000	8.7469e-07	1.6002	3.4407e-09	2.6003	8.8950e-10	2.6000

We point out that scaling B-spline functions also have the computational advantages for the collocation method. In the following, we consider a variable-coefficient evolution equation

$$u_t - (k_1 x^{2-\beta} {}_0D_x^{2-\beta} u + k_2 (1 - x)^{2-\beta} {}_xD_1^{2-\beta} u) = f, (x, t) \in \Omega \times (0, T] \quad \textbf{(1.51)}$$

with the right-hand term

$$f(x, t) = -12\exp(-t)\Big\{ x^2(1 - x)^2 + \frac{1}{6}[k_1 x^2 + k_2(1 - x)^2]$$

$$- \frac{1}{\beta+1}[k_1 x^3 + k_2(1 - x)^3] + \frac{2}{(\beta+1)(\beta+2)}[k_1 x^4 + k_2(1 - x)^4] \Big\}, \quad x \in \Omega$$

and the initial condition $u(x, 0) = x^2(1 - x)^2, x \in \Omega$.

For $x \in \Omega$, the analytical solution is $u(x,t) = \exp(-t)x^2(1-x)^2$. We choose the approximation space $S_J = span\{\Phi_J, d = 4\}$ (see Appendix A2), and use the collocation points $\{1/2^{J+1}, k/2^J|_{k=1}^{2^J-1}, 1 - 1/2^{J+1}\}$. In the implementation, one needs to calculate the left and right fractional differential matrices $A_L := {}_0D_x^{2-\beta}\Phi_J|_{x_l}$ and $A_R := {}_xD_1^{2-\beta}\Phi_J|_{x_l}$. The former can be got by first using the results in Appendix A2 to get the formula of ${}_0D_x^{2-\beta}\Phi_J$, then obtain A_L directly or use the fact that the sub-matrix produced by $2^{J/2}{}_0D_x^{2-\beta}\phi(2^Jx - k)|_{l/2^J}$, $k, l = 2,3,\cdots,2^J - 2$ has a Toeplitz structure (similar to the previous proof); and the latter can be easily obtained by using the relationship $A_R = A_L(end: -1: 1, end: -1: 1)$ (see Appendix A3).

The Crank-Nicolson scheme is used to get the full discretization approximation of (1.51) with the time stepsize $1/2^{2J}$ and $T = 1/2$.

Table 1.3. Convergence performance of the cubic spline collocation method with $k_1 = k_2 = 1$.

J	$\beta = 2/10$		$\beta = 8/10$		$\beta = 0$	
	L_∞-Err	L_∞-Rate	L_∞-Err	L_∞-Rate	L_∞-Err	L_∞-Rate
5	5.8773e-05	—	1.8431e-06	—	1.6167e-04	—
6	1.2862e-05	2.1920	2.6580e-07	2.7937	4.0533e-05	1.9958
7	2.8036e-06	2.1977	3.8223e-08	2.7978	1.0441e-05	1.9990

Table **1.3** shows the expected convergence order $2 + \beta$ of collocation method for fractional PDE, agreeing with the classical conclusion when $\beta = 0$. Though the superconvergence can be obtained for classical PDE by carefully averaging the derivative values gotten in collocation points, it seems that this result maynot be directly extended to fractional PDE. For convenience, let's consider the frequently used Hermite spline collocation method. Take the collocation space

$$V_J := span\{\pi_2(2^Jx + 1)|_\Omega, \pi_1(2^Jx - k), \pi_2(2^Jx - k), \pi_2(2^Jx - 2^J + 1))|_\Omega\},$$

where $|_\Omega$ denotes the restriction in Ω, $k = 0,1,\cdots,2^J - 2$, and π_1, π_2 are the cubic Hermite compactly supported functions given as

$$\pi_1(x) = \begin{cases} -x^2(2x - 3) & 0 \le x < 1, \\ (x - 2)^2(2x - 1) & 1 \le x \le 2, \end{cases}$$

$$\pi_2(x) = \begin{cases} x^2(x-1) & 0 \le x < 1, \\ (x-2)^2(x-1) & 1 \le x \le 2. \end{cases}$$

To determine the unknown coefficients, one needs total 2^{J+1} collocation points. As well known for classic PDE ($\beta = 0$), when the general collocation points such as the third-quarter points of every interval $[i/2^J, (i+1)/2^J], i = 0, 1, \cdots, 2^J - 1$ are used, the convergence order is 2. But if the Gauss nodes are used, one arrives at the superconvergence result of order 4. Unfortunately, in both case the convergence order are $2 + \beta$ for the fractional PDE, except the approximation accuracy maybe improved. The numerical results are presented in Table **1.4**, where the abbr 'Equi' and 'Gauss' denote the two types of collocation points mentioned above.

Table 1.4. Convergence performance of the cubic Hermite collocation method with $k_1 = 1$ and $k_2 = 0$.

J	$\beta = 5/10$, Equi		$\beta = 5/10$, Gauss		$\beta = 0$, Equi		$\beta = 0$, Gauss	
	L_∞-Err	L_∞-Rate	L_∞-Err	L_∞-Rate	L_∞-Err	L_∞-Rate	L_∞-Err	L_∞-Rate
5	8.2952e-06	—	1.9362e-07	—	3.5693e-05	—	3.5875e-08	—
6	1.4663e-06	2.5001	3.2737e-08	2.5642	8.9270e-06	1.9994	2.2506e-09	3.9946
7	2.5912e-07	2.5005	5.6891e-09	2.5247	2.2316e-06	2.0001	1.4098e-10	3.9968

Example 2.

Now, we focus on the wavelet preconditioning schemes for solving the fractional PDEs. The presented numerical results with $d = 2$, and in this case the coefficient matrix has a full Toeplitz structure. The Toeplitz matrix-vector product is performed directly by FFT. For the other bases, the computational procedure is almost the same after a slight modification, e.g., when $d = 3$, $A_1 c_j := \left({_0 D_x^{-\beta/2}} D\Phi_{J,x} D_1^{-\beta/2} D\Phi_J \right) c_j$ can be decomposed into several blocks with H being the Toeplitz matrix:

$$(A_1 c_j)(1) = \left[a_1, r(\mathbf{a_2})^T, 0 \right] c_j,$$
$$(A_1 c_j)(2 : \text{end} - 1) = c_j(1) \mathbf{a_1} + H c_j(2 : \text{end} - 1) + c_j(\text{end}) \mathbf{a_2},$$
$$(A_1 c_j)(\text{end}) = \left[a_2, r(\mathbf{a_1})^T, a_1 \right] c_j.$$

We first reveal that the preconditioning brings a uniform matrix condition number and an improved spectral distribution. Considering (1.5) with $\kappa_\beta = 1, p = 1$ and $\kappa_\beta = 1, p = 1/2$, the condition numbers for different β are presented in Table **1.5**; one can see that without preconditioning, the condition number of the stiffness matrix behaves like $\mathcal{O}(2^{J(2-\beta)})$, which means the conditional number increases fast with the refinement especially when β is small.

Table 1.5. Primal condition numbers of (1.5) with $q = 0, \kappa_\beta = 1$, and $d = 2$.

J	$p = 1, \beta = 1/2$		$p = 1/2, \beta = 1/2$		$p = 1, \beta = 1/5$		$p = 1/2, \beta = 1/5$	
	Con-Num	Rate	Con-Num	Rate	Con-Num	Rate	Con-Num	Rate
8	1.4763e+03	—	1.8304e+03	—	8.7494e+03	—	9.1119e+03	—
9	4.1754e+03	1.5000	5.1784e+03	1.5003	3.0467e+04	1.8000	3.1732e+04	1.8001
10	1.1810e+04	1.5000	1.4648e+04	1.5002	1.0609e+05	1.8000	1.1050e+05	1.8000

After preconditioning, the uniformly bounded condition numbers with different wavelet preconditioners are obtained; see Table **1.6**, where 'inte-', 'Semi-', and 'Bior-' denote the interpolation wavelet, semiorthogonal wavelet, and biorthogonal wavelet ($d = \tilde{d} = 2$), respectively, having been introduced in Section 1.2. Note that when performing the decomposition by semiorthogonal and biorthogonal wavelets, the interpolation wavelet has been used for S_1 and S_2.

Table 1.6. Preconditioned condition numbers of (1.5) with $q = 0, \kappa_\beta = 1, d = 2$, and $J_0 = 0$.

p	J	$\beta = 1/2$			$\beta = 1/5$		
		Inte-	Semi-	Bior-	Inte-	Semi-	Bior-
	8	3.0970	5.8363	13.3957	1.5953	10.2897	12.2315
$p = 1$	9	3.2286	6.1158	14.4784	1.6269	10.5561	12.9767
	10	3.3457	6.3622	15.4103	1.6540	10.7779	13.5788
	8	3.2614	8.0344	12.7702	1.5935	11.1094	12.2830
$p = 1/2$	9	3.4745	8.1648	13.6624	1.6296	11.4094	13.0063
	10	3.6686	8.2634	14.4026	1.6608	11.6511	13.5854

We also display the matrix eigenvalue distribution for $\beta = 1/5$ in Figs. (**1.1** and **1.2**); they show the preconditioning benefits of a more concentrated eigenvalue distribution.

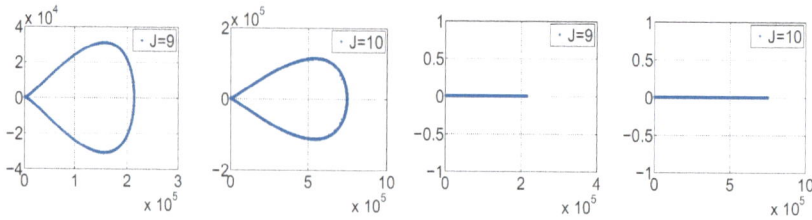

Fig. (1.1). Eigenvalue distribution of the non-preconditioned systems with $p = 1$ (first two) and $p = 1/2$ (last two).

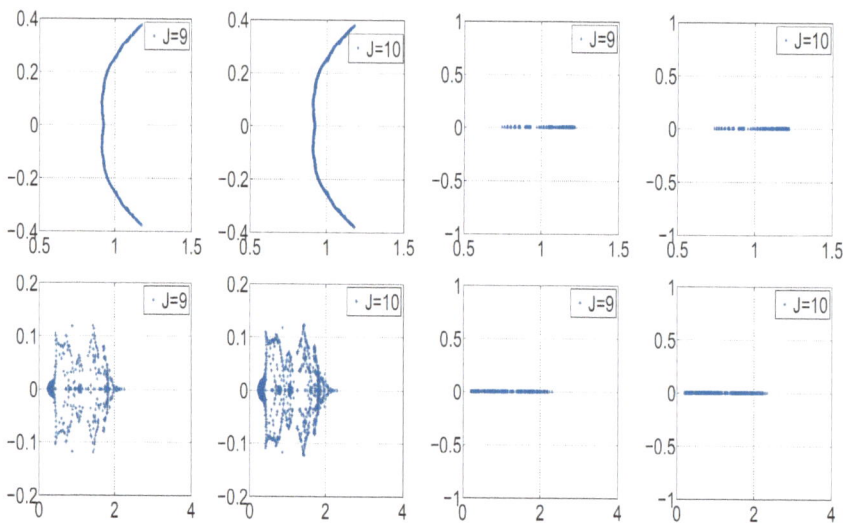

Fig. (1.2). Eigenvalue distribution of the preconditioned systems with the interpolation wavelet (first line) and the semiorthogonal wavelet (second line), respectively (the first two columns are for $p = 1$ and the last two columns for $p = 1/2$).

To explore the effectiveness of this preconditioned system, we numerically solve (1.5) with

$$f = \frac{1}{\Gamma(1+\beta)}\left(-2x^\beta + \beta x^{\beta-1}\right), \quad x \in \Omega$$

and

$$f = \frac{1}{2\Gamma(1+\beta)}\left(-2x^\beta + \beta x^{\beta-1} - 2(1-x)^\beta + \beta(1-x)^{\beta-1}\right), \quad x \in \Omega$$

for $p = 1$ and $p = 1/2$, respectively. We use GMRES and Bi-CGSTAB to solve the algebraic system before and after preconditioning, and the numerical results are given in Tables **1.7** and **1.8**, respectively. The comparisons for the two methods are made almost with the same L_2 approximation error, not listed in the tables. The stopping criterion for solving the linear systems is

$$\frac{\|r(k)\|_{l_2}}{\|r(0)\|_{l_2}} \le 1e - 8,$$

with $r(k)$ being the residual vector of linear systems after k iterations. It should be noted that the GMRES method for $p = 1$ without preconditioning stops before reaching this criterion.

Table 1.7. Numerical results of (1.5), solved by GMRES and Bi-CGSTAB, with $q = 0, \kappa_\beta = 1, \beta = 1/5$, and $d = 2$.

J	$p = 1$, GMRES		$p = 1/2$, GMRES		$p = 1$, Bi-CGSTAB		$p = 1/2$, Bi-CGSTAB	
	Iter	CPU(s)	Iter	CPU(s)	Iter	CPU(s)	Iter	CPU(s)
8	2.5500e+02	0.3443	1.1800e+02	0.0915	2.6350e+02	0.0672	1.1700e+02	0.0303
9	5.1100e+02	1.5217	2.2000e+02	0.3230	5.3550e+02	0.2408	2.0950e+02	0.1012
10	1.0230e+03	7.4783	4.1200e+02	1.3219	1.1665e+03	0.6731	3.9150e+02	0.2280

Table 1.8. Numerical results of (1.5), solved by the preconditioned GMRES and Bi-CGSTAB, with $q = 0, \kappa_\beta = 1, \beta = 1/5, d = 2$, and $J_0 = 0$.

p	J	GMRES, Inte-		GMRES, Semi-		Bi-CGSTAB, Inte-		Bi-CGSTAB, Semi-	
		Iter	CPU(s)	Iter	CPU(s)	Inter	CPU(s)	Iter	CPU(s)
	8	13.0	0.0094	27.0	0.0203	8.0	0.0077	19.0	0.0198
$p = 1$	9	13.0	0.0115	28.0	0.0258	9.5	0.0133	20.0	0.0272
	10	13.0	0.0209	28.0	0.0452	9.5	0.0149	22.0	0.0376
	8	9.0	0.0075	25.0	0.0227	6.5	0.0064	18.0	0.0201
$p = 1/2$	9	9.0	0.0084	26.0	0.0248	7.5	0.0087	20.0	0.0267
	10	9.0	0.0163	26.0	0.0370	8.0	0.0126	21.0	0.0363

Example 3.

In this example, we focus on the previously proposed ad-hoc wavelet adaptive algorithms for the fractional PDEs. (1.5) is solved by the biorthogonal wavelet bases Ψ^J ($d = 3$ and $\tilde{d} = 3$), the regularity of its exact solution, $u(x) = (1 - x)^{11/10} - (1 - x)$ ($x \in \Omega$), is weak at the area close to 1; and the parameters $\kappa_\beta = 1, p = 0$, and the source term

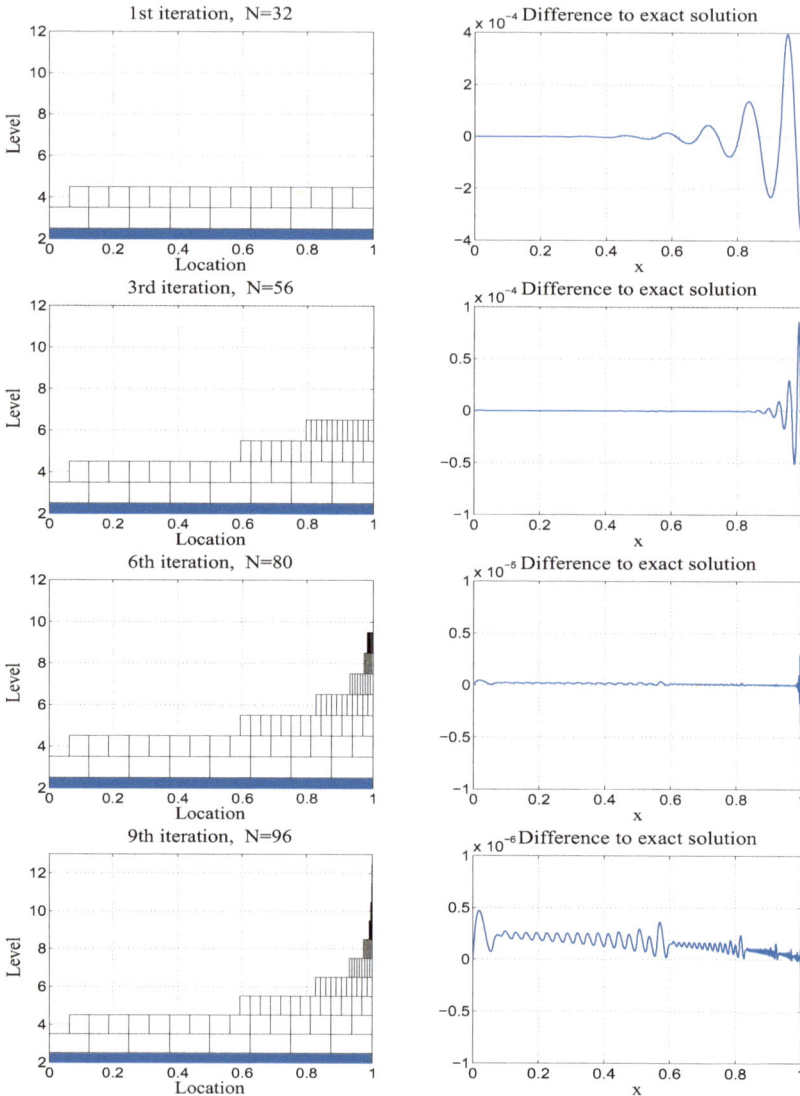

Fig. (1.3). Distribution of adaptive wavelet bases and curve of the approximation error gotten by Algorithm 2.

$$f(x) = -\frac{\Gamma(21/10)(1-x)^{\beta-\frac{9}{10}}}{\Gamma\left(\beta+\frac{1}{10}\right)} + \frac{(1-x)^{\beta-1}}{\Gamma(\beta)}, \quad x \in \Omega.$$

In the algorithm 2, we take $J_0 = 3, \varepsilon(j) = 1e - 5$. For every iteration step, the finally extended irregular indexes are obtained by firstly adding the children of all the significant indexes and then including two neighbors, *i.e.*, the right and left neighbors, of each index of the extended irregular indexes. When $\beta = 1/2$, the sets of wavelet indices that corresponding to the adaptively chosen wavelets and the corresponding error $u - \hat{u}_{J_0+m}$ are presented in Fig. (**1.3**), where the blue bar denotes that we have used all the scaling bases in the coarsest level J_0. One can see that the algorithm in fact automatically recognizes the whereabouts of the boundary layer of the solution u, and adds wavelets locally to there. It also reveals that the newly added computational costs are spend in the most needed place, and the large peaks of the errors are successively reduced. Moreover, for different β, from the decreasing of the L_2 approximation error of the adaptive and uniform Galerkin schemes with the increasing of the freedom N (the loglog coordinate) in Fig. (**1.4**), one can see that the adaptive MGM is remarkably superior to the uniform MGM.

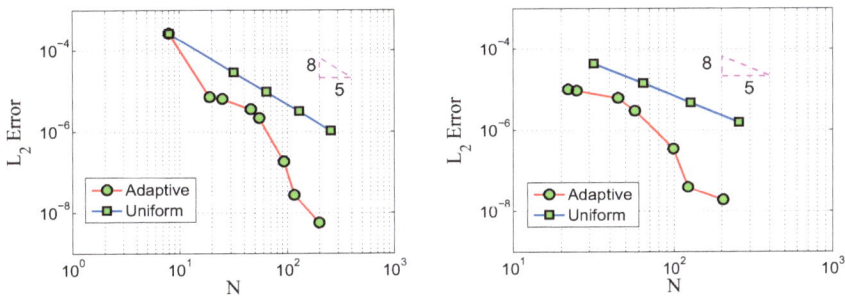

Fig. (1.4). L_2 errors versus freedom N for the adaptive and the uniform Galerkin approximations with $\beta = 1/2$ (left) and $\beta = 4/5$ (right), respectively.

1.6. CONCLUSION AND DISCUSSION

This chapter focuses on digging out the potential benefits, and providing the techniques and extensive numerical experiments in solving the fractional PDEs by wavelet numerical methods. The scaling B-spline bases display their powerfulness in saving computational cost when generating stiffness matrix, *i.e.*, by using the scaling bases, the stiffness matrix has the quasi-Toeplitz structure; and then the effective preconditioners are provided by wavelets. We also numerically show that the heuristic wavelet adaptive scheme works very well for fractional PDEs; in particular, it is still easy to get the local regularity indicator even for the fractional problem; and the algorithm descriptions are provided.

After finishing this work, one of the directions of our further research appears, *i.e.*, applying the wavelet compression property to fractional operator. A key difference between the fractional and classical operators is that the former is non-local, and then both the matrixes generated by the scaling and the multiscale bases are no longer sparse. Fortunately, the wavelet compression not only allows one to design adaptive algorithm, but it seems also effective for the fractional operators. Considering the discretization of the operator:

$$\mathscr{B}u = -D\left(\frac{2}{3}\,_0D_x^{1-\beta} - \frac{1}{3}\,_xD_1^{1-\beta}\right)u$$

in the approximation space S_J with $J = 10$, we first compute the matrix A_J or $\hat{A}_J$ (here the multiscale wavelet bases also have been normalized with D, proposed in Section 1.3). Then we get the compressed matrix by setting all entries of A_J or $\hat{A}_J$ with modulus less than $\varepsilon = 10^{-4} \times 2^{-J}$ to zeros. The comparison results are displayed in Table **1.9** and Fig. **(1.5)** , where $(\cdot\ \%)$ denotes the percentage of the non-zero entries of the compressed matrix. It can be seen that many entries in $\hat{A}_J$ are so small that they can be omitted to retrieve the famous finger structure, whereas essentially all entries in A_J are significant. These may be allow one to design the paralleled sparse approximate inverse preconditioner or to develop more efficient adaptive algorithms (see [27] for the PDEs).

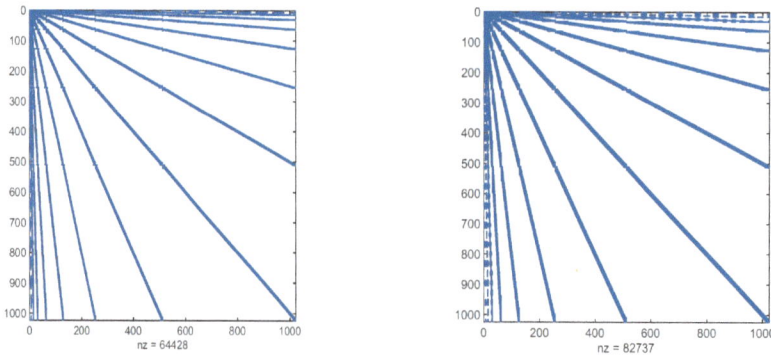

Fig. (1.5). Sparsity patterns of the compression matrix $\hat{A}_J$ by using the semiorthogonal wavelet with $\beta = 5/10$ and $d = 2$ (left) and the biorthogonal wavelet $^{2,4}\psi$ with $d = 2$ and $\tilde{d} = 4$ bases (right).

Table 1.9. Compression capacity of the different bases for operator $\mathscr{B}$.

ψ	$\beta = 8/10$		$\beta = 5/10$		$\beta = 2/10$		Note	
	A_J	$\hat{A}_J$	A_J	$\hat{A}_J$	A_J	$\hat{A}_J$	wavelet	compression
$d = 2, Inte -$	99.80%	99.07%	99.80%	80.38	99.80%	48.04%	interpolation	No/Yes
$d = 2, Semi -$	99.80%	6.66%	99.80%	6.14%	99.80%	5.31%	semiorthogonal	Yes
$d = 2, \tilde{d} = 4$	99.80%	8.15%	99.80%	7.89%	99.80%	7.15%	biorthogonal	Yes
$d = 3, \tilde{d} = 3$	100%	10.18%	100%	9.52%	100%	8.30%	biorthogonal	Yes

APPENDIX

A1. The expressions of the primal biorthogonal wavelets Ψ_J

For $d = 2, \tilde{d} = 2$ and $J \geq 2$, define

$$\psi^{2,0}(x) = \tfrac{5}{4}N_2(2x) - \tfrac{3}{2}N_2(2x - 1) - \tfrac{1}{2}N_2(2x - 2) + \tfrac{1}{2}N_2(2x - 3) + \tfrac{1}{4}N_2(2x - 4),$$

$$\psi^2(x) = -\tfrac{1}{4}N_2(2x + 2) - \tfrac{1}{2}N_2(2x + 1) + \tfrac{3}{2}N_2(2x) - \tfrac{1}{2}N_2(2x - 1) - \tfrac{1}{4}N_2(2x - 2).$$

Then $\Psi_j = 2^{\frac{j}{2}} \left\{ \psi^{2,0}(2^j x), \psi^2(2^j x - k) \Big|_{k=1}^{2^j - 2}, \psi^{2,0}(2^j(1 - x)) \right\}$.

For $d = 3, \tilde{d} = 3$ and $J \geq 3$, define

$$\psi^{3,0}(x) = -\tfrac{315}{256}b_1(x) + \tfrac{1365}{1024}N_3(2x) - \tfrac{81}{1024}N_3(2x - 1) - \tfrac{279}{512}N_3(2x - 2)$$

$$- \tfrac{33}{512}N_3(2x - 3) + \tfrac{135}{1024}N_3(2x - 4) + \tfrac{45}{1024}N_3(2x - 5),$$

$$\psi^{3,1}(x) = -\tfrac{15}{32}b_1(x) - \tfrac{15}{128}N_3(2x) + \tfrac{195}{128}N_3(2x - 1) - \tfrac{75}{64}N_3(2x - 2)$$

$$- \tfrac{13}{64}N_3(2x - 3) + \tfrac{27}{128}N_3(2x - 4) + \tfrac{9}{128}N_3(2x - 5),$$

$$\psi^3(x) = -\tfrac{3}{32}N_3(2x + 6) - \tfrac{9}{32}N_3(2x + 5) + \tfrac{7}{32}N_3(2x + 4) + \tfrac{45}{32}N_3(2x + 3)$$

$$-\frac{45}{32}N_3(2x+2) - \frac{7}{32}N_3(2x+1) + \frac{9}{32}N_3(2x) + \frac{3}{32}N_3(2x-1),$$

where $b_1(x)$ are the boundary scaling function given in Section 1.2. Then

$$\Psi_j = 2^{\frac{j}{2}}\left\{\psi^{3,0}(2^j x), \psi^{3,1}(3^j x), \psi^3(2^j x - k)\big|_{k=3}^{2^j-2}, \psi^{3,1}(2^j(1-x)), \psi^{3,0}(2^j(1-x))\right\}.$$

A2. The Riemann-Liouville fractional derivative of Φ_J

In the following, we only give the left fractional derivative of Φ_J for $d = 4$. Similarly for the other d. Define

$$b_1(x) := 3x_+ - \frac{9}{2}x_+^2 + \frac{7}{4}x_+^3 - 2(x-1)_+^3 + \frac{1}{4}(x-2)_+^3,$$

$$b_2(x) := \frac{3}{2}x_+^2 - \frac{11}{12}x_+^3 + \frac{3}{2}(x-1)_+^3 - \frac{3}{4}(x-2)_+^3 + \frac{(x-3)_+^3}{6},$$

Then $\Phi_J = 2^{\frac{J}{2}}\left\{b_1(2^J x), b_2(2^J x), N_4(2^J x - k)\big|_{k=0}^{2^J-4}, b_2(2^J(1-x)), b_1(2^J(1-x))\right\}.$

For $x_0 \geq 0, v \in \mathbb{N}$ and $k \in \mathbb{N}^+$, it is easy to check that

$$_0D_x^{1-\beta}(H(x-x_0)v(x)) = H(x-x_0)_{x_0}D_x^{1-\beta}v(x),$$

$$_aD_x^{1-\beta}(x-a)^v = \frac{\Gamma(v+1)(x-a)^{v+\beta-1}}{\Gamma(v+\beta)},$$

$$(b-ax)_+^k = (b-ax)^k + (-1)^{k-1}(ax-b)_+^k,$$

where $H(x)$ denotes the Heaviside function. Then for $b/a \geq 0$, one has

$$_0D_x^{1-\beta}(ax-b)_+^k = a^{1-\beta}\frac{\Gamma(k+1)}{\Gamma(k+\beta)}(ax-b)_+^{k+\beta-1},$$
$$_0D_x^{1-\beta}(b-ax)_+^k = (-1)^{k-1}{_0D_x^{1-\beta}}(ax-b)_+^k + \sum_{m=0}^k \binom{k}{m}a^k b^{k-m}\frac{m!(-1)^m}{\Gamma(m+\beta)}x_+^{m+\beta-1}. \quad \textbf{(1.52)}$$

Define

$$M_1(x) := {_0D_x^{1-\beta}}b_2(x) = \frac{1}{\Gamma(\beta+2)}\left(3x_+^{\beta+1} - \frac{11}{2(\beta+2)}x_+^{\beta+2}\right)$$

$$+\frac{1}{2\Gamma(\beta+3)}\left(18(x-1)_+^{\beta+2} - 9(x-2)_+^{\beta+2} + 2(x-3)_+^{\beta+2}\right),$$

$$M_2(x) := {}_0D_x^{1-\beta}N_4(x) = \frac{1}{\Gamma(\beta+3)}\sum_{i=0}^{4}\binom{4}{i}(-1)^i(x-i)_+^{\beta+2},$$

$$M_3(x,l) := {}_0D_x^{1-\beta}b_2(l-x) = \frac{-1}{\Gamma(\beta+2)}\left(3(x-l)_+^{\beta+1} + \frac{11}{2(\beta+2)}(x-l)_+^{\beta+2}\right)$$

$$+\frac{1}{2\Gamma(\beta+3)}\left(18(x-l+1)_+^{\beta+2} - 9(x-l+2)_+^{\beta+2} + 2(x-l+3)_+^{\beta+2}\right).$$

Then using formulaes (1.52) again, it holds that

$${}_0D_x^{1-\beta}\left(2^{J/2}\phi_{b_2}(2^Jx)\right) = 2^{J(3/2-\beta)}M_1(2^Jx),$$

$${}_0D_x^{1-\beta}\left(2^{J/2}N_4(2^Jx-k)\right) = 2^{J(3/2-\beta)}M_2(2^Jx-k),$$

$${}_0D_x^{1-\beta}\left(2^{J/2}b_2(2^J(1-x))\right) = 2^{J(3/2-\beta)}M_3(2^Jx,2^J).$$

The similar formulaes can also be derived for $2^{J/2}b_1(2^Jx)$ and $2^{J/2}b_1(2^J(x-1))$.

A3. The proof of $A_R = A_L(end: -1: 1, end: -1: 1)$ in Example 1

In fact, for $v(x)$ satisfying $v(0) = v(1) = 0$, there exists

$${}_0D_x^{2-\beta}v(x) = {}_0D_x^{-\beta}v''(x) + \frac{v'(0)}{\Gamma(\beta)}x^{\beta-1},$$

$${}_xD_1^{2-\beta}v(x) = {}_xD_1^{-\beta}v''(x) - \frac{v'(1)}{\Gamma(\beta)}(1-x)^{\beta-1}.$$

For $k = 0, 1, \cdots, 2^J - 4,$

$${}_0D_x^{(2-\beta)}N_4(2^Jx-k) = \frac{2^{2J}}{\Gamma(\beta)}\int_0^x (x-\xi)^{\beta-1}(N_4)''(2^J\xi-k)\mathrm{d}\xi$$

$$= \frac{2^{J(2-\beta)}}{\Gamma(\beta)}\int_0^{2^Jx-k}(2^Jx-\xi-k)^{\beta-1}(N_4)''(\xi)\mathrm{d}\xi,$$

$${}_xD_1^{2-\beta}N_4\left(2^Jx - (2^J-4-k)\right)$$

$$= \frac{2^{2J}}{\Gamma(\beta)}\int_x^1(\xi-x)^{\beta-1}(N_4)''(2^J-2^J\xi-k)\mathrm{d}\xi$$

$$= \frac{2^{J(2-\beta)}}{\Gamma(\beta)}\int_0^{2^J-2^Jx-k}(2^J-k-\xi-2^Jx)^{\beta-1}(N_4)''(\xi)\mathrm{d}\xi,$$

then

$$_0D_y^{2-\beta}N_4(2^Jy-k)|_{y=x} = {}_yD_1^{2-\beta}N_4(2^Jy-(2^J-4-k))|_{y=1-x}, x,y \in (0,1),$$

where the properties $N_4(x) = N_4(4-x)$ and $supp\, N_4(x) = (0,4)$ are used. Noticing that $supp\, b_2(x) = (0,3), b_2(0) = (b_2)'(0) = 0, supp\, b_1(x) = (0,2)$ and $b_1(0) = 0$, one can obtain

$$_0D_y^{2-\beta}b_2(2^Jy)\,|_{y=x} = {}_yD_1^{2-\beta}b_2\left(2^J(1-y)\right)|_{y=1-x}, \quad x,y \in (0,1),$$

$$_0D_y^{2-\beta}b_1(2^Jy)\,|_{y=x} = {}_yD_1^{2-\beta}b_1\left(2^J(1-y)\right)|_{y=1-x}, \quad x,y \in (0,1).$$

CONSENT FOR PUBLICATION

Not applicable.

CONFLICT OF INTEREST

The authors declare no conflict of interest, financial or otherwise.

ACKNOWLEDGEMENTS

The authors thank Xudong Wang and Lijing Zhao for their help in modifying the paper. This work was supported by the National Natural Science Foundation of China under Grant No. 11271173 and 11671182.

REFERENCES

[1]   B. Alpert, G. Beylkin, R. Coifman, and V. Rokhlin, "Wavelet-like bases for the fast solution of second-kind integral equations," *SIAM J. Sci. Comput.*, vol. 14, no. 1, pp. 159–184, Jan. 1993.

[2]   E. Barkai, R. Metzler, and J. Klafter, "From continuous time random walks to the fractional Fokker-Planck equation," *Rhys. Rev. E*, vol. 61, pp. 132–138, Jan. 2000.

[3]   W. Cai, and J.Z. Wang, "Adaptive multiresolution collocation methods for initial-boundary value problems of nonlinear PDEs," *SIAM J. Numer. Anal.*, vol. 33, no. 3, pp. 937–970, Jun. 1996.

[4]   H.F. Chan, and X.Q. Jin, *An Introduction to Iterative Toeplitz Solvers*. SIAM: Philadelphia, 2007.

[5]   C.K. Chui, and E. Quak, "Wavelets on a bounded interval," In: *Numerical Methods in Approximation Theory*, Basel: Birkhäuser, 1992, pp. 53–75.

[6]   A. Cohen, "Wavelet methods in numerical analysis," In: *Handbook of Numerical Analysis*, Amsterdam: Elsevier, 2000, pp. 417–711.

[7]   W.H. Deng, "Finite element method for the space and time fractional Fokker-Planck equation," *SIAM J. Numer. Anal.*, vol. 47, no. 1, pp. 204–226, Oct. 2008.

[8]   B. Dybiec, E. Gudowska-Nowak, and P. hänggi, "Lévy-Brownan motion on finite intervals: Mean first passage time analysis," *Phys. Rev. E*, vol. 73, no. 4, p. 046104, Apr. 2006.

[9]   V.J. Ervin and J.P. Roop, "Variational formulation for the stationary fractional advection dispersion equation," *Numer. Methods Partial Differential Equations*, vol. 22, no. 3, pp. 558–576, May 2006.

[10]   M. Holmström, "Solving hyperbolic PDEs using interpolating wavelets," *SIAM J. Sci. Comput.*, vol. 21, no. 2, pp. 405–420, Sep. 1999.

[11]   H. Jafari, S.A. Yousefi, M.A. Firoozjaee, S. Momani, and C.M. Khalique, "Application of Legendre wavelets for solving fractional differential equations," *Comput. Math. Appl.*, vol. 62, no. 3, pp. 1038–1045, Aug. 2011.

[12]   B.T. Jin, R. Lazarov, J. Pasciak, and W. Runadell, "Variational formulation of problems involving fractional order differential operators," *Math. Comp.*, vol. 84, pp. 2665–2700, Apr. 2015.

[13]   X. Li and C. Xu, "Existence and uniqueness of the weak solution of the space-time fractional diffusion equation and a spectral method approximation," *Commun. Comput. Phys.*, vol. 8, no. 5, pp. 1016–1051, Nov. 2010.

[14]   S. Lei and H. Sun, " A circulant preconditioner for fractional diffusion equations," *J. Comput. Phys.*, vol. 242, pp. 715–725, Jun. 2013.

[15]   F.R. Lin, S.W. Yang, and X.Q. Jin, "Preconditioned iterative methods for fractional diffusion equation," *J. Comput. Phys.*, vol. 256, pp. 109–117, Jan. 2014.

[16]   F. Mainardi, *Fractional Calculus and Waves in Linear Viscoelasticity: An Introduction to Mathematical Models*. Imperial College Press: London, 2010.

[17]   M.M. Meerschaert, and C. Tadjeran, "Finite difference approximations for fractional advection-dispersion flow equations," *J. Comput. Appl. Math.*, vol. 172, pp. 65–77, Nov. 2004.

[18]   M.M. Meerschaert, and A. Sikorskii, *Stochastic Models for Fractional Calculus*. De Gruyter: Berlin, 2012.

[19]   T. Moroney, and Q.Q. Yang, "A banded preconditioner for the two-sided, nonlinear space-fractional diffusion equation," *Comput. Math. Appl.*, vol. 66, no. 5, pp. 659–667, Sep. 2013.

[20]   H.K. Pang and H.W. Sun, "Multigrid method for fractional diffusion equations," *J. Comput. Phys.*, vol. 231, no. 2, pp. 693–703, Jan. 2012.

[21]   M. Primbs, *"Stabile biorthogonale Spline-Waveletbasen auf dem Intervall,"* PhD thesis, Universität Duisburg-Essen, Germany, 2006.

[22]   S. Dahlke, M. Fornasier, M. Primbs, T. Raasch, and M. Werner, "Nonlinear and adaptive frame approximation schemes for elliptic PDEs: Theory and numerical experiments," *Numer. Methods Partial Differential Equations*, vol. 25, no. 6, pp. 1366–1401, Nov. 2009.

[23]   M. Primbs, "New stable biorthogonal spline-wavelets on the interval," *Results. Math.*, vol. 57, no. 1, pp. 121-162, Mar. 2010.

[24]   A. Saadatmandia, M. Dehghanb, and M.R. Azizi, "The Sinc-Legendre collocation method for a class of fractional convection-diffusion equations with variable coefficients," *Commun. Nonlinear Sci. Numer. Simul.*, vol. 17, no. 11, pp. 4125- 4136, Nov. 2012.

[25]   U. Saeed and M. Rehman, "Haar wavelet-quasilinearization technique for fractional nonlinear differential equations," *Appl. Math. Comput.*, vol. 220, pp. 630–648, Sep. 2013.

[26]   W.Y. Tian, H. Zhou, and W.H. Deng, "A class of second order difference approximation for solving space fractional diffusion equations," *Math. Comp.*, vol. 84, pp. 1703–1727, Jan. 2015.

[27]   K. Urban, *Wavelet Methods for Elliptic Partial Differential Equations*. Oxford University Press: Oxford, 2009.

[28]   H. Wang, D.P. Yang, and S.F. Zhu, "Inhomogeneous dirichlet boundary-value problems of space-fractional diffusion equations and their finite element approximations," *SIAM J. Numer. Anal.*, vol. 52, no. 3, pp. 1292–1310, Jun. 2014.

[29]   L.F. Wang, Y.P. Ma, and Z.J. Meng, "Harr wavelet method for solving fractional partial differential equations numerically," *Appl. Math. Comput.*, vol. 227, pp. 66–76, Jan. 2014.

[30]   H. Wang, K.X. Wang, and T. Sircar, "A direct $O(N \log^2 N)$ finite difference method for fractional diffusion equations," *J. Comput. Phys.*, vol. 229, no. 21, pp. 8095–8104, Oct. 2010.

[31] B. J. West, "Fractional calculus in bioengineering," *Stat. Phys.,* Vol. 126, no. 6, pp. 1285–1286, Mar. 2007.

[32] Q.W. Xu and J.S. Hesthaven, "Discontinuous Galerkin method for fractional convection-diffusion equations," *SIAM J. Numer. Anal.,* vol. 52, no. 1, pp. 405–423, Feb. 2014.

[33] Q. Yang, I. Turner, F. Liu, and M. llic', "Novel numerical methods for solving the time-space fractional diffusion equation in two dimensions," *SIAM J. Sci. Comput.,* vol. 33, no. 3, pp. 1159–1180, May 2011.

[34] M. Zaslavsky, "Chaos, fractional kinetics, and anomalous transport," *Phys. Rep.,* vol. 371, no. 6, pp. 461–580, Dec. 2002.

[35] M. Zayernouri, and G.E. Karniadakis, "Fractional spectral collocation method," *SIAM J. Sci. Comput.,* vol. 36, no. 1, pp. A40–A62, Jan. 2014.

[36] Y.N. Zhang, Z.Z. Sun, and H.L. Liao, "Finite difference methods for the time fractional diffusion equation on non-uniform meshes," *J. Comput. Phys.,* vol. 265, pp. 195–210, May 2014.

A Survey on the Recent Results Regarding Maximum Principles for the Time-Fractional Diffusion Equations

Yuri Luchko[1,*] and Masahiro Yamamoto[2,3]

[1]Department of Mathematics, TU of Applied Sciences Berlin, Germany
[2]Graduate School of Mathematical Sciences, The University of Tokyo, Japan
[3]Research Center of Nonlinear Problems of Mathematical Physics, Peoples' Friendship University of Russia, Moscow, Russia

Abstract: In this chapter, a survey on the recent results regarding the maximum principles for the time-fractional diffusion equations is presented. We formulate the maximum principles for the time-fractional partial differential equations with both the Caputo fractional derivative and the Riemann-Liouville fractional derivative. Along with the single-term time-fractional differential equations, the multi-term equations and the equations of the distributed order are considered. We also discuss some important applications of the maximum principles for the time-fractional diffusion equations including a priori estimates for solutions of the initial-boundary-value problems for these equations and uniqueness of their solutions.

Keywords: Fractional derivatives, Time-fractional diffusion equation, Initial-boundary-value problems, Maximum principle, A priori estimates, Uniqueness of solutions.

AMS Subject Classification: 26A33, 35A05, 35B30, 35B50, 35C05, 35E05, 35L05, 45K05, 60E99.

2.1. INTRODUCTION

The maximum principles is a well known and widely applied tool in the theory of partial differential equations of elliptic and parabolic type (see *e.g.* the books [28] and [29] entirely devoted to the maximum principles and their applications). The maximum principles have a clear and straightforward physical background and provide important a priori information regarding solutions to the boundary- or initial-boundary-value problems for the partial differential equation of elliptic or parabolic type, respectively.

The fractional diffusion equation with the time-fractional derivative of order α between zero and one interpolates between the PDEs of elliptic and parabolic

*Corresponding author Yuri Luchko: Department of Mathematics, TU of Applied Sciences Berlin, Germany; Tel: +49 30 4504 5295; Fax: +49 30 4504 2011; E-mail: luchko@beuth-hochschule.de

Sachin Bhalekar (Ed.)

type. Thus it would be natural to expect that an appropriate maximum principle would be valid for this equation, too. Because the fractional derivatives are non-local operators, the standard proof technique for the maximum principles does not work and thus the maximum principles for the fractional differential equations remained unproved until very recently. The prehistory of the maximum principles for the fractional differential equations started with the papers [11] and [13], where some arguments related to a kind of a maximum principle have been used for analysis of the fractional diffusion equation. To the best of the authors' knowledge, a (weak) maximum principle for the time-fractional diffusion equation was formulated for the first time in the paper [21]. In the paper [21], the maximum principle was proved for a time-fractional diffusion equation that is obtained from the conventional diffusion equation by replacing the first-order time derivative by the Caputo fractional derivative of the order α ($0 < \alpha \leq 1$). After its publication, the maximum principles for the fractional differential equations and their applications became a popular topic under very intensive development. In the meantime, several papers devoted to the maximum principles for the fractional partial differential equations with different fractional derivatives have been published. It is worth mentioning that the maximum principles for the fractional differential equations have a series of important applications including a priori estimates for solutions of the corresponding initial-boundary-value problems and the uniqueness of their solutions. Say, in [20] the maximum principle proved in the paper [21] was applied to establish uniqueness and existence results for solutions of the initial-boundary-value problems for the time-fractional diffusion equation with the Caputo fractional derivative. In the paper [5], a maximum principle for the time-fractional diffusion equation with the Riemann-Liouville fractional derivative was derived and employed to show uniqueness of solutions to the initial-boundary-value problems both for linear and non-linear one-dimensional fractional diffusion equations with the Riemann-Liouville time-fractional derivative. In the linear case, these solutions were constructed in form of the Fourier series with respect to the eigenfunctions of the corresponding Sturm-Liouville eigenvalue problems.

In the theory of the fractional partial differential equations, not only single-term equations, but also multi-term equations and the distributed order equations are studied. A maximum principle for the multi-term time-fractional diffusion equation was proved in [18] for the case of the Caputo fractional derivatives and in [4] for the case of the Riemann-Liouville fractional derivatives. As to the time-fractional diffusion equations of the distributed order, the maximum principles for them were proved in [22] for the equations with the Caputo derivatives and in [3] for the equations with the Riemann-Liouville derivatives, respectively.

One of the most recent research topics in the theory of the fractional differential equations is the general time-fractional diffusion equation, which generalizes the single- and the multi-term time-fractional diffusion equations as well as the time-fractional diffusion equation of the distributed order. The general time-fractional diffusion equation contains a fractional derivative with a general (not necessarily power-law) kernel that was introduced and investigated in [12]. In [26], a maximum principle for the time-fractional diffusion equation with the general fractional derivative of the Caputo type was formulated and proved. As an application of the maximum principle, uniqueness of both the strong and the weak solutions to the initial-boundary-value problem for this equation was established. Existence of a suitably defined generalized solution to the initial-boundary-value problem for the general time-fractional diffusion equation with the homogeneous boundary conditions was proved in [26], too.

The publications mentioned above dealt with the weak maximum principles for the time-fractional diffusion equations. A kind of a strong maximum principle for the single-term time-fractional diffusion equation was proved in [15]. In a very recent paper [25], a strong maximum principle for this equation was proved for a weak solution and under weaker conditions compared to those formulated in [15]. A strong maximum principle for the time-fractional diffusion equations with multiple Caputo derivatives was established in [14]. As an application, uniqueness of solution to the problem of determining the temporal component of the source term was proved in [14].

In this chapter, we give a short survey of the results regarding the maximum principles for the time-fractional diffusion equations that were obtained by the authors of the chapter and their co-authors. The rest of the chapter is organized as follows: In the 2nd section, the maximum principles for the time-fractional diffusion equations with the Caputo fractional derivative are introduced and applied for analysis of the initial-boundary-value problems for these equations. The 3rd section is devoted to the maximum principles for the time-fractional diffusion equations with the Riemann-Liouville fractional derivatives and their applications. In the 4th section, we give an overview of some resent results regarding the maximum principle for the time-fractional diffusion equation with the general fractional derivative. Finally, in the last section, some conclusions and open problems are discussed.

2.2. TIME-FRACTIONAL DIFFUSION EQUATIONS WITH THE CAPUTO FRACTIONAL DERIVATIVE

In this section, we present some of the recent results regarding the maximum principles for the time-fractional diffusion equation with the Caputo fractional derivative. For the details and complete proofs we refer the reader to the papers [17-25].

We start with a single-term time-fractional diffusion equation over an open bounded domain $\Omega \times (0,T), \Omega \subset \mathbb{R}^n$. This equation is obtained from the conventional diffusion equation by replacing the first-order time derivative by the fractional Caputo derivative of the order α $(0 < \alpha \leq 1)$:

$$(D_t^\alpha u)(t) = L_x(u) + F(x,t), \tag{2.1}$$

$$0 < \alpha \leq 1, (x,t) \in \Omega_T := \Omega \times (0,T), \Omega \subset \mathbb{R}^n,$$

where

$$L_x(u) = \text{div}\,(p(x)\,\text{grad}\,u) - q(x)u, \tag{2.2}$$

$$p \in C^1(\overline{\Omega}), q \in C(\overline{\Omega}), p(x) > 0, q(x) \geq 0, x \in \overline{\Omega}, \tag{2.3}$$

the fractional derivative D_t^α is defined in the Caputo sense

$$(D_t^\alpha f)(t) = (I^{1-\alpha} f')(t), 0 < \alpha \leq 1, \tag{2.4}$$

I^α being the fractional Riemann-Liouville integral

$$(I^\alpha f)(t) := \begin{cases} \frac{1}{\Gamma(\alpha)} \int_0^t (t-\tau)^{\alpha-1} f(\tau)\,d\tau, 0 < \alpha < 1, \\ f(t), \alpha = 0, \end{cases} \tag{2.5}$$

and the domain Ω with the boundary $\partial\Omega$ is open and bounded in $\mathbb{R}^n$.

If $\alpha = 1$, the equation (2.1) coincides with the conventional diffusion equation, so that in the further discussions we focus on the case $0 < \alpha < 1$. Following [21], for the equation (2.1) we consider the following initial-boundary-value problem with the initial condition

$$u|_{t=0} = u_0(x), x \in \overline{\Omega} \tag{2.6}$$

and the Dirichlet boundary condition

$$u|_{\partial\Omega} = v(x,t), (x,t) \in \partial\Omega \times [0,T], \tag{2.7}$$

where $\partial\Omega$ denotes as usual the boundary of the domain Ω and $\overline{\Omega}$ is its closure.

The (weak) maximum principle for the equation (2.1) will be formulated for the strong solutions that are defined as follows:

Definition 2.1. *A strong solution of the initial-boundary-value problem (2.1), (2.6), (2.7) is defined as a function $u = u(x,t)$ defined in the domain $\bar{\Omega}_T := \bar{\Omega} \times [0,T]$ that belongs to the space $CW_T(\Omega) := C(\bar{\Omega}_T) \cap W_t^1((0,T)) \cap C_x^2(\Omega)$ and satisfies both the equation (2.1) and the initial and boundary conditions (2.6)-(2.7). By $W_t^1((0,T))$ the space of functions f that satisfy the inclusions $f \in C^1((0,T])$ and $f' \in L((0,T))$ is denoted.*

The proof of the maximum principle for the equation (2.1) presented in the paper [21] is based on an extremum principle for the Caputo fractional derivative (2.4).

Theorem 2.1. *Let a function $f \in W_t^1((0,T)) \cap C([0,T])$ attain its maximum over the interval $[0,T]$ at the point $\tau = t_0, t_0 \in (0,T]$. Then the Caputo fractional derivative of the function f is non-negative at the point t_0 for any $\alpha, 0 < \alpha < 1$, i.e.,*

$$(D_t^\alpha f)(t_0) \geq 0, 0 < \alpha < 1. \tag{2.8}$$

Let us note that the inequality (2.8) was proved for the first time in the paper [21]. Later on, a stronger inequality, namely,

$$(D_t^\alpha f)(t_0) \geq \frac{t_0^{-\alpha}}{\Gamma(1-\alpha)}(f(t_0) - f(0)), 0 < \alpha < 1 \tag{2.9}$$

was derived in [2] (see also [5]). However, for the proof of the maximum principle for the time-fractional diffusion equation (2.1), the inequality (2.8) is sufficient.

In the paper [21], the following weak maximum principle for the generalized time-fractional diffusion equation (2.1) was proved:

Theorem 2.2. *Let a function $u \in CW_T(\Omega)$ be a solution of the time-fractional diffusion equation (2.1) in the domain Ω_T and $F(x,t) \leq 0, (x,t) \in \Omega_T$.*
Then either $u(x,t) \leq 0, (x,t) \in \bar{\Omega}_T$ or the function u attains its positive maximum on the bottom or back-side parts $S_\Omega^T := (\bar{\Omega} \times \{0\}) \cup (\partial\Omega \times [0,T])$ of the boundary of the domain Ω_T, i.e.,

$$u(x,t) \leq \max\{0, \max_{(x,t)\in S_\Omega^T} u(x,t)\}, \forall(x,t) \in \bar{\Omega}_T. \tag{2.10}$$

Substituting $-u$ instead of u in the results stated above, the minimum principle can be obtained:

Theorem 2.3. *Let a function $u \in CW_T(\Omega)$ be a solution of the time-fractional diffusion equation (1) in the domain Ω_T and $F(x,t) \geq 0, (x,t) \in \Omega_T$.*

Then either $u(x,t) \geq 0, (x,t) \in \bar{\Omega}_T$ or the function u attains its negative minimum on the part S_Ω^T of the boundary of the domain Ω_T, i.e.,

$$u(x,t) \geq \min\{0, \min_{(x,t)\in S_\Omega^T} u(x,t)\}, \forall (x,t) \in \bar{\Omega}_T. \tag{2.11}$$

One of the most important applications of the maximum principle is uniqueness of solution of the initial-boundary-value problem (2.1), (2.6)-(2.7). To prove the uniqueness, some a priori estimates of the solution norm are needed.

Theorem 2.4. *Let u be a strong solution to the initial-boundary-value problem (2.1), (2.6)-(2.7) and F belong to the space $C(\bar{\Omega}_T)$ and $M := \| F \|_{C(\bar{\Omega}_T)}$.*

Then the following estimate of the solution norm holds true:

$$\| u \|_{C(\bar{\Omega}_T)} \leq \max\{M_0, M_1\} + \frac{T^\alpha}{\Gamma(1+\alpha)} M, \tag{2.12}$$

where

$$M_0 := \| u_0 \|_{C(\bar{\Omega})}, M_1 := \| v \|_{C(\partial\Omega\times[0,T])}. \tag{2.13}$$

The results formulated in Theorem 2.4 are then used to prove the following important theorem:

Theorem 2.5. *The initial-boundary-value problem (2.1), (2.6)-(2.7) possesses at most one strong solution. This solution continuously depends on the data given in the problem in the sense that if*

$$\| F - \tilde{F} \|_{C(\bar{\Omega}_T)} \leq \varepsilon,$$

$$\| u_0 - \tilde{u}_0 \|_{C(\bar{\Omega})} \leq \varepsilon_0, \| v - \tilde{v} \|_{C(\partial\Omega\times[0,T])} \leq \varepsilon_1,$$

then the estimate

$$\| u - \tilde{u} \|_{C(\bar{\Omega}_T)} \leq \max\{\varepsilon_0, \varepsilon_1\} + \frac{T^\alpha}{\Gamma(1+\alpha)} \varepsilon \tag{2.14}$$

holds true for the corresponding strong solutions u and $\tilde{u}$.

Uniqueness of the strong solution immediately follows from the fact that the homogeneous problem (2.1), (2.6)-(2.7), *i.e.*, the problem with $F \equiv 0$, $u_0 \equiv 0$, and $v \equiv 0$ has only one strong solution, namely, $u(x,t) \equiv 0, (x,t) \in \overline{\Omega}_T$. The last statement is a simple consequence of the norm estimate (2.12) established in Theorem 2.4. The same estimate is used to prove the inequality (2.14). This time, it is applied to the function $u - \tilde{u}$ that is a strong solution to the initial-boundary-value problem (2.1), (2.6)-(2.7) with the functions $F - \tilde{F}$, $u_0 - \tilde{u}_0$, and $v - \tilde{v}$ instead of the functions F, u_0, and v, respectively.

As to the question regarding existence of a solution to the initial-boundary-value problem (2.1), (2.6)-(2.7), it was answered in [20] for the case of a generalized solution in the Vladimirov sense ([32]) and in [17] for the case of the strong solution.

Recently, the initial-boundary-value problems for the time-fractional diffusion equations were considered in the fractional Sobolev spaces (see [9]). In the paper [25], a maximum principle for a suitably defined weak solution from the fractional Sobolev space was established. Moreover, in [25], the maximum principle was proved for a time-fractional diffusion equation without any restrictions on the sign of the coefficient by the unknown function. In what follows, we shorty present some of the results obtained in [25].

Let $0 < \alpha < 1$. We consider the following initial-boundary-value problem for the single-term time-fractional diffusion equation

$$(D_t^\alpha u)(t) = \sum_{i,j=1}^n \partial_i \left(a_{ij}(x) \partial_j u(x,t) \right) + c(x)u(x,t) + F(x,t),$$
$$x \in \Omega \subset \mathbb{R}^n, t > 0, \tag{2.15}$$

$$u(x,t) = 0, \ x \in \partial\Omega, t > 0, \tag{2.16}$$

$$u(x,0) = a(x), \ x \in \Omega \tag{2.17}$$

in a bounded domain Ω with a smooth boundary $\partial\Omega$. In what follows, we always suppose that $a_{ij} \equiv a_{ji} \in C^1(\overline{\Omega})$, $1 \leq i,j \leq n$, $c \in C(\overline{\Omega})$, and there exists a constant $\mu_0 > 0$ such that $\sum_{i,j=1}^n a_{ij}(x)\xi_i\xi_j \geq \mu_0 \sum_{i=1}^n \xi_i^2$ for all $x \in \overline{\Omega}$ and $\xi_1, \ldots, \xi_n \in \mathbb{R}$, *i.e.*, that the spatial differential operator in equation (2.15) is uniformly elliptic one. As to the fractional derivative D_t^α, it is an extension of the Caputo fractional derivative to the closure $H_\alpha(0,T)$ of $_0C^1[0,T] := \{u \in C^1[0,T]; u(0) = 0\}$ in the fractional Sobolev space $H^\alpha(\Omega)$ (see [9] for details).

Thus we interpret the problem (2.15)-(2.17) as the fractional diffusion equation (2.15) subject to the inclusions

$$\begin{cases} u(\cdot,t) \in H_0^1(\Omega), t > 0, \\ u(x,\cdot) - a(x) \in H_\alpha(0,T), x \in \Omega. \end{cases} \tag{2.18}$$

According to the results presented in [9], for any initial condition $a \in L^2(\Omega)$ and any source function $F \in L^2(\Omega \times (0,T))$, there exists a unique weak solution $u_{a,F} \in L^2(0,T;H^2(\Omega) \cap H_0^1(\Omega))$ to the problem (2.15)-(2.17) satisfying the condition $u_{a,F} - a \in H_\alpha(0,T; L^2(\Omega))$. For $\frac{1}{2} < \alpha < 1$, in view of the Sobolev embedding, the solution u belongs to the function space $C([0,T]; L^2(\Omega))$ and satisfies the initial condition (2.17) in L^2-sense.

The weak maximum principle for the equation (2.15) says that the inequalities $F(x,t) \geq 0, (x,t) \in \Omega \times (0,T)$ and $a(x) \geq 0, x \in \Omega$ yield   the   inequality $u(x,t) \geq 0, (x,t) \in \Omega \times (0,T)$, where u is a weak solution to the initial-boundary value problem (2.15)-(2.17) defined as in [9].

A maximum principle for the strong solution to the initial-boundary value problem (2.15)-(2.17) can be proved following the method employed in the paper [21] under the assumption that

$$c(x) \leq 0, x \in \overline{\Omega}.$$

In the paper [21], the case of an inhomogeneous Dirichlet boundary condition

$$u|_{\partial\Omega\times(0,T)} = b(x,t)$$

was also considered, but in [25], for the sake of technical simplicity, the homogeneous boundary condition $u|_{\partial\Omega\times(0,T)} = 0$ was assumed (although the method from [25] can be applied in the case of an inhomogeneous boundary condition, too).

The main result of the paper [25] is the weak maximum principle for the equation (2.15) with any $c \in C(\overline{\Omega})$ without the non-negativity condition $c(x) \leq 0, x \in \overline{\Omega}$. As it is known, the weak maximum principle for the partial differential equations of the parabolic type is valid without any condition on the sign of the coefficient $c = c(x)$ (see *e.g.*, [28] or [33]). The proof of this fact uses the properties of the exponential function and reduces the case of a bounded coefficient $c = c(x)$ to the case of a non-positive coefficient. This technique does not work in the case of the fractional diffusion equation (2.15) and thus in [25], a different proof method was employed.

Let us now denote the solution to the initial-boundary value problem (2.15)-(2.17) defined as in [9] by $u_{a,F}$ and formulate the main results derived in [25].

Theorem 2.6. *Let $a \in L^2(\Omega)$ and $F \in L^2(\Omega \times (0,T))$. If $F(x,t) \geq 0$ a.e. (almost everywhere) in $\Omega \times (0,T)$ and $a(x) \geq 0$ a.e. in Ω, then $u_{a,F}(x,t) \geq 0$ a.e. in $\Omega \times (0,T)$.*

Let us mention that in the paper [21] the maximum principle for the strong solution was stated pointwise (*i.e.*, for all points from $\overline{\Omega} \times [0,T]$) under the assumption that $c(x) \leq 0, x \in \Omega$. In Theorem 2.6, the maximum principle is formulated for the weak solution. Its proof presented in [25] is based on the fixed point theorem and the property that the solution mapping $\{a,F\} \to u_{a,F}$ preserves its sign on the set of the weak solutions.

Theorem 2.6 immediately yields the following comparison property:

Corollary 2.1. *Let $a_1, a_2 \in L^2(\Omega)$ and $F_1, F_2 \in L^2(\Omega \times (0,T))$ satisfy the inequalities $a_1(x) \geq a_2(x)$ a.e. in Ω and $F_1(x,t) \geq F_2(x,t)$ a.e. in $\Omega \times (0,T)$, respectively. Then $u_{a_1,F_1}(x,t) \geq u_{a_2,F_2}(x,t)$ a.e. in $\Omega \times (0,T)$.*

Corollary 2.1 can be employed among other things to remove the condition $c(x) \leq 0$ from the formulation of the strong maximum principle for the fractional diffusion equation that was derived in [15].

Let us now fix a source function $F = F(x,t) \geq 0$ and an initial condition $a = a(x) \geq 0$ and denote by $u_c = u_c(x,t)$ the weak solution to the initial-boundary-value problem (2.15)-(2.17) with the coefficient $c = c(x)$. Then the following comparison property with respect to the coefficient $c(x)$ is valid:

Theorem 2.7. *Let $c_1, c_2 \in C(\overline{\Omega})$ satisfy the inequality $c_1(x) \geq c_2(x)$ in Ω. Then $u_{c_1}(x,t) \geq u_{c_2}(x,t)$ in $\Omega \times (0,T)$.*

One of the useful consequences from Theorem 2.7 is given in the following statement (strong maximum principle):

Corollary 2.2. *Let $n \leq 3$ ($\Omega \subset \mathbb{R}^n$), the initial condition $a \in L^2(\Omega)$ satisfy the inequality $a(x) \geq 0$ a.e. in Ω, a do not vanish identically and the source function be identically equal to zero, i.e., $F(x,t) \equiv 0, x \in \Omega, t > 0$.*

Then the weak solution u to the initial-boundary-value problem (2.15)-(2.17) satisfies the inclusion $u \in C((0,T]; C(\overline{\Omega}))$ and for each $x \in \Omega$ the set $\{t > 0; u(x,t) \leq 0\}$ is at most a finite set.

For the proofs of Theorems 2.6 and 2.7 and Corollaries 2.1 and 2.2, we refer the reader to the paper [25].

As mentioned in the introduction, the maximum principles for both the multi-term fractional diffusion equation and the fractional diffusion equation of the distributed order have been also derived. In the paper [18], a weak maximum principle of the type formulated in Theorem 2.2 was proved for the multi-term fractional diffusion equation

$$P(D_t)u = L_x(u) + F(x,t), (x,t) \in \Omega_T := \Omega \times (0,T), \Omega \subset \mathbb{R}^n, \quad \textbf{(2.19)}$$

where the operator L_x is defined by (2.2)-(2.3) and the multi-term fractional derivative $P(D_t)$ by the equation

$$P(D_t) = D_t^\alpha + \sum_{i=1}^m \lambda_i D_t^{\alpha_i}, \quad \textbf{(2.20)}$$

$$0 < \alpha_m < \cdots < \alpha_1 < \alpha \leq 1, 0 \leq \lambda_i, i = 1, \cdots, m, m \in \mathbb{N}$$

with the fractional derivatives in the Caputo sense. In [14], a version of a strong maximum principle of the type formulated in Corollary 2.2 was derived for the multi-term fractional diffusion equation

$$P(D_t)u = \sum_{i,j=1}^n \partial_i(a_{ij}(x) \partial_j u(x,t)) + c(x)u(x,t) + F(x,t), x \in \Omega \subset \mathbb{R}^n, t > 0 \quad \textbf{(2.21)}$$

with the homogeneous boundary condition

$$u(x,t) = 0, \ x \in \partial\Omega, t > 0, \quad \textbf{(2.22)}$$

and an inhomogeneous initial condition

$$u(x,0) = a(x), \ x \in \Omega. \quad \textbf{(2.23)}$$

In the equation (2.21), the operator $P(D_t)$ is defined by (2.20), $0 < \alpha < 1$, Ω is an open bounded domain with a smooth boundary $\partial\Omega$ and the spatial differential operator in equation (2.21) is uniformly elliptic one. Moreover, the condition $c(x) \leq 0, x \in \overline{\Omega}$ was supposed to be valid.

To finalize this section, let us mention that a weak maximum principle of the type formulated in Theorem 2.2 was proved in [22] for the fractional diffusion equation of the distributed order

$$\mathbb{D}_t^{w(\alpha)} u = L_x(u) + F(x,t), (x,t) \in \Omega_T := \Omega \times (0,T), \Omega \subset \mathbb{R}^n, \qquad (2.24)$$

where the operator L_x is given by (2.2)-(2.3). The distributed order derivative $\mathbb{D}_t^{w(\alpha)}$ is defined by the equation

$$\mathbb{D}_t^{w(\alpha)} f(t) = \int_0^1 (D_t^\alpha f)(t) \, w(\alpha) \, d\alpha, \qquad (2.25)$$

where D_t^α denotes the Caputo fractional derivative (2.4), $w: [0,1] \to \mathbb{R}$ is a continuous function, not identically equal to zero on the interval $[0,1]$, the conditions

$$0 \leq w(\alpha), w \not\equiv 0, \alpha \in [0,1], \int_0^1 w(\alpha) \, d\alpha = W > 0 \qquad (2.26)$$

hold true, and the domain Ω with the boundary $\partial\Omega$ is open and bounded in $\mathbb{R}^n$.

In the 4th Section, a maximum principle for a fractional diffusion equation with the general fractional derivative of the Caputo type will be presented. This equation contains the single- and the multi-term fractional diffusion equations as well as the diffusion equation of the distributed order as its particular cases.

2.3. TIME-FRACTIONAL DIFFUSION EQUATIONS WITH THE RIEMANN-LIOUVILLE FRACTIONAL DERIVATIVE

The maximum principles for the single- and multi-term time-fractional diffusion equations as well as for the time-fractional equation of the distributed order with the Riemann-Liouville fractional derivatives were derived in [3-5]. In this section, we present a short overview of these results. For the details and complete proofs, we refer the readers to the original papers [3-5].

The Riemann-Louville fractional derivative is defined as follows:

$$(D_{0+}^\alpha f)(t) = \frac{d}{dt} (I^{1-\alpha} f)(t), 0 < \alpha \leq 1, \qquad (2.27)$$

I^α being the Riemann-Liouville fractional integral (2.5). For the theory of the Riemann-Liouville fractional integrals and derivatives, the reader is referred to *e.g.*, [30]. It is well known that for $0 < \alpha < 1$ and $f \in C^1[0,T]$ the Riemann-Liouville fractional derivative (2.5) is connected with the Caputo fractional derivative (2.4) by the relation (see *e.g.*, [23])

$$(D_{0+}^\alpha f)(t) = \frac{t^{-\alpha}}{\Gamma(1-\alpha)} f(0) + (D_t^\alpha f)(t). \tag{2.28}$$

Still the theory of the fractional differential equations with the Caputo fractional derivatives is very different from one for the equations with the Riemann-Liouville fractional derivatives and it is not always possible to "translate" the results obtained for the equations with the Caputo derivatives to the case of the equations with the Riemann-Liouville derivatives and vice versa. In particular, the maximum principles for the fractional differential equations with the Riemann-Liouville fractional derivatives were derived independently of ones for the equations with the Caputo fractional derivatives.

The main component of the proofs of the maximum principles for the time-fractional diffusion equations with the Riemann-Liouville fractional derivatives is an estimate for the Riemann-Liouville fractional derivative of a function at its maximum point that was established in [2] (see also [3]-[5]) and is given in the following theorem:

Theorem 2.8. *Let a function* $f \in C^1[0,T]$ *attain its maximum at a point* $t_0 \in (0,T]$ *and* $0 < \alpha < 1$. *Then the inequality*

$$(D_{0+}^\alpha f)(t_0) \geq \frac{t_0^{-\alpha}}{\Gamma(1-\alpha)} f(t_0) \tag{2.29}$$

holds true.

Let us start with some results for a family of the one-dimensional time-fractional diffusion equations

$$(D_{0+}^\alpha u)(t) = L(u) + f(x,t,u), 0 < \alpha \leq 1, (x,t) \in \Omega_T = (0,\ell) \times (0,T], \tag{2.30}$$

where L is a second-order differential operator defined by

$$L(u) = a(x,t)u_{xx} + b(x,t)u_x + c(x,t)u \tag{2.31}$$

and the fractional derivative D_{0+}^α is the Riemann-Liouville derivative (2.27). In what follows, we assume that the functions $a = a(x,t), b = b(x,t)$, and $c =$

$c(x,t)$ from the formula (2.31) are continuous functions on $\overline{\Omega_T} = [0, \ell] \times [0, T]$ and $a(x,t) > 0, (x,t) \in \overline{\Omega_T}$. The function $f = f(x, t, u)$ is supposed to be continuous with respect to the variables x and t and smooth with respect to the variable u.

The equation (2.30) is a non-linear equation. If the source term f does not depend on the unknown function u, the equation becomes linear.

Let us introduce the notation

$$P_\alpha(u) = (D_{0+}^\alpha u)(t) - L(u), 0 < \alpha < 1. \tag{2.32}$$

The the equation (2.30) can be rewritten in the form

$$P_\alpha(u) = f(x, t, u), (x, t) \in \Omega_T. \tag{2.33}$$

The operator P_α can be interpreted as a fractional differential operator of the parabolic type. In what follows, we formulate a weak and a strong maximum principles for this operator that are analogous to those for the partial differential operators of the parabolic type.

A weak maximum principle for the operator P_α is formulated in the following theorem:

Theorem 2.9. *Let a function* $u \in C([0, T]; C^2[0, \ell]) \cap C^1([0, T]; C[0, \ell])$ *satisfy the inequality* $P_\alpha(u(x, t)) \leq 0, (x, t) \in \Omega_T$ *and* $c(x, t) \leq 0, (x, t) \in \Omega_T$. *Then the inequality*

$$\max_{(x,t)\in\overline{\Omega_T}} u(x, t) \leq \max_{(x,t)\in S_1 \cup S_2 \cup S_3} \{u(x, t), 0\} \tag{2.34}$$

holds true, where

$$S_1 = \{x = 0, 0 \leq t \leq T\}, S_2 = \{0 \leq x \leq \ell, t = 0\}, \tag{2.35}$$

$$S_3 = \{x = \ell, 0 \leq t \leq T\}, S_4 = \{0 \leq x \leq \ell, t = T\}. \tag{2.36}$$

In the proof of the weak maximum principle given in [5], a stronger statement than the inequality (2.34) was in fact deduced, namely, that a function u that fulfills the conditions of Theorem 2.9 cannot attain its positive maximum at a point $(x_0, t_0) \in \Omega_T$.

As a consequence, a kind of a strong maximum principle for the fractional differential operator P_α was formulated and proved in [5].

Theorem 2.10. *Let a function* $u \in C([0,T]; C^2[0,\ell]) \cap C^1([0,T]; C[0,\ell])$ *satisfy the equation* $P_\alpha(u) = 0, (x,t) \in \Omega_T$ *and* $c(x,t) \le 0, (x,t) \in \Omega_T$.

If the function u *attains its maximum and its minimum at some points that belong to* Ω_T, *then the function* u *is a constant, more precisely* $u(x,t) = 0, (x,t) \in \overline{\Omega_T}$.

Let us now consider some applications of the maximum principles to analysis of the initial-boundary-value problems for linear and non-liner fractional diffusion equations with the Riemann-Liouville fractional derivatives. We start with the liner fractional diffusion equation in the form

$$P_\alpha(u) = (D_{0+}^\alpha u)(t) - L(u) = 0, 0 < \alpha < 1, (x,t) \in \Omega_T \qquad (2.37)$$

along with with the initial and boundary conditions

$$u(x,0) = g(x), 0 \le x \le \ell, \qquad (2.38)$$

$$u(x,t) = h(x,t), (x,t) \in S_1 \cup S_3, \qquad (2.39)$$

where S_1 and S_3 are defined as in (2.35)-(2.36).

The following result is a direct consequence of the weak maximum principle (Theorem 2.9):

Theorem 2.11. *Let* $u \in C([0,T]; C^2[0,\ell]) \cap C^1([0,T]; C[0,\ell])$ *fulfill the inequality* $P_\alpha(u) \le 0, (x,t) \in \Omega_T$ *and* $c(x,t) \le 0, (x,t) \in \Omega_T$. *If* u *satisfies the initial and boundary conditions (2.38) and (2.39),* $g(x) \le 0, 0 \le x \le \ell$, *and* $h(x,t) \le 0, (x,t) \in S_1 \cup S_3$, *then* $u(x,t) \le 0, (x,t) \in \overline{\Omega_T}$.

Let us now consider a non-linear fractional diffusion equation in the form

$$Q_\alpha(u) = (D_{0+}^\alpha u)(t) - (a(x,t)u_{xx} + b(x,t)u_x) = f(x,t,u), 0 < \alpha < 1, \quad (2.40)$$

where $(x,t) \in \Omega_T$, $a(x,t) > 0, (x,t) \in \Omega_T$ and D_{0+}^α stands for the Riemann-Liouville fractional derivative.

Under some suitable conditions on the non-linear term f, the maximum principle for the operator P_α leads to an uniqueness result for the initial-boundary-value problem (2.40), (2.38)-(2.39) for the non-linear fractional diffusion equation.

Theorem 2.12. *Let $f = f(x,t,u)$ be a smooth and non-increasing function with respect to the variable u. Then the initial-boundary-value problem (2.40), (2.38)-(2.39) for the non-linear fractional diffusion equation possesses at most one solution $u = u(x,t), (x,t) \in \overline{\Omega_T}$ in the function space*

$$C([0,T]; C^2[0,\ell]) \cap C^1([0,T]; C[0,\ell]).$$

A uniqueness result for the initial-boundary-value problem (2.37), (2.38)-(2.39) for the linear fractional diffusion equation is a simple consequence from Theorem 2.12. More precisely, let us consider the initial-boundary-value problem (2.40), (2.38)-(2.39):

$$P_\alpha(u) = (D_{0+}^\alpha u)(t) - (a(x,t)u_{xx} + b(x,t)u_x + c(x,t)u) = g(x,t). \quad (2.41)$$

Because the function $f(x,t,u) = c(x,t)u + g(x,t)$ is a smooth and non-increasing function with respect to the variable u, we can apply Theorem 2.12 to obtain the following result:

Corollary 2.3. *The initial-boundary-value problem (2.41), (2.38)-(2.39) for the linear fractional diffusion equation possesses at most one solution in the function space $C([0,T]; C^2[0,\ell]) \cap C^1([0,T]; C[0,\ell])$.*

Another direct consequence of the weak maximum principle is the following stability result for the solutions of the initial-boundary-value problems (2.41), (2.38)-(2.39) for the linear fractional diffusion equation:

Theorem 2.13. *Let $u_1, u_2 \in C([0,T]; C^2[0,\ell]) \cap C^1([0,T]; C[0,\ell])$ be two solutions of the fractional diffusion equation (2.41) that satisfy the same boundary condition (2.39) and the initial conditions $u_1(x,0) = g_1(x)$, $u_2(x,0) = g_2(x)$, $0 \le x \le \ell$, respectively, and $c(x,t) \le 0, (x,t) \in \Omega_T$. Then the inequality*

$$\max_{(x,t)\in\Omega_T} |u_1(x,t) - u_2(x,t)| \le \max_{x\in[0,\ell]} |g_1(x) - g_2(x)|$$

holds true.

Let us now consider a family of the multi-term time-fractional diffusion equations in the form

$$\aleph(D_t)u = L_x(u) + f(x,t,u), (x,t) \in \Omega \times (0,T], \tag{2.42}$$

where Ω is a bounded open domain in $\mathbb{R}^n$ with a smooth boundary $\partial\Omega$ (for example, of C^2 class), L_x is a uniformly elliptic operator defined by

$$L_x(u) = \sum_{i,j=1}^n a_{ij}(x,t)\frac{\partial^2 u}{\partial x_i \partial x_j} + \sum_{i=1}^n b_i(x,t)\frac{\partial u}{\partial x_i}, \tag{2.43}$$

$$\aleph(D_t) = D_{0+}^\alpha + \sum_{i=1}^m \lambda_i D_{0+}^{\alpha_i}, 0 < \alpha_m < \cdots < \alpha_1 < \alpha \le 1, 0 \le \lambda_i, i = 1, \cdots, m, m \in \mathbb{N}, \tag{2.44}$$

and the fractional derivative D_{0+}^α is the Riemann-Liouville derivative (2.27). The function f is assumed to be continuous with respect to the variables x and t, and smooth with respect to the variable u.

To prove a maximum principle for the equation (2.42), we need an estimate for the Riemann-Liouville fractional derivative of a function at its maximum point, which is similar to the result stated in Theorem 2.8. However, the condition $f \in C^1[0,T]$ from Theorem 2.8 is too restrictive and in fact not necessary for existence of the Riemann-Liouville fractional derivative. Say, the function $f(t) = \sqrt{t} - t$ attains its maximum at the point $t_0 = \frac{1}{4} \in (0,1]$ and is α-differentiable on the interval $(0,1]$ for $0 < \alpha < 1$, but f does not belong to the space $C^1[0,1]$. It is a natural question whether the inequality from Theorem 2.8 holds true for such functions, *i.e.*, for the functions from the space $CW^1[0,T] := C[0,T] \cap W^1(0,T]$. We recall that by $W^1(0,T]$ we denote the space of functions $f \in C^1(0,T]$ such that $f' \in L^1[0,T]$. An answer to this question is given in the following two lemmas.

Lemma 2.1. *Let $f \in CW^1[0,T]$ attain its maximum at a point $t_0 \in (0,T)$ and $0 < \alpha < 1$. Then the inequality*

$$(D_{0+}^\alpha f)(t_0) \ge \frac{t_0^{-\alpha}}{\Gamma(1-\alpha)} f(t_0)$$

holds true.

Lemma 2.2. *Let $f \in CW^1[0,T]$. Then $(D_{0+}^\alpha f)(t)$ is continuous for $t \in (0,T]$.*

Combining the results of Lemmas 2.1 and 2.2 we have the following statement.

Theorem 2.14. *Let $f \in CW^1[0,T]$ attain its maximum at a point $t_0 \in (0,T]$ and $0 < \alpha < 1$. Then the inequality*

$$(D_{0+}^\alpha f)(t_0) \geq \frac{t_0^{-\alpha}}{\Gamma(1-\alpha)} f(t_0) \qquad (2.45)$$

holds true.

The inequality (2.45) can be employed among other things to derive a weak and a strong maximum principles for the fractional differential operator P_α of the parabolic type that is defined by the formula

$$P_\alpha(u) = \aleph(D_t)u(x,t) - L_x(u) - c(x,t)u$$

$$= \aleph(D_t)u(x,t) - \sum_{i,j=1}^n a_{ij}(x,t)\frac{\partial^2 u}{\partial x_i \partial x_j} - \sum_{i=1}^n b_i(x,t)\frac{\partial u}{\partial x_i} - c(x,t)u, \quad (2.46)$$

where the function $c = c(x,t)$ is bounded on $\bar{\Omega} \times [0,T]$.

We start with a weak maximum principle that is formulated in the following theorem.

Theorem 2.15. *Let a function $u \in C([0,T]; C^2(\bar{\Omega})) \cap W^1((0,T]; L^2(\Omega))$ satisfy the inequality $P_\alpha(u(x,t)) \leq 0, (x,t) \in \Omega \times (0,T]$ and $c(x,t) \leq 0, (x,t) \in \Omega \times (0,T]$. Then the inequality*

$$\max_{(x,t)\in \bar{\Omega}\times[0,T]} u(x,t) \leq \max\{\max_{x\in\bar{\Omega}} u(x,0), \max_{(x,t)\in \partial\Omega\times[0,T]} u(x,t), 0\} \qquad (2.47)$$

holds true.

A version of the strong maximum principle for the fractional differential operator P_α is given in the next theorem.

Theorem 2.16. *Let a function $u \in C([0,T]; C^2(\bar{\Omega})) \cap W^1((0,T]; L^2(\Omega))$ satisfy the equation $P_\alpha(u) = 0, (x,t) \in \Omega \times (0,T]$ and $c(x,t) \leq 0, (x,t) \in \Omega \times (0,T]$.*

If the function u attains its maximum and its minimum at some points that belong to $\Omega \times (0,T]$ then the function u is a constant, more precisely, $u(x,t) = 0, (x,t) \in \Omega \times (0,T]$.

The maximum principles for the multi-term fractional differential operator P_α defined by (2.46) can be applied for analysis of the initial-boundary-value problems for the multi-term time-fractional diffusion equations. Let us start with the initial-boundary-value problem for the linear equation of the form

$$P_\alpha(u) = \aleph(D_t)u(x,t) - L_x(u) - c(x,t)u = 0, (x,t) \in \Omega \times (0,T], \quad \textbf{(2.48)}$$

$$u(x,0) = g(x), x \in \Omega, \quad \textbf{(2.49)}$$

$$u(x,t) = h(x,t), (x,t) \in \partial\Omega \times (0,T], \quad \textbf{(2.50)}$$

where the function $c = c(x,t)$ is bounded on $\bar{\Omega} \times [0,T]$ and the operators L_x and $\aleph(D_t)$ are defined by (2.43) and (2.44), respectively.

The weak maximum principle formulated in Theorem 2.15 automatically leads to the following result:

Theorem 2.17. *Let a function* $u \in C([0,T]; C^2(\bar{\Omega})) \cap W^1((0,T]; L^2(\Omega))$ *fulfill the inequality* $P_\alpha(u) \leq 0, (x,t) \in \Omega \times (0,T]$ *and* $c(x,t) \leq 0, (x,t) \in \Omega \times (0,T]$. *If* u *satisfies the initial and boundary conditions (2.49) and (2.50), respectively,* $g(x) \leq 0, x \in \Omega$, *and* $h(x,t) \leq 0, (x,t) \in \partial\Omega \times (0,T]$, *then* $u(x,t) \leq 0, (x,t) \in \bar{\Omega} \times [0,T]$.

Theorem 2.17 is used, among other things, for analysis of a non-linear multi-term time-fractional reaction-diffusion equation in the form

$$Q_\alpha(u) = \aleph(D_t)u(x,t) - L_x(u) = f(x,t,u), (x,t) \in \Omega \times (0,T]. \quad \textbf{(2.51)}$$

Indeed, under some suitable conditions on the non-linear term f, the maximum principle for the operator P_α or, more precisely, its implication given in Theorem 2.17, leads to a uniqueness result for the initial-boundary-value problem (2.51), (2.49)-(2.50) for the multi-term fractional diffusion equation that is formulated in the following theorem.

Theorem 2.18. *Let* $f = f(x,t,u)$ *be a smooth and non-increasing function with respect to the variable* u.
Then the initial-boundary-value problem (2.51), (2.49)-(2.50) for the multi-term fractional diffusion equation possesses at most one solution $u = u(x,t), (x,t) \in \bar{\Omega} \times [0,T]$ *in the function space* $C([0,T]; C^2(\bar{\Omega})) \cap W^1((0,T]; L^2(\Omega))$.

Theorem 2.18 applied to the function $f(x,t,u) = c(x,t)u + g(x,t)$ (if $c(x,t) \leq 0, (x,t) \in \Omega \times (0,T]$, the function f is a smooth and non-increasing function with respect to the variable u) immediately leads to the following result:

Theorem 2.19. *Let* $c(x,t) \leq 0, (x,t) \in \Omega \times (0,T]$. *The initial-boundary-value problem (2.51), (2.49)-(2.50) for the linear multi-term fractional diffusion equation*

$$\aleph(D_t)u(x,t) - L_x(u) - c(x,t)u = g(x,t) \tag{2.52}$$

possesses at most one solution in the function space $C([0,T]; C^2(\bar{\Omega})) \cap W^1((0,T]; L^2(\Omega))$.

A stability result of the type formulated in Theorem 2.13 is valid also for the initial-boundary-value problem (2.51), (2.49)-(2.50) for the linear multi-term fractional diffusion equation (see [4] for details).

Another important point related to the weak maximum principle is the condition $c(x,t) \leq 0, (x,t) \in \Omega \times (0,T]$ posed on the function $c = c(x,t)$ that is required in Theorem 2.17. In the case of the partial differential equations of parabolic type ($\alpha = 1$ and $\lambda_i = 0, i = 1, \cdots, m$ in (2.46)), the corresponding weak maximum principle is valid without any conditions on the sign of the function $c = c(x,t)$ (of course, $c = c(x,t)$ has to be bounded on $\Omega \times (0,T]$). It turns out that also in the case of the multi-term fractional diffusion equation the condition $c(x,t) \leq 0, (x,t) \in \Omega \times (0,T]$ can be removed from the formulation of the weak maximum principle under an additional condition posed on the function $a = a(x,t)$. In particular, we refer to [4], where this result was proved for the following initial-boundary-value problem for the one-dimensional multi-term fractional diffusion equation

$$P_\alpha^* u = \aleph(D_t)u(x,t) - (a(x,t)u_{xx} + b(x,t)u_x + c(x,t)u), (x,t) \in (0,\ell) \times (0,T], \tag{2.53}$$

$$u(x,0) = g(x), x \in (0,\ell) \tag{2.54}$$

$$u(0,t) = h_1(t), t \in (0,T] \tag{2.55}$$

$$u(\ell,t) = h_2(t), t \in (0,T], \tag{2.56}$$

where $a(x,t) > 0, (x,t) \in [0,\ell] \times [0,T]$.

Theorem 2.20. *Let a function* $u \in C([0,T]; C^2([0,\ell])) \cap W^1((0,T]; L^2(0,\ell))$ *satisfy the inequality* $P_\alpha^*(u) \le 0, (x,t) \in (0,\ell) \times (0,T]$ *and the initial and boundary conditions (2.54) and (2.55), respectively, the function* $c = c(x,t)$ *be bounded on* $(0,\ell) \times (0,T]$, *and the inequalities* $g(x) \le 0, 0 \le x \le \ell$ *and* $h_1(t) \le 0, h_2(t) \le 0, t \in (0,T]$ *hold true. Moreover, let* $b_1 \le \min_{(x,t)\in[0,\ell]\times[0,T]} b(x,t)$, $c_1 \le \sup_{(x,t)\in[0,\ell]\times[0,T]} c(x,t)$ *and the inequality*

$$a(x,t) + b_1 x + \frac{1}{2}c_1 x^2 > 0 \tag{2.57}$$

hold true for all $(x,t) \in [0,\ell] \times [0,T]$, *where* $\ell < \frac{1}{\sqrt{\beta^*}}$ *and*

$$\beta^* = \frac{1}{2} \sup_{(x,t)\in[0,\ell]\times[0,T]} \left\{ \frac{c(x,t)}{a(x,t)+b_1 x+\frac{1}{2}c_1 x^2} \right\}. \tag{2.58}$$

Then the inequality $u(x,t) \le 0, (x,t) \in [0,\ell] \times [0,T]$ *holds true.*

Finally, let us present some results for the fractional diffusion equations of the distributed order with the Riemann-Liouville fractional derivatives in the form (see [3] for details):

$$D_t^{\omega(\alpha)} u = L_x(u) - q(x)u + g(x,t), (x,t) \in \Omega \times (0,T], \tag{2.59}$$

where Ω is a bounded open domain in $\mathbb{R}^n$ with a smooth boundary $\partial\Omega$, L_x is a uniformly elliptic operator defined by (2.43), and $q \in C(\overline{\Omega}), q(x) \ge 0, x \in \overline{\Omega}$.

The distributed order derivative $D_t^{\omega(\alpha)}$ is defined as follows:

$$\left(D_t^{\omega(\alpha)} f \right)(t) = \int_0^1 (D_{0+}^\alpha f)(t)\omega(\alpha)d\alpha, t > 0, \tag{2.60}$$

where D_{0+}^α is the Riemann-Liouville fractional derivative (2.27) and $\omega = \omega(\alpha)$ is a non-negative weight function that is continuous and not identically equal to zero on the interval $[0,1]$, *i.e.*,

$$\omega \in C([0,1]), 0 \le \omega(\alpha), \alpha \in [0,1], \int_0^1 \omega(\alpha)d\alpha = W > 0.$$

As in the case of the multi-term fractional diffusion equation, the inequality (2.45) given in Theorem 2.14 is employed to derive a weak and a strong maximum principles for the fractional differential operator of parabolic type that is defined by the formula

$$P_{\omega(\alpha)}(u) = D_t^{\omega(\alpha)}u - L_x(u) + q(x)u$$

$$= D_t^{\omega(\alpha)}u - \sum_{i,j=1}^{n} a_{ij}(x,t)\frac{\partial^2 u}{\partial x_i \partial x_j} - \sum_{i=1}^{n} b_i(x,t)\frac{\partial u}{\partial x_i} + q(x)u. \quad \textbf{(2.61)}$$

The formulations of the weak and the strong maximum principles are given in the following theorems:

Theorem 2.21. *Let a function* $u \in C([0,T]; C^2(\bar{\Omega})) \cap W^1((0,T]; L^2(\Omega))$ *satisfy the inequality*

$$P_{\omega(\alpha)}(u(x,t)) \leq 0, (x,t) \in \Omega \times (0,T].$$

Then the inequality

$$\max_{(x,t)\in \bar{\Omega}\times[0,T]} u(x,t) \leq \max\{\max_{x\in\bar{\Omega}} u(x,0), \max_{(x,t)\in \partial\Omega\times[0,T]} u(x,t), 0\} \quad \textbf{(2.62)}$$

holds true.

Theorem 2.22. *Let a function* $u \in C([0,T]; C^2(\bar{\Omega})) \cap W^1((0,T]; L^2(\Omega))$ *satisfy the equation* $P_{\omega(\alpha)}(u(x,t)) = 0, (x,t) \in \Omega \times (0,T].$

If the function u *attains its maximum and its minimum at some points that belong to* $\Omega \times (0,T]$ *then the function* u *is a constant, more precisely,* $u(x,t) \equiv 0, (x,t) \in \Omega \times (0,T].$

The maximum principles from Theorems 2.21 and 2.22 can be used for analysis of the initial-boundary-value problems for the linear and non-linear time-fractional diffusion equations of distributed order. Let us start with the linear equation in the form

$$Q_{\omega(\alpha)}(u) = D_t^{\omega(\alpha)}u - L_x(u) + q(x)u - g(x,t) = 0, (x,t) \in \Omega \times (0,T], \quad \textbf{(2.63)}$$

$$u(x,0) = h(x), x \in \Omega, \quad \textbf{(2.64)}$$

$$u(x,t) = r(x,t), (x,t) \in \partial\Omega \times [0,T]. \quad \textbf{(2.65)}$$

In what follows, we suppose that the functions $g = g(x,t)$, $h = h(x)$, and $r = r(x,t)$ are continuous on $\overline{\Omega}_T = \overline{\Omega} \times [0,T]$, $\overline{\Omega}$, and $\partial\Omega \times [0,T]$, respectively.

The weak maximum principle allows to obtain some a priori estimates for the solution u:

Theorem 2.23. *Let a function* $u \in C([0,T]; C^2(\overline{\Omega})) \cap W^1((0,T]; L^2(\Omega))$ *be a solution to the initial-boundary-value problem (2.63)-(2.65). Then the solution norm estimate*

$$||u||_{C(\overline{\Omega}_T)} \le \max\{R, H\} + \frac{G\, T_M}{W_\Gamma}$$

holds true with $R = ||r||_{C(\partial\Omega \times [0,T])}$, $H = ||h||_{C(\overline{\Omega})}$, $G = ||g||_{C(\overline{\Omega}_T)}$, $W_\Gamma = \int_0^1 \frac{w(\alpha)}{\Gamma(1-\alpha)}\, d\alpha > 0$, *and* $T_M = \max\{T, 1\}$.

The following important uniqueness and stability results are direct consequences of Theorem 2.23.

Theorem 2.24. *The initial-boundary-value problem (2.63)-(2.65) for the distributed order fractional diffusion equation possesses at most one solution in the function space* $C([0,T]; C^2(\overline{\Omega})) \cap W^1((0,T]; L^2(\Omega))$.

Theorem 2.25. *Let* $u_1, u_2 \in C([0,T]; C^2(\overline{\Omega})) \cap W^1((0,T]; L^2(\Omega))$ *be two solutions to the distributed order fractional diffusion equation (2.63) that satisfy the same boundary condition (2.65) and the initial conditions* $u_1(x,0) = h_1(x), u_2(x,0) = h_2(x), x \in \Omega$, *respectively. Then the norm inequality*

$$||u_1 - u_2||_{C(\overline{\Omega}_T)} \le ||h_1 - h_2||_{C(\overline{\Omega})} \tag{2.66}$$

holds true.

Finally, let us consider the initial-boundary-value problems for the nonlinear fractional diffusion equation of the distributed order

$$NQ_{\omega(\alpha)}(u) = D_t^{\omega(\alpha)} u - L_x(u) + q(x)u - F(x,t,u) = 0 \tag{2.67}$$

with the initial and boundary conditions (2.64)-(2.65). The operator $NQ_{\omega(\alpha)}$ can be represented in the form

$$NQ_{\omega(\alpha)}(u) = P_{\omega(\alpha)}(u) - F(x,t,u), \tag{2.68}$$

where $P_{\omega(\alpha)}$ is given by (2.61). The weak maximum principle for the operator $NQ_{\omega(\alpha)}$ is formulated in the following theorem:

Theorem 2.26. *Let a function $u \in C([0,T]; C^2(\bar{\Omega})) \cap W^1((0,T]; L^2(\Omega))$ fulfill the inequality $NQ_{\omega(\alpha)}(u) \leq 0, (x,t) \in \Omega \times (0,T]$ and $F(x,t,u) \leq 0, (x,t) \in \Omega \times (0,T]$. If the function u satisfies the initial and boundary conditions (2.64) and (2.65), respectively, with $h(x) \leq 0, x \in \Omega$, and $r(x,t) \leq 0, (x,t) \in \Omega \times (0,T]$, then $u(x,t) \leq 0, (x,t) \in \bar{\Omega}_T$.*

The next results are concerned with stability, uniqueness, and comparison of the solutions to the initial-boundary-value problem (2.67), (2.64)-(2.65) under some suitable conditions posed on the non-linear part F.

Theorem 2.27. *Let $u_1, u_2 \in C([0,T]; C^2(\bar{\Omega})) \cap W^1((0,T]; L^2(\Omega))$ be two solutions to the distributed order fractional diffusion equation (2.67) that satisfy the same boundary condition (2.65) and the initial conditions $u_1(x,0) = h_1(x), u_2(x,0) = h_2(x), x \in \Omega$, respectively. If $F = F(x,t,u)$ is a smooth and non-increasing function with respect to the variable u, then the inequality*

$$||u_1 - u_2||_{C(\bar{\Omega}_T)} \leq ||h_1 - h_2||_{C(\bar{\Omega})} \tag{2.69}$$

holds true.

Theorem 2.28. *Let $F = F(x,t,u)$ be a smooth and non-increasing function with respect to the variable u. Then the initial-boundary value problem (2.67), (2.64)-(2.65) for the non-linear distributed order fractional diffusion equation possesses at most one solution $u = u(x,t), (x,t) \in \bar{\Omega}_T$ in the function space $C([0,T]; C^2(\bar{\Omega})) \cap W^1((0,T]; L^2(\Omega))$.*

Theorem 2.29. *Let $u_1, u_2 \in C([0,T]; C^2(\bar{\Omega})) \cap W^1((0,T]; L^2(\Omega))$ be two solutions to the initial-boundary-value problem (2.67), (2.64)-(2.65) for the distributed order fractional diffusion equation with the non-linear parts $F = F_1(x,t,u)$ and $F = F_2(x,t,u)$, respectively.*

Let $F_1 = F_1(x,t,u)$ be a smooth and non-increasing function with respect to the variable u and the inequality $F_1(x,t,u) \leq F_2(x,t,u)$ hold true for all $(x,t) \in \bar{\Omega}_T$ and $u \in C([0,T]; C^2(\bar{\Omega})) \cap W^1((0,T]; L^2(\Omega))$. Then

1. $u_1(x,t) \le u_2(x,t), (x,t) \in \overline{\Omega}_T$.

2. In the case $T_M \le W_\Gamma$, the norm inequality

$$||u_1 - u_2||_{C(\overline{\Omega}_T)} \le ||F_1 - F_2||_{C(\overline{\Omega})} \qquad (2.70)$$

holds true, where $T_M = max\{T, 1\}$ and $W_\Gamma = \int_0^1 \frac{w(\alpha)}{\Gamma(1-\alpha)} d\alpha > 0$.

2.4. TIME-FRACTIONAL DIFFUSION EQUATIONS WITH THE GENERAL FRACTIONAL DERIVATIVE

In this section, we present some results regarding a maximum principle for a general time-fractional diffusion equation. This equation generalizes the single- and the multi-term time-fractional diffusion equations as well as the time-fractional diffusion equation of the distributed order (see the very recent paper [26] for details).

Following [12], we define the general fractional derivative of the Caputo type in the form

$$(\mathbb{D}^C_{(k)} f)(t) = \int_0^t k(t - \tau) f'(\tau) \, d\tau \qquad (2.71)$$

and the general fractional derivative of the Riemann-Liouville type in the form

$$(\mathbb{D}^{RL}_{(k)} f)(t) = \frac{d}{dt} \int_0^t k(t - \tau) f(\tau) \, d\tau, \qquad (2.72)$$

where k is a nonnegative locally integrable function. For an absolutely continuous function f with the inclusion $f' \in L_1^{loc}(\mathbb{R}_+)$, we easily obtain the relation

$$(\mathbb{D}^C_{(k)} f)(t) = \frac{d}{dt} \int_0^t k(t - \tau) f(\tau) \, d\tau - k(t) f(0) = (\mathbb{D}^{RL}_{(k)} f)(t) - k(t) f(0) \qquad (2.73)$$

between the Caputo and Riemann-Liouville types of the general fractional derivatives. Let us mention that in the paper [12] the general fractional derivative was introduced in form (2.73) that is valid for a wider class of functions (in particular, for absolutely continuous functions) compared to the definition (2.71) that additionally requires the inclusion $f' \in L_1^{loc}(\mathbb{R}_+)$.

Setting

$$k(\tau) = \frac{\tau^{-\alpha}}{\Gamma(1-\alpha)}, 0 < \alpha < 1 \tag{2.74}$$

in the formulas (2.71) and (2.72), we obtain the conventional Caputo and Riemann-Liouville fractional derivatives, respectively. Other important particular cases of (2.71) and (2.72) are given by

$$k(\tau) = \sum_{k=1}^{n} a_k \frac{\tau^{-\alpha_k}}{\Gamma(1-\alpha_k)}, 0 < \alpha_1 < \cdots < \alpha_n < 1 \tag{2.75}$$

and

$$k(\tau) = \int_0^1 \frac{\tau^{-\alpha}}{\Gamma(1-\alpha)} \, d\rho(\alpha), \tag{2.76}$$

where ρ is a Borel measure on $[0,1]$. They correspond to the multi-term derivatives and the derivative of the distributed order, respectively.

In the paper [12], Kochubei introduced a special class of general fractional derivatives in the form (2.73) with the kernel functions k satisfying the following conditions:

C1) The Laplace transform $\tilde{k}$ of k,

$$\tilde{k}(p) := \int_0^{\infty} k(t) \, e^{-pt} \, dt,$$

exists for all $p > 0$,

C2) $\tilde{k}(p)$ is a Stiltjes function,

C3) $\tilde{k}(p) \to 0$ and $p\tilde{k}(p) \to \infty$ as $p \to \infty$,

C4) $\tilde{k}(p) \to \infty$ and $p\tilde{k}(p) \to 0$ as $p \to 0$.

For the general fractional derivatives with the kernels satisfying the conditions C1)-C4) the following results were derived in the paper [12]:

(A) For any $\lambda > 0$, the initial value problem for the relaxation equation

$$(\mathbb{D}_{(k)}^C u)(t) = -\lambda u(t), t > 0 \tag{2.77}$$

with the initial condition

$$u(0) = 1 \tag{2.78}$$

has a unique solution $u_\lambda = u_\lambda(t)$ that belongs to the class $C^\infty(\mathbb{R}_+)$ and is a completely monotone function, *i.e.*,

$$(-1)^n u_\lambda^{(n)}(t) \geq 0, t > 0, n = 0,1,2 \ldots . \tag{2.79}$$

(B) There exists a completely monotone function $\kappa = \kappa(t)$ with the property

$$(k * \kappa)(t) = \int_0^t k(t - \tau)\kappa(\tau)\, d\tau \equiv 1, t > 0. \tag{2.80}$$

(C) For $f \in L_1^{loc}(\mathbb{R}_+)$, the relation

$$(\mathbb{D}_{(k)}^C \mathbb{I}_{(k)} f)(t) = f(t) \tag{2.81}$$

holds true, where the general fractional integral $\mathbb{I}_{(k)}$ is defined by the formula

$$(\mathbb{I}_{(k)} f)(t) = \int_0^t \kappa(t - \tau) f(\tau)\, d\tau. \tag{2.82}$$

In what follows, we deal with a maximum principle and its applications to analysis of the fractional diffusion equation with the general fractional derivative of the Caputo type in the form

$$(\mathbb{D}_{(k)}^C u(x,\cdot))(t) = D_2(u) + D_1(u) - q(x)u(x,t) + F(x,t), (x,t) \in \Omega \times (0,T], \tag{2.83}$$

where $q \in C(\bar{\Omega})$ and $q(x) \geq 0$ for $x \in \bar{\Omega}$, D_1 is a first-order and D_2 is a second-order spatial differential operators, respectively:

$$D_1(u) = \sum_{i=1}^n b_i(x)\frac{\partial u}{\partial x_i}, D_2(u) = \sum_{i,j=1}^n a_{ij}(x)\frac{\partial^2 u}{\partial x_i\, \partial x_j}. \tag{2.84}$$

Moreover we assume that D_2 is a uniformly elliptic differential operator.

In this section, we consider the initial-boundary-value problems for the equation (2.83) with the initial condition

$$u(x,t)|_{t=0} = u_0(x), x \in \bar{\Omega} \tag{2.85}$$

and the boundary conditions

$$u(x,t)|_{(x,t)\in\partial\Omega\times(0,T]} = v(x,t). \tag{2.86}$$

A solution to this problem in the strong sense is defined as follows:

Definition 2.2. *The space of functions* $u = u(x,t), (x,t) \in \bar{\Omega} \times [0,T]$ *that satisfy the inclusions* $u \in C(\bar{\Omega} \times [0,T])$, $u(\cdot,t) \in C^2(\Omega)$ *for any* $t > 0$, *and* $\partial_t u(x,\cdot) \in C(0,T] \cap L_1(0,T)$ *for any* $x \in \Omega$ *will be denoted by* $S(\Omega,T)$.

Definition 2.3. *A function* $u \in S(\Omega,T)$ *satisfying both the equation (2.83) and the initial and boundary conditions (2.85)-(2.86) is called a strong solution of the initial-boundary-value problem (2.83), (2.85)-(2.86) for the general time-fractional diffusion equation.*

To derive some a priori estimates in the uniform norm of the strong solution to the initial-boundary-value problem (2.83), (2.85)-(2.86) and then to show its uniqueness, the following assumptions posed on the kernel k of the general fractional derivatives (2.71) and (2.72) are supposed to hold true:

K1) $k \in C^1(\mathbb{R}_+) \cap L_1^{loc}(\mathbb{R}_+)$,

K2) $k(\tau) > 0$ and $k'(\tau) < 0$ for $\tau > 0$,

K3) $k(\tau) = o(\tau^{-1}), \tau \to 0$.

Let us note that the conditions C1)-C4) mentioned before are not needed for validity of the weak maximum principle for the general diffusion equation (2.83). However if the condition C3) is fulfilled, then it follows from the Feller-Karamata Tauberian theorem for the Laplace transform (see [8]) that the condition K3) is satisfied, too.

As in the previous sections, to prove the maximum principle for the general diffusion equation (2.83), some estimates of the general fractional derivative of a function f at its maximum point are needed. They are given in the following theorem.

Theorem 2.30. *Let the conditions K1)-K3) be fulfilled, a function* $f \in C([0,T])$ *attain its maximum over the interval* $[0,T]$ *at the point* $t_0, t_0 \in (0,T]$, *and* $f' \in C((0,T]) \cap L_1(0,T)$.
Then the following inequalities hold true:

$$(\mathbb{D}_{(k)}^{RL} f)(t_0) \geq k(t_0) f(t_0), \tag{2.87}$$

$$(\mathbb{D}^C_{(k)}f)(t_0) \geq k(t_0)(f(t_0) - f(0)) \geq 0. \tag{2.88}$$

Let us note here that for the conventional Riemann-Liouville and Caputo fractional derivatives, the inequalities (2.87) and (2.88) take the form

$$(\mathbb{D}^\alpha_{0+}f)(t_0) \geq \frac{t_0^{-\alpha}}{\Gamma(1-\alpha)} f(t_0) \tag{2.89}$$

and

$$(\mathbb{D}^\alpha f)(t_0) \geq \frac{t_0^{-\alpha}}{\Gamma(1-\alpha)} (f(t_0) - f(0)) \geq 0. \tag{2.90}$$

Now we present a formulation of a weak maximum principle for the operator

$$\mathbb{P}_{(k)}(u) := (\mathbb{D}^C_{(k)}u)(t) - D_2(u) - D_1(u) + q(x)u(x,t). \tag{2.91}$$

Theorem 2.31. *Let the conditions K1)-K3) be fulfilled and a function* $u \in S(\Omega, T)$ *satisfy the inequality*

$$\mathbb{P}_{(k)}(u) \leq 0, (x,t) \in \Omega \times (0,T]. \tag{2.92}$$

Then the following weak maximum principle holds true:

$$\max_{(x,t)\in \bar{\Omega}\times[0,T]} u(x,t) \leq \max\{\max_{x\in\bar{\Omega}} u(x,0), \max_{(x,t)\in \partial\Omega\times[0,T]} u(x,t), 0\}. \tag{2.93}$$

The weak maximum principle can be applied, among other things, to prove some a priori estimates for the strong solutions of the initial-boundary-value problem (2.83), (2.85)-(2.86) for the general time-fractional diffusion equation.

Theorem 2.32. *Let the conditions C1)-C4) and K1)-K3) be fulfilled and* u *be a strong solution of the initial-boundary-value problem (2.83), (2.85)-(2.86).*
Then the following estimate in the uniform norm holds true:

$$\| u \|_{C(\bar{\Omega}\times[0,T])} \leq \max\{M_0, M_1\} + M f(T), \tag{2.94}$$

where

$$M_0 = \| u_0 \|_{C(\bar{\Omega})}, M_1 = \| v \|_{C(\partial\Omega\times[0,T])}, M = \| F \|_{C(\Omega\times[0,T])}, \tag{2.95}$$

and

$$f(t) = \int_0^t \kappa(\tau)\, d\tau, \tag{2.96}$$

the function κ being defined by (2.80).

The norm estimates given in Theorem 2.32 can be applied to prove the following important result concerning uniqueness of the strong solution and its continuous dependence on problem data.

Theorem 2.33. *The initial-boundary-value problem (2.83), (2.85)-(2.86) possesses at most one strong solution.*

This solution continuously depends on the problem data in the sense that if u and $\tilde{u}$ are strong solutions to the problems with the sources functions F and $\tilde{F}$ and the initial and boundary conditions u_0 and $\tilde{u}_0$ and v and $\tilde{v}$, respectively, and

$$\| F - \tilde{F} \|_{C(\bar{\Omega}\times[0,T])} \le \varepsilon,$$

$$\| u_0 - \tilde{u}_0 \|_{C(\bar{\Omega})} \le \varepsilon_0, \| v - \tilde{v} \|_{C(\partial\Omega\times[0,T])} \le \varepsilon_1,$$

then the following norm estimate holds true:

$$\| u - \tilde{u} \|_{C(\bar{\Omega}\times[0,T])} \le \max\{\varepsilon_0, \varepsilon_1\} + \varepsilon\, f(T), \tag{2.97}$$

the function f being defined by (2.96).

In the rest of this section, we deal with a weak solution to the initial-boundary-value problem (2.83), (2.85)-(2.86) (generalized solution in the sense of Vladimirov [32]) that is defined as follows:

Definition 2.4. *Let* $F_k \in C(\bar{\Omega} \times [0,T])$, $u_{0k} \in C(\bar{\Omega})$ *and* $v_k \in C(\partial\Omega \times [0,T])$, $k = 1,2,\ldots$ *be three sequences of functions that satisfy the following conditions:*

W1) There exist the functions F, u_0, and v, such that

$$\| F_k - F \|_{C(\bar{\Omega}\times[0,T])} \to 0 \ \ as \ k \to \infty, \tag{2.98}$$

$$\| u_{0k} - u_0 \|_{C(\bar{\Omega})} \to 0 \ \ as \ k \to \infty, \tag{2.99}$$

$$\| v_k - v \|_{C(\partial\Omega\times[0,T])} \to 0 \ \ as \ k \to \infty. \tag{2.100}$$

W2) For any $k = 1, 2, \ldots$ *there exists a strong solution* $u_k = u_k(x, t)$ *to the initial-boundary-value problem*

$$u_k|_{t=0} = u_{0k}(x), x \in \bar{\Omega}, \tag{2.101}$$

$$u_k|_{\partial\Omega\times(0,T]} = v_k(x, t) \tag{2.102}$$

for the general time-fractional diffusion equation

$$\left(\mathbb{D}^C_{(k)} u_k(x, \cdot)\right)(t) = D_2(u_k) + D_1(u_k) - q(x)u_k(x, t) + F_k(x, t),$$
$$(x, t) \in \Omega \times (0, T]. \tag{2.103}$$

Then the function $u \in C(\bar{\Omega} \times [0, T])$ *defined by*

$$\| u_k - u \|_{C(\bar{\Omega}\times[0,T])} \to 0 \ as \ k \to \infty \tag{2.104}$$

is called a weak solution to the initial-boundary-value problem (2.83), (2.85)-(2.86) *for the general time-fractional diffusion equation.*

In [26], the correctness of Definition 2.4 was shown. As a limit of a sequence of the continuous functions in the uniform norm, a weak solution to the problem (2.83), (2.85)-(2.86) is a continuous function, not a distribution. However, the weak solution is not required to be a smooth function from the function space $S(\Omega, T)$ that the strong solution has to belong to.

On the other hand, any strong solution is evidently also a weak solution. As we have seen above, a strong solution, if it exists, is unique. Let us consider the question if a weak solution is unique, too.

If the problem (2.83), (2.85)-(2.86) possesses a weak solution, then the functions F, u_0 and v from the problem formulation have to belong to the spaces $C(\bar{\Omega} \times [0, T]), C(\bar{\Omega})$ and $C(\partial\Omega \times [0, T])$, respectively, as the limits of sequences of the continuous functions in the uniform norm.

Moreover, the estimate (2.94) for the strong solutions of the problem (2.83), (2.85)-(2.86) holds true for the weak solutions, too. To show this, we let k tend to $+\infty$ in the inequality

$$\| u_k \|_{C(\bar{\Omega}_T)} \leq \max\{M_{0k}, M_{1k}\} + M_k f(T), \tag{2.105}$$

$$M_{0k} := \| u_{0k} \|_{C(\bar{\Omega})}, M_{1k} := \| v_k \|_{C(\partial\Omega \times [0,T])}, M_k := \| F_k \|_{C(\bar{\Omega} \times [0,T])}$$

that is valid for each $k = 1, 2 \ldots$.

The estimate (2.94) for the weak solutions is then employed to prove the following important theorem.

Theorem 2.34. *The initial-boundary-value problem (2.83), (2.85)-(2.86) for the general time-fractional diffusion equation possesses at most one weak solution in the sense of Definition 2.4. The weak solution - if it exists - continuously depends on the data given in the problem in the sense of the estimate (2.97).*

2.5. TIME-SPACE FRACTIONAL DIFFUSION EQUATIONS

Finally let us mention some recent publications dealing with the maximum principles for other kinds of the fractional partial differential equations. A maximum principle for the multi-term time-space fractional diffusion equations with the modified Riesz space-fractional derivative in the Caputo sense was introduced and employed in [34]. In [16], a maximum principle for the multi-term time-space variable-order fractional differential equations with the Riesz-Caputo fractional derivatives was proved and applied for analysis of these equations. In [1], a maximum principle for the time-space fractional diffusion equation with the fractional Laplace operator and the Riemann-Liouville time-fractional derivative was derived. The maximum principles for a time-space fractional diffusion equation with the Riemann-Liouville time-fractional derivative and the fractional Laplace operator were derived in [10] both for the strong and for the weak solutions. For results regarding the maximum principles for the space-fractional elliptic and parabolic partial differential equations with the fractional Laplace operator, we refer the reader to [6] and to the literature mentioned there.

We conclude this section with a sketch of our latest paper [24]. In our discussions, a framework of the theory of semigroups (see, *e.g.*, [27] and [31]) will be employed. Let X be a Hilbert space over $\mathbb{R}$ with the scalar product $(\cdot, \cdot)$. For $0 < \alpha, \beta < 1$, we consider the following evolution equation in the Hilbert space X:

$$D_t^{\alpha} u(t) = -(-A)^{\beta} u \ in \ X, t > 0 \qquad (2.106)$$

and

$$u(0) = a \in X. \qquad (2.107)$$

Here, for simplicity, we assume that the operator A is self-adjoint, has compact resolvent, and $(-\infty, 0] \subset \rho(-A)$, $\rho(-A)$ being the resolvent of $-A$. For example, these conditions are fulfilled for $X = L^2(\Omega)$, $A = \Delta$ and $D(A) = H^2(\Omega) \cap H_0^1(\Omega)$ with a bounded smooth domain $\Omega \subset \mathbb{R}^n$, $n \geq 1$.

We note that the equation (2.106) implies that $u(\cdot, t) := u(t) \in D((-A)^\beta)$ for $t > 0$ and thus a boundary condition is incorporated into the equation (2.106). Say, in the case $\Omega \subset \mathbb{R}^n$ is a bounded domain and $A = \Delta$ with $D(A) = H^2(\Omega) \cap H_0^1(\Omega)$, we can verify that (2.106) yields $u(t) \in H_0^1(\Omega)$ for $t > 0$ if $\beta > \frac{1}{4}$.

Let us denote a solution to (2.106) by $u_{\alpha, \beta}(t)$. The weak maximum principle for the evolution equation (2.106) is stated as follows:

Theorem 2.35. *Let $0 < \alpha, \beta < 1$. If $a \geq 0$ in Ω, then $u_{\alpha, \beta}(\cdot, t) \geq 0$ in Ω for $t \geq 0$.*

Let us discuss a short sketch of a proof of Theorem 2.35. The main idea is first to prove non-negativity of $u_{\alpha, \beta}$ in the case $\alpha = 1$ and then to extend this result to the general case. We start with the following important result:

Lemma 2.3. $u_{1, \beta}(\cdot, t) \geq 0$ *in Ω for $t \geq 0$ if $a \geq 0$ in Ω.*

Proof of Lemma 2.3. First we employ the representation

$$((-A)^\beta + 1)^{-1} a = \frac{\sin \pi \beta}{\pi} \int_0^\infty \frac{\mu^\beta (-A + \mu)^{-1} a}{\mu^{2\beta} + 2\mu^\beta \cos \pi \beta + 1} d\mu, \ a \in X \qquad (2.108)$$

(see, *e.g.*, the formula (2.32) on p. 37 in [31]). By the maximum principle for A, for $\mu \geq 0$, we have $(-A + \mu)^{-1} a \geq 0$ if $a \geq 0$ in Ω. Moreover, for $0 < \beta < 1$, we can directly verify that $\mu^{2\beta} + 2\mu^\beta \cos \pi \beta + 1 > 0$ for all $\mu \geq 0$. Hence by (2.108) we get the inequality

$$(1 + (-A)^\beta)^{-1} a \geq 0 \ \text{if } a \geq 0 \text{ in } \Omega. \qquad (2.109)$$

By the product formula (see, *e.g.*, the formula (5.16) on p. 92 in [27]), we have

$$e^{-(-A)^\beta t} a = \lim_{\ell \to \infty} \left(1 + \frac{t}{\ell}(-A)^\beta\right)^{-\ell} a, \ a \in X. \qquad (2.110)$$

Putting (2.109) and (2.110) together completes the proof of the lemma.

Now we establish a transformation formula between the solutions $u_{1,\beta}$ and $u_{\alpha,\beta}$. Let us denote by Φ_α a particular case of the Wright function in the form:

$$\Phi_\alpha(\eta) = \sum_{\ell=0}^\infty \frac{(-\eta)^\ell}{\ell! \Gamma(-\alpha\ell+1-\alpha)}. \tag{2.111}$$

Lemma 2.4. *Let* $0 < \alpha, \beta < 1$. *Then*

$$u_{\alpha,\beta}(x,t) = \int_0^\infty \Phi_\alpha(\eta) u_{1,\beta}(x, t^\alpha \eta) d\eta, x \in \Omega, t > 0. \tag{2.112}$$

Proof of Lemma 2.4. Let us order the eigenvalues of the operator $-A$

$$0 < \lambda_1 \le \lambda_2 \le \cdots$$

and denote by φ_ℓ an eigenfunction for λ_ℓ such that $\{\varphi_\ell\}_{\ell \in \mathbb{N}}$ is an orthonormal basis of X. We note that $\lambda_\ell > 0$, $\ell \in \mathbb{N}$ because of the condition $\rho(-A) \supset (-\infty, 0]$. Then

$$u_{1,\beta}(x,t) = \sum_{\ell=1}^\infty e^{-\lambda_\ell^\beta t}(a, \varphi_\ell)\varphi_\ell \tag{2.113}$$

and

$$u_{\alpha,\beta}(x,t) = \sum_{\ell=1}^\infty E_{\alpha,1}(-\lambda_\ell^\beta t^\alpha)(a, \varphi_\ell)\varphi_\ell. \tag{2.114}$$

It can be verified that both series are convergent in $C([0,T]; X)$. Now we use the known relation

$$\int_0^\infty \Phi_\alpha(\eta) e^{z\eta} d\eta = E_{\alpha,1}(z), z \in \mathbb{C} \tag{2.115}$$

and the inequality

$$\Phi_\alpha(\eta) \ge 0, \eta > 0 \tag{2.116}$$

(see, *e.g.*, (1.31) and (1.32) on p. 14 in [7])). By (2.115) we have

$$\int_0^\infty \Phi_\alpha(\eta) e^{-(\lambda_\ell^\beta t^\alpha)\eta} d\eta = E_{\alpha,1}(-\lambda_\ell^\beta t^\alpha)$$

and hence

$$\int_0^\infty \Phi_\alpha(\eta) u_{1,\beta}(x, t^\alpha \eta) d\eta = \int_0^\infty \Phi_\alpha(\eta) \sum_{\ell=1}^\infty e^{-(\lambda_\ell^\beta t^\alpha)\eta} (a, \varphi_\ell) \varphi_\ell d\eta$$

$$= \sum_{\ell=1}^\infty E_{\alpha,1}(-\lambda_\ell^\beta t^\alpha)(a, \varphi_\ell)\varphi_\ell = u_{\alpha,\beta}(x, t), x \in \Omega, t > 0$$

that completes the proof of the lemma.

Theorem 2.35 follows directly from Lemmas 2.3, 2.4 and the inequality (2.116). Indeed, the condition $a \geq 0$ and Lemma 2.3 yield the inequality $u_{1,\beta}(x, \xi) \geq 0$ for $x \in \Omega$ and $\xi \geq 0$. Hence the inequality (2.116) and Lemma 2.4 yield the inequality $u_{\alpha,\beta}(x, t) \geq 0$ for $x \in \Omega$ and $t \geq 0$.

2.6. DISCUSSIONS AND OPEN PROBLEMS

In this chapter, some selected results regarding the maximum principles for the time-fractional diffusion equations of different types were presented. In particular, we considered the single- and the multi-term fractional diffusion equations as well as the equations of the distributed order both with the Caputo and the Riemann-Liouville fractional derivatives. Moreover, we presented also the maximum principles for the strong and the weak solutions for the fractional diffusion equations with the general fractional derivatives.

Our discussions were mainly concerned with the weak maximum principles that have the same form as in the case of the partial differential equations of the parabolic type. As to the strong maximum principles, their formulations are much weaker for the fractional diffusion equations compared to those known for the parabolic partial differential equations. Moreover, the strong maximum principles were not yet proved for all types of the time-fractional diffusion equations and thus this topic is worth to be investigated in detail.

Another important direction of further research is connected with the maximum principles for the appropriate defined weak solutions to the initial-boundary-value problems for the fractional diffusion equations without a restriction on the sign of the coefficient by the unknown function. Until now, this topic was treated only for the single-term fractional diffusion equation with the Caputo derivative (see [25]). To solve this problem, a new technique, namely, reduction of the differential equation to an integral equation and then application of the fixed point theorem for investigation of the integral equation was suggested in [25].

This new technique can be used to derive several other important results. In particular, a maximum principle for a coupled system of the time-fractional diffusion equations with the fractional derivatives of the same order $\alpha, 0 < \alpha < 1$ can be proved without a restriction on the sign of the coefficient by the unknown function. More general fractional differential equations such as multi-term time-fractional diffusion equations, diffusion equations of the distribute order, and the general diffusion equations studied in [12] and [25] can be analyzed with this technique, too.

In this chapter, some of applications of the maximum principles for the fractional differential equations were discussed. However, many potentially important topics like *e.g.*, applications of the maximum principles to the method of the upper and lower solutions of the non-linear fractional diffusion equations are still not worked out. Another topic that is worth to be investigated is concerned with applications of the maximum principles to numerical schemes for the initial-boundary-value problems for the fractional diffusion equations of different types. The maximum principles for the partial differential equations are known to have discrete analogues that lead to stable and accurate numerical algorithms. This topic seems to be not yet investigated for the fractional partial differential equations.

CONSENT FOR PUBLICATION

Not applicable.

CONFLICT OF INTEREST

The authors declare no conflict of interest, financial or otherwise.

ACKNOWLEDGEMENTS

The second named author was partly supported by Grants-in-Aid for Scientific Research (S) 15H05740 and (S) 26220702, and by A3 Foresight Program "Modeling and Computation of Applied Inverse Problems" by Japan Society for the Promotion of Science, and by the Ministry of Education and Science of the Russian Federation (the Agreement number No.02 a03. 21. 0008).

REFERENCES

[1]   A. Alsaedi, B. Ahmad, and M. Kirane, "Maximum principle for certain generalized time and space fractional diffusion equations," *Quart. Appl. Math.*, vol. 73, pp. 163–175, 2015.

[2]   M. Al-Refai, "On the fractional derivative at extreme points," *Elect. J. of Qualitative Theory of Diff. Eqn.*, vol. 55, pp. 1–5, 2012.

[3]   M. Al-Refai, and Yu. Luchko, "Analysis of fractional diffusion equations of distributed order: Maximum principles and their applications," *Analysis*, vol. 36, pp. 123–133, May 2016.

[4]   M. Al-Refai, and Yu. Luchko, "Maximum principle for the multi-term time-fractional diffusion equations with the Riemann-Liouville fractional derivatives," *Appl. Math. Comput.*, vol. 257, pp. 40–51, Apr. 2015.

[5]   M. Al-Refai, and Yu. Luchko, "Maximum principles for the fractional diffusion equations with the Riemann-Liouville fractional derivative and their applications," *Fract. Calc. Appl. Anal.*, vol. 17, pp. 483–498, Jun. 2014.

[6]   B. Barrios, and M. Medina, "Strong maximum principles for fractional elliptic and parabolic problems with mixed boundary conditions," arXiv: 1607.01505, Analysis of PDEs (math.AP), 2016.

[7]   E. Bazhlekova, *"Fractional Evolution Equations in Banach Space"*, Ph.D. thesis, Technische Universiteit Eindhoven, the Netherlands, 2001.

[8]   W. Feller, *An Introduction to Probability Theory and its Applications*, Vol. 2, Wiley: New York, 1966.

[9]   R. Gorenflo, Yu. Luchko, and M. Yamamoto, "Time-fractional diffusion equation in the fractional Sobolev spaces," *Fract. Calc. Appl. Anal*, vol. 18, pp. 799–820, Jun. 2015.

[10]   J. Jia, and K. Li, "Maximum principles for a time-space fractional diffusion equation," *Appl. Math. Lett.*, vol. 62, pp. 23–28, Dec. 2016.

[11]   S.D. Eidelman, and A.N. Kochubei, "Cauchy problem for fractional diffusion equations," *J. Diff. Equat.*, vol. 199, pp. 211–255, May 2004.

[12]   A.N. Kochubei, "General fractional calculus, evolution equations, and renewal processes," *Integr. Equa. Operator Theory*, vol. 71, pp. 583–600, Dec. 2011.

[13]   A.N. Kochubei, "Fractional-order diffusion," *Differential Equations*, vol. 26, pp. 485–492, Apr. 1990.

[14]   Y. Liu, "Strong maximum principle for multi-term time-fractional diffusion equations and its application to an inverse source problem," *Comput. Math. Appl.*, vol. 73, pp. 96–108, Jan. 2017.

[15]   Y. Liu, W. Rundell, and M. Yamamoto, "Strong maximum principle for fractional diffusion equations and an application to an inverse source problem," *Fract. Calc. Appl. Anal.*, vol. 19, pp. 888–906, Aug. 2016.

[16]   Z. Liu, Sh. Zeng, and Y. Bai, "Maximum principles for multi-term space-time variable-order fractional diffusion equations and their applications," *Fract. Calc. Appl. Anal.*, vol. 19, pp. 188–211, Feb. 2016.

[17]   Yu. Luchko, "Initial-boundary-value problems for the one-dimensional time-fractional diffusion equation," *Fract. Calc. Appl. Anal.*, vol. 15, pp. 141–160, Mar. 2012.

[18]   Yu. Luchko, "Initial-boundary-value problems for the generalized multi-term time-fractional diffusion equation," *J. Math. Anal. Appl.*, vol. 374, pp. 538–548, Feb. 2011.

[19]   Yu. Luchko, "Maximum principle and its application for the time-fractional diffusion equations," *Fract. Calc. Appl. Anal.*, vol. 14, pp. 110–124, Mar. 2011.

[20]   Yu. Luchko, "Some uniqueness and existence results for the initial-boundary value problems for the generalized time-fractional diffusion equation," *Comput. Math. Appl.*, vol. 59, pp. 1766–1772, Mar. 2010.

[21]   Yu. Luchko, "Maximum principle for the generalized time-fractional diffusion equation," *J. Math. Anal. Appl.*, vol. 351, pp. 218–223, Mar. 2009.

[22]   Yu. Luchko, "Boundary value problems for the generalized time-fractional diffusion equation of distributed order," *Fract. Calc. Appl. Anal.*, vol. 12, pp. 409–422, 2009.

[23]   Yu. Luchko, "Operational method in fractional calculus," *Fract. Calc. Appl. Anal.*, vol. 2, pp. 463–489, 1999.

[24] Yu. Luchko, W. Rundell, and M. Yamamoto, "Maximum principle for time-space fractional diffusion equations," preprint.

[25] Yu. Luchko, and M. Yamamoto, "On the maximum principle for a time-fractional diffusion equation," *Fract. Calc. Appl. Anal*, vol. 20, pp. 1131-1145, 2017.

[26] Yu. Luchko, and M. Yamamoto, "General time-fractional diffusion equation: Some uniqueness and existence results for the initial-boundary-value problems," *Fract. Calc. Appl. Anal.*, vol. 19, pp. 676–695, Jun. 2016.

[27] A. Pazy, *Semigroups of Linear Operators and Applications to Partial Differential Equations*. Springer: Berlin, 1983.

[28] M.H. Protter, and H.F. Weinberger, *Maximum Principles in Differential Equations*. Springer: Berlin, 1999.

[29] P. Pucci, and J.B. Serrin, *The Maximum Principle*. Birkhäuser: Basel, 2007.

[30] S.G. Samko, A.A. Kilbas, and O.I. Marichev, *Fractional Integrals and Derivatives: Theory and Applications*. Gordon and Breach: New York, 1993.

[31] H. Tanabe, *Equations of Evolution*. Pitman: London, 1979.

[32] V.S. Vladimirov, *Equations of Mathematical Physics*. Nauka: Moscow, 1971.

[33] W. Walter, "On the strong maximum principle for parabolic differential equations," *Proc. Edinb. Math. Soc.*, vol. 29, pp. 93–96, Feb. 1986.

[34] H. Ye, F. Liu, V. Anh, and I. Turner, "Maximum principle and numerical method for the multi-term time-space Riesz-Caputo fractional differential equations," *Appl. Math. Comput.*, vol. 227, pp. 531–540, Jan. 2014.

CHAPTER 3

Integral-Balance Solution to Nonlinear Subdiffusion Equation

Jordan Hristov[*]

Dept. Chem. Eng, UCTM, Sofia 1756, 8 Kl.Ohridsky Blvd, Bulgaria

Abstract: Improved double-integration technique to approximate integral-balance solutions of non-linear fractional subdiffusion equations has been conceived. The time-fraction subdiffusion equation with Dirichlet boundary condition and a power-law fractional diffusivity has been chosen as a test example. Problems pertinent to approximation of time-fractional Riemann-Liouville derivative when the distribution is expressed as a parabolic profile with unspecified exponent and accuracy of the solutions have been analyzed. The final solution is a closed-form can be presented with either a similarity variable of a fractional order as independent variable or by an effective similarity variable incorporating the effects of both the fractional order and the nonlinearity of the diffusion coefficient. Optimization problem pertinent to determination of the optimal exponent of the parabolic profile, dependent on both the fractional order and the nonlinearity parameter of the diffusion coefficient, has been developed by a modified technique transforming the time-varying domain of integration into one with fixed boundaries. It was clearly defined that the approximate profile can exhibit a concave behaviour, typical for subdiffusion relaxation processes when the non-linearity of the diffusion coefficient is low and the fractional order is high. Otherwise, the increase in the nonlinearity of the diffusion coefficient results in convex profiles typical for the degenerate diffusion behaviour.

Keywords: Subdiffusion, degenerate diffusion, integral method, approximate solution.

AMS Subject Classification: 26A33, 35K05, 40C10.

3.1. INTRODUCTION

Anomalous diffusion in real physical systems such as media with fractal geometry [1], thick biological membranes [2], drug transmission through skins and polymers [3], highly branched porous structures [4-6] and complex systems discussed in [7-13] and the references therein is commonly modelled by subdiffusion equation models.

---
[*]**Corresponding author Jordan Hristov:** Dept. Chem. Eng, UCTM, Sofia 1756, 8 Kl.Ohridsky Blvd, Bulgaria; Tel: +359 885 82 77 12; E-mail: jordan.hristov@mail.bg

The extensive applied approaches for solving fractional-order equations are purely mathematical [14-22]. The integral-balance method is among the approximate analytical methods providing closed form solutions [23-25], *etc.*

The present chapter focuses on an approximate solution of fractional-time diffusion equation with non-linear coefficient.

$$\frac{\partial^\mu u(x,t)}{\partial t^\mu} = \frac{\partial}{\partial x}\left(f(u)\frac{\partial u(x,t)}{\partial x}\right) \tag{3.1}$$

This equation has been analyzed by coefficient of diffusion in many cases approximated as either power-law $f(u) = D_{\mu0}u^m$ or exponential $f(u) = D_{\mu0}e^{mu(x,t)}$, functions; $D_{\mu0}$ is a sort of fractional diffusion coefficient of dimensions $[D_{\mu0}] = [m^2/s^\mu]$ corresponding to linear problems with $m = 0$ solved in [26] . With the power law coefficient $D_{\mu0}u^m$ and $m > 0$ equation (3.1) models the so-called *slow-diffusion*, while with $m < 0$ it is related to the *fast diffusion problems*. In the integer order ($\mu = 1$) models this problem with ($m > 0$) is well-investigated [27-30]. The model (3.1) with $\mu = 1$ and $f(u) = D_{\mu0}u^m$, for instance, in contrast to the linear diffusion equation ($m = 0$), is uniformly parabolic in any region where $u(x, t)$ is not zero, but degenerates in the vicinity of any point where $u(x,t) = 0$ [27]. The main performance of this degeneracy is that *any disturbance propagates at finite speed giving rise to a front or interface in the solution*. Therefore, due to the non-linearity of the diffusivity coefficient there exist solutions propagating with well-defined fronts separating the disturbed $u \neq 0$ and the undisturbed medium $u = 0$ [28-30].

The model (3.1) has been investigated for fast diffusion with $m < 0$ [19, 31, 32], while the large time behaviour of both global and non-global solutions of the fast diffusion model has been investigated in [33]. Further, approximate analytical solutions by the Adomian decomposition method (ADM) [34] and homotopy perturbation method (HAM) [35] have been developed by Das *et al.* [19, 32] for both $m > 0$ and $m < 0$ (with Caputo fractional derivatives). In addition, for the case of slow diffusion ($m > 0$) Sun *et al.* [36] has been developed by a numerical solution by a non-Boltzmann scaling of the governing equation, while Plociczak and Okrasinska [31] have developed self-similar solution of problem (3.1).

The present chapter reports approximate closed-form solutions of the model (3.1) by the integral-balance method applying a double-integration technique, already applied to the linear case ($m = 0$) [26]. The simpler version of the integral-balance approach known as Heat-balance Integral Method (HBIM) [38] has been successfully applied to develop approximate closed-form solutions to fractional subdiffusion equations [23-25, 39] and further applied to other complex models involving time-fractional derivatives [40-44]. The main issue of the present chapter is the double-integration technique allowing solving nonlinear problems (slow diffusion) modelled by the subdiffusion equation (3.1). To the literature background a few studies [32, 37] and approximate solutions (not in closed-forms) to this problem have been reported so far. The efficiency of the double-integration technique is exemplified by a solution of the Dirichlet problem.

This chapter presents a solution amalgamating the principle results of integral-balance solutions of two principle cases of the model (3.1) : the case with $m = 0$ and $0 < \mu < 1$ [26] and the case with $m > 0$ and $\mu = 1$ [45, 46].

3.2. PROBLEM FORMULATION

Let us consider a subdiffusion of temperature (concentration) $u(x, t)$ in a semi-infinite subdiffusive material described by one-dimensional fractional equation, with boundary and initial conditions

$$\frac{\partial^\mu u(x,t)}{\partial t^\mu} = \frac{\partial}{\partial x}\left(D_{\mu 0} u^m \frac{\partial u(x,t)}{\partial x}\right) \tag{3.2}$$

with boundary and initial conditions

$$u(0, t) = u_s(t), t \geq 0; u(x, t) = u_\infty = 0; x > 0; \frac{\partial u(x,t)}{\partial t} = 0, x > 0 \tag{3.3}$$

The time-fractional derivative would be presented as a left Riemann-Liouville fractional derivative of with respect to the time [47].

$$\frac{\partial^\mu u}{\partial t^\mu} =_{RL} D_t^\mu u(x, t) = \frac{1}{\Gamma(1-\mu)} \frac{d}{dt} \int_0^t \frac{u(x,t)}{(t-\tau)^\mu} d\tau \tag{3.4}$$

or as a left Caputo derivative

$$\frac{\partial^\mu u}{\partial t^\mu} =_C D_t^\mu u(x, t) = \frac{1}{\Gamma(1-\mu)} \int_0^t \frac{1}{(t-\tau)^\mu} \frac{du(x,t)}{dt} d\tau \tag{3.5}$$

Further, in cases where the type of the fractional derivative is not specified we will use the notation $\frac{\partial^\mu u}{\partial t^\mu}$.

The integral-balance approach suggests a finite depth of penetration δ (a sharp front) which evolves in time, *i.e.* $\delta(t)$. Beyond the point $x = \delta(t)$, the medium is undisturbed and the following boundary conditions (the so-called Goodman's boundary conditions) are imposed instead (3.3 [38])

$$u_a(x, \delta) = 0, x > \delta; \frac{\partial u_a(\delta,t)}{\partial x} = 0; \delta(0) = 0, t = 0 \tag{3.6}$$

In the context of subdiffusion, the concept of a final penetration depth is motivated be experimental facts of almost sharp fronts of penetration of the diffusion substances [2, 3]. In accordance with the heat-balance concept, with the single-step integration (HBIM) [23-25], at any time the integral of both sides (3.1) along δ , when $m = 0$ (the linear case) should be

$$\int_0^\delta \frac{\partial^\mu u}{\partial t^\mu} = -D_0 \left(\frac{\partial u}{\partial x}\right)_{x=0} \tag{3.7}$$

Then, replacing u by an assumed profile u_a it is possible to derive an ordinary differential equation describing the propagation of the penetration depth $\delta(t)$ (see details in section 3 of this chapter). This drawback, for integer-order equations can be avoided by the double-integration method (DIM) usefully solving non-linear problems [26, 45, 46]. Now, we will demonstrate how the same principle can be applied to the fractional-time subdiffusion equation with a non-linear diffusion equation: an approach applied before only to the case with $m = 0$ [26].

3.3.   INTEGRAL   METHOD   TO   THE   NON-LINEAR   DIFFUSION EQUATION

3.3.1. Preliminary Treatment of the Diffusion Term

The successful application of the integral method developed in this chapter requires a preliminary treatment of the diffusion term in the right-hand side of (3.2), namely [45, 46]

$$D_{\mu 0} u^m \frac{\partial u}{\partial x} = \frac{D_{\mu 0}}{m+1} \frac{\partial u^{m+1}}{\partial x^2} \tag{3.8}$$

Hence, the model (3.2) can be expressed as

$$\frac{\partial^\mu u}{\partial t^\mu} = \frac{D_{\mu 0}}{m+1} \frac{\partial^2 u^{m+1}}{\partial x^2} \tag{3.9}$$

3.3.1.1. Single-integration Approach (HBIM)

This technique of integration is well-known as the Heat-Balance integral (HBIM) of Goodman [38]. The HBIM techniques requires only one step: integration of (3.2) from 0 to δ, namely

$$\int_0^\delta \frac{\partial^\mu u}{\partial t^\mu} dx = -\frac{D_{\mu 0}}{m+1} \left(\frac{\partial u^{m+1}}{\partial x} \right)_{x=0} \tag{3.10}$$

Equation (3.10) is the principle integral relations of HBIM

3.3.1.2. Double-Integration Approach (DIM)

The technique of DIM [26, 45, 46] considers two consequent integrations of the diffusion equations. The first step of DIM is the integration from 0 to x. The second principle step of DIM is integration from 0 to δ. Applying this technique to (3.1) we have [26, 45]

$$\int_0^\delta \int_0^x \frac{\partial^\mu u(x,t)}{\partial t^\mu} dx dx = \int_0^\delta \int_0^x \left(f(u) \frac{\partial u(x,t)}{\partial x} \right) dx dx \tag{3.11}$$

Moreover, the relationship (3.11) can be presented in an alternative form, more suitable for application to equations with fractional derivatives [26, 43]

$$\int_0^\delta \int_x^\delta \frac{\partial^\mu u(x,t)}{\partial t^\mu} dx dx = \int_0^\delta \int_x^\delta \left(f(u) \frac{\partial u(x,t)}{\partial x} \right) dx dx \tag{3.12}$$

Applying this technique to (3.2) we have

$$\int_0^\delta \int_x^\delta \frac{\partial^\mu u(x,t)}{\partial t^\mu} dx dx = \int_0^\delta \int_x^\delta \frac{\partial}{\partial x} \left(D_{\mu 0} u^m \frac{\partial u(x,t)}{\partial x} \right) dx dx \tag{3.13}$$

With the transformed right-hand side of (3.2), as it was expressed by (3.9), we have

$$\int_0^\delta \int_x^\delta \frac{\partial^\mu u(x,t)}{\partial t^\mu} dx dx = \int_0^\delta \int_x^\delta \frac{D_{\mu 0}}{m+1} \frac{\partial u^{m+1}}{\partial x^2} dx dx \tag{3.14}$$

This finally reads as

$$\int_0^\delta \int_x^\delta \frac{\partial^\mu u(x,t)}{\partial t^\mu} dx dx = \frac{D_{\mu 0}}{m+1} u^{m+1}(0,t) \tag{3.15}$$

Equation (3.14) is the principle integral relation of the double-integration method when the differential equation is of a fractional order. If, for instance, we may apply the Leibniz rule for differentiation under the integral sign (when $\mu = 1$), then the integer order counterpart of (3.12) is [45, 48].

$$\frac{d}{dt}\int_0^\delta xudx = -\frac{D_0}{m+1}u^{m+1}(0,t) \qquad (3.16)$$

The left side of (3.14) (and (3.15), too) is termed the fractional-time Double-balance Integral (FT-DBI) [26].

3.4. SOLUTION OF THE PROBLEM BY DIM

The integral-balance method suggests approximation of the distribution $u(x,t)$ by an approximate profile $u_a(x,t)$ depending only on the space coordinate x and coefficients dependent on $\delta(t)$. The final outcome is an ordinary differential equation defining the time-evolution of $\delta(t)$.

The approximate profile u_a used here is assumed to a be parabolic with unspecified exponent [26, 39, 40, 42, 43, 44, 46, 48, 49]

$$u_a = u_s\left(1-\frac{x}{\delta}\right)^n \qquad (3.17)$$

The profile (3.17) satisfies all boundary conditions imposed at the front for any n [48, 50], that is $u_a(\delta,t) = \partial u_a(\delta,t)/\partial x = 0$. For the sake of simplicity, allowing to demonstrate DIM by a classical example, we consider $u(0,t) = 1$ that immediately leads to $u_s = 1$; this is equivalent to solve the problem with a dimensionless profile $u_a/u_s = \bar{u} = (1 - x/\delta)^n$. Besides, we assume zero initial conditions which allows successfully applying both the Riemann-Liouville and the Caputo time-fractional derivatives.

3.4.1. Approximation of the Fractional-time Double-balance Integral (FT-DBI)

3.4.1.1. FT-DBI with Riemann-Liouville Time-fractional Derivative

We will briefly demonstrate the approximation of the left-hand side of eq. (3.15) developed in details in the preceding works [26, 43]. Following eq. (3.15) and replacing u by u_a defined by (3.17) we have, as a first step, to integrate (3.2) from x to δ, then again from 0 to δ, namely

$$\int_0^\delta \int_{x\ RL}^\delta D_t^\mu u_a(x,t)dxdx = \int_0^\delta \int_x^\delta \left(\frac{1}{\Gamma(1-\mu)} \frac{d}{dt} \int_0^t \frac{1}{(t-\tau)^\mu}\left(1-\frac{x}{\delta}\right)^n d\tau\right) dxdx \quad (3.18)$$

The integration with respect to x in (3.18) yields [26]:

$$\int_0^\delta \int_{x\ RL}^\delta D_t^\mu u_a(x,t)dxdx = \frac{1}{\Gamma(1-\mu)} \frac{d}{dt} \int_0^t \frac{1}{(t-\tau)^\mu} \frac{\delta^2}{(n+1)(n+2)} d\tau = N_{RL}D_t^\mu\delta^2 \quad (3.19)$$

where $N_{RL} = ((n+1)(n+2))^{-1}$

3.4.1.2. FT-DBI with Caputo Time-fractional Derivative

The integration in LHS of eq.(3.15) with the Caputo derivative and the assumed profile (3.17) we have

$$\int_0^\delta \int_{x\ C}^\delta D_t^\mu u_a(x,t)dxdx = \int_0^\delta \int_x^\delta \frac{1}{\Gamma(1-\mu)} \int_0^t \left(\frac{1}{(t-\tau)^\mu} \frac{du_a(x,t)}{dt} d\tau\right) dxdx \quad (3.20)$$

with

$$\frac{du_a(x,t)}{dt} = \frac{x}{\delta^2} n \left(1-\frac{x}{\delta}\right)^{n-1} \frac{d\delta}{dt}$$

one obtains from (3.19)

$$\int_0^\delta \int_{x\ C}^\delta D_t^\mu u_a(x,t)dxdx = \frac{1}{\Gamma(1-\mu)} \int_0^t \frac{1}{(t-\tau)^\mu} \left(\frac{2}{(n+1)(n+2)}\right) \delta \frac{d\delta}{dt} d\tau = N_C D_t^\mu\delta^2 \quad (3.21)$$

where $N_C = N_{RL} = ((n+1)(n+2))^{-1}$

3.4.2. Penetration Depth

In order to elucidate the benefit of using the double-integration technique instead the simple HBIM approach, we will develop next the penetration depth equations by both methods.

3.4.2.1. DIM Solution

Applying the principle equation (3.15) with the results (3.19) and (3.21) we obtain two fractional equations about the penetration depth

$$_{RL}D_t^\mu\delta^2 = D_{\mu 0} \frac{(n+1)(n+2)}{m+1} \quad (3.22)$$

$$_c D_t^\mu \delta^2 = D_{\mu 0} \frac{(n+1)(n+2)}{m+1} \tag{3.23}$$

When the integer order counterpart of the model (3.2) is at issue, for instance, with $\mu = 1$ and $D_0 = D_{\mu=1}$ the principle equation defining the penetration depth in accordance with the double-integration approach is

$$\frac{d\delta^2}{dt} = D_0 \frac{(n+1)(n+2)}{m+1} \Rightarrow \delta(t) = \sqrt{D_0 t} \sqrt{\frac{(n+1)(n+2)}{m+1}} \tag{3.24}$$

The right-hand sides of (3.22) and (3.23) are constants. Therefore the fractional integrations with the Riemann-Liouville derivative ([47]-p.139) results in

$$_{RL}\delta^2 - \left(_{RL}D_t^{\mu-1}\delta^2 \right)_{t=0} \frac{t^{\mu-1}}{\Gamma(\mu)} =_{RL} J^\mu \left(D_{\mu 0} \frac{(n+1)(n+2)}{m+1} \right) \tag{3.25}$$

The integral-balance method imposes a physical condition $\delta(t = 0) = 0$ (that is the diffusion layer at $t = 0$ if of zero thickness) and therefore $\lim_{t \to 0} (_{RL}D^{\mu-1-k}\delta^2(t)) = 0$. Consequently from (3.22) we have

$$_{RL}\delta^2 = D_{\mu 0} \frac{(n+1)(n+2)}{m+1} \frac{t^\mu}{\Gamma(1+\mu)} \Rightarrow_{RL} \delta(t) = \sqrt{D_{\mu 0} t^\mu} \sqrt{\frac{(n+1)(n+2)}{\Gamma(1+\mu)(m+1)}} \tag{3.26}$$

For $\mu = 1$ and $m = 0$, for instance, we get the integer order relationship (see 3.24) obtained by DIM [48].

Alternately, looking for the solution of (3.21) by applying the Laplace transform we have [47]

$$L\left(_{RL}D_t^\mu \delta^2(t), s \right) = s^\mu \delta^2(s) - \sum_{p=0}^{K-1} s^k \left(_{RL}D_t^{\mu-k-1}\delta^2(t) \right)_{t=0} = \frac{1}{s} \frac{D_{\mu 0}(n+1)(n+2)}{(m+1)} \tag{3.27}$$

where $K - 1 < \mu < K$

The condition $\delta(t = 0) = 0$ leads to

$$_{RI,}\delta^2(s) = \frac{1}{s^{\mu+1}} D_{\mu 0} \left(\frac{(n+1)(n+2)}{(m+1)} \right)$$

and applying the inverse Laplace transform we obtain the same result as (3.26), namely

$$_{RL}\delta^2(t) = \frac{t^\mu}{\Gamma(1+\mu)} D_{\mu 0} \frac{(n+1)(n+2)}{(m+1)} \tag{3.28}$$

With the Caputo derivative, applying the Laplace transform to (3.23) we have [47]

$$L\left(_C D_t^\mu \delta^2(t)\right) = s_C^\mu \delta^2(s) \sum_{k=0}^{K-1} s^{\mu-k-1} \frac{d^{(d)}}{dt}(\delta^2(0)) = \frac{1}{s} D_{\mu 0} \frac{(n+1)(n+2)}{(m+1)} \tag{3.29}$$

where $K - 1 < \mu < K$

The physical condition $\delta(t = 0) = 0$ leads to

$$_C\delta^2 = D_{\mu 0}(n+1)(n+2)\frac{t^\mu}{\Gamma(1+\mu)} \Rightarrow_C \delta = \sqrt{D_{\mu 0} t^\mu}\sqrt{\frac{(n+1)(n+2)}{\Gamma(1+\mu)(m+1)}} \tag{3.30}$$

Therefore, the penetration depth equations are equal since $_{RL}D_t^\mu \delta^2(t) =_C D_t^\mu \delta^2(t)$; this is an expected result when the initial condition is $\delta(t = 0) = 0$ [47].

3.4.2.2. Penetration Depth: HBIM Solution

To this end, we have focused the attention on the double integration method in derivation of the equation of the penetration depth. For comparison only we will demonstrate briefly how the simple HBIM works and what are the differences in the evaluation of the penetration depths. Here, we have to take into account that

$$\frac{D_{\mu 0}}{m+1}\left(\frac{\partial u_a^{m+1}}{\partial n}\right)_{x=0} = \frac{D_{\mu 0}}{m+1}\left(-(m+1)\frac{n}{\delta}\left(1-\frac{x}{\delta}\right)^{(m+1)n-1}\right)_{x=0} \tag{3.31}$$

Hence,

$$\frac{D_{\mu 0}}{m+1}\left(\frac{\partial u_a^{m+1}}{\partial n}\right)_{x=0} = -D_{\mu 0}\frac{n}{\delta} \tag{3.32}$$

Hence, applying equation (3.10) we obtain:

With Riemann-Liouville derivative

$$\int_0^\delta {}_{RL}D_t^\mu u_a(x,t)dx = \int_0^\delta \left(\frac{1}{\Gamma(1-\mu)}\frac{d}{dt}\int_0^t \frac{1}{(\tau-t)^\mu}\left(\frac{\delta}{n+1}\right)d\tau\right)dx = \frac{n}{\delta} \tag{3.33}$$

Then, following the definition (3.4) we get

$$_{RL}D_t^\mu \frac{\delta}{n+1} = \frac{n}{\delta} \Rightarrow _{RL} \delta(_{RL}D_t^\mu) = n(n+1) \tag{3.34}$$

Applying the Laplace transform to (3.34) we get

$$\delta^2(s) = \frac{1}{s^{\mu+1}} n(n+1) \tag{3.35}$$

The inverse Laplace transform applied to (3.35) yields

$$_{RL}\delta(t)_{(HBIM)} = \sqrt{D_{\mu 0}n(n+1)} \frac{t^{\frac{\mu}{2}}}{\sqrt{\Gamma(1+\mu)}} \tag{3.36}$$

Hence, with HBIM *we lost the effect of the non-linearity* expressed by the parameter m, which disappears through the integration of the right-hand side of the diffusion equation. This effect of the HBIM technique was already reported in [45] for integer-order models.

With Caputo derivative

Similarly to the previous operations we have

$$\int_0^\delta {}_C D_t^\mu u_a(x,t)dx = \int_0^\delta \left(\frac{1}{\Gamma(1-\mu)} \int_0^t \frac{\frac{x}{\delta^2} n \left(1 - \frac{x}{\delta}\right)^{n-1} \frac{d\delta}{dt}}{(t-\tau)^\mu} d\tau \right) dx = \frac{_C D_t^\mu \delta(t)}{n+1} \tag{3.37}$$

Then, from eq. (3.10) we get

$$_C D_t^\mu \delta = D_{\mu 0} \frac{1}{\delta} n(n+1) \tag{3.38}$$

Applying the Laplace transform solution to (3.38) we obtain

$$s^\mu(\delta(s))^2 = \frac{1}{s} n(n+1) \Rightarrow_C \delta(t)_{(HBIM)} = \sqrt{D_{\mu 0}n(n+1)} \frac{t^{\frac{\mu}{2}}}{\sqrt{\Gamma(1+\mu)}} \tag{3.39}$$

Hence, we also lost the effect of the parameter m.

If $\mu = 1$, for instance, then eq. (3.39) becomes $\delta d\delta/dt = D_{\mu 0}n(n+1)$ and the penetration depth is $\delta = \sqrt{D_0 t}\sqrt{2n(n+1)}$ as in the classical HBIM solution [38, 50]. In general, *the HBIM solution cannot encounter the effect of the nonlinearity through the exponent* m, in contrast to the DIM solution (in the case of integer-order model [45], too).

It is obvious that the HBIM and DIM results in different expressions of the penetration depth. Comparing the penetration depths we get the ratio.

$$\frac{c\delta_{(DIM)}}{c\delta_{(HBIM)}} = \frac{n+2}{n(m+1)}$$

However, with DIM we do not need $(\partial u / \partial x)_{x=0}$ to be evaluated through the assumed profile, that is advantage of the method.

3.4.2.3. *Alternative Frozen Front Approach (FFA)*

This approach has been applied to the first integral-balance (HBIM) solutions of the fractional diffusion equation [23, 24, 25, 39, 51]. In these early studies it was more physically motivated rather than mathematically proved. The main physical assumption behind the FFA is based on a conjecture that the rate of the penetration front is extremely slow with respect the short memory time of the fractional operator, that is for a certain time range the value of δ could be considered as a constant. However, *this assumption is valid only in the evaluation of the convolution integral.*

The memory term in the time-fractional derivative, in fact, accounts for a delay in the propagation speed of the disturbances imposed at the boundary $x = 0$, with respect to the speed provided by the classical diffusion equation ($\mu = 1$). The concept of a finite penetration depth $\delta(t)$ comes from the integer-order parabolic equations and does the same; that is, it corrects the *unphysical infinite propagation speed* which is inherent to this model.

Therefore, applying the integral-balance concept to a model containing a memory term expressed as a Volterra integral, we overlap two techniques relevant to the concept of finite propagation speed. The FFA, however, separates the objects, *i.e.* the terms of the model to which these two concepts work. That is, the concept of a final depth works on all terms without memories, while in the convolution integral is assumed as a constant due to the short-time effect of this term as it was conjectured in [23, 25, 39, 51].

The FFA conjecture is that the assumed profile (3.48) expressed as $u_{a(FFA)} = (1 - x/\delta_{FFA})^n$ is *time-independent* (since δ does not vary within the short relaxation process modelled by the memory integral) *when approximating the convolution integral* in the Riemann-Liouville fractional derivative through the definition (3.3) and solving the convolution integral with respect to t. Hence, we may write (with $D_\mu(\partial u_{a(FFA)}/\partial x))_{x=0} = \frac{n}{m+1}\frac{1}{\delta_{FFA}}$ and omitting the subscript FFA)

$$_{RL}D_t^\mu u_a(x,t) = \frac{1}{\Gamma(1-\mu)}\frac{d}{dt}\int_0^t \frac{u_a}{(t-\tau)^\mu} = \frac{1}{(1-\mu)\Gamma(1-\mu)}\frac{d}{dt}(u_a t^{1-\mu}) \qquad (3.40)$$

Here u_a, we mention again, is considered as constant due the conjecture explained above. *This assumption is valid for the approximation of the fractional derivative of the Riemann-Liouville sense only.*

Now, integrating (3.2) from 0 to δ in accordance with HBIM we get

$$\frac{d}{dt}(\delta t^{1-\mu}) = D_{\mu 0}\frac{N_{FFA}(\mu,n)}{\delta}, N_{FFA}(\mu,n) = n(n+1)(1-\mu)\Gamma(1-\mu) \quad (3.41)$$

Setting $\delta t^{1-\mu} = Z(t)$ we obtain

$$\frac{d}{dt}Z^2 = 2D_{\mu 0}M(\mu,n)t^{1-\mu} \Rightarrow Z^2 = D_{\mu 0}\frac{2M(\mu,n)}{2-\mu}t^{2-\mu} + C_1, C_1 = 0 \quad (3.42)$$

With $\delta(t=0) = 0$ we get

$$\delta_{\mu(HBIM)FFA} = \sqrt{D_{\mu 0}t^\mu}\sqrt{2n(n+1)\frac{\Gamma(2-\mu)}{(2-\mu)}} \qquad (3.43)$$

Similarly with the DIM technique we get

$$\delta_{\mu(DIM)FFA} = \sqrt{D_{\mu 0}t^\mu}\sqrt{\frac{(n+1)(n+2)}{m+1}\frac{\Gamma(2-\mu)}{(2-\mu)}} \qquad (3.44)$$

The expressions (3.43) and (3.44) preserve the non-Boltzmann scaling $\sqrt{D_{\mu 0}t^\mu}$ as it was developed in the mathematically correct approximation of u_a (see (3.36, (3.30) and (3.39)). Besides, for $\mu = 1$ and $m = 0$, eq. (3.44) reduces to the classical DIM solution $\delta_{\mu=1(DIM)} = \sqrt{(n+1)(n+2)}$, which is a well-known result of the DIM [52].

Therefore, violating the mathematical rules and using the physical idea of the frozen front in the approximation of the time-fractional Riemann-Liouville derivative we obtained expressions for the penetration depths which resemble the ones developed by the correct approach. The non-Boltzmann scaling $D_{\mu 0}\sqrt{t^\mu}$ is preserved, the forms $N_{HBIM}(n)$ and $N_{DIM}(n)$, too . The only differences appear in the factors depending on the fractional order μ. In general, we did *mathematically incorrect, but physically motivated*, calculations and got analogous results, avoiding solutions of fractional differential equations with respect to δ. Therefore, it is mandatory, to demonstrate, what is the error in derivation of penetration depth when FFA is applied. To do that we create the ratios $\Delta_{\mu(HBIM)(FFA)}/\delta_{\mu(HBIM)}$ and $\Delta_{\mu(DIM)(FFA)}/\delta_{\mu(DIM)}$, namely

$$\Delta_{\mu(HBIM)(FFA)} = \sqrt{2}\sqrt{\Gamma(1+\mu}\qquad\qquad(3.45)$$

$$\Delta_{\mu(DIM)(FFA)} = \sqrt{\Gamma(1+\mu)\Gamma(2-\mu)}\qquad\qquad(3.46)$$

The expressions (3.45) and (3.46) are valid if the exponent n is stipulated (*i.e.*, equal for all integration techniques applied), that is, *the exponent is not optimized*! The plots of $\Delta_{\mu(HBIM)(FFA)}$ and $\Delta_{\mu(DIM)(FFA)}$ are shown in Fig. (**3.1**) a,b for $0 \leq \mu \leq 1$. For the DIM solution the curve is almost symmetrical and the maximum deviation is for about $\mu = 0.5$, while the asymmetric curve corresponding to the HBIM solution indicates a range $0 < \mu < 0.5$ where the difference is less than 0.02.

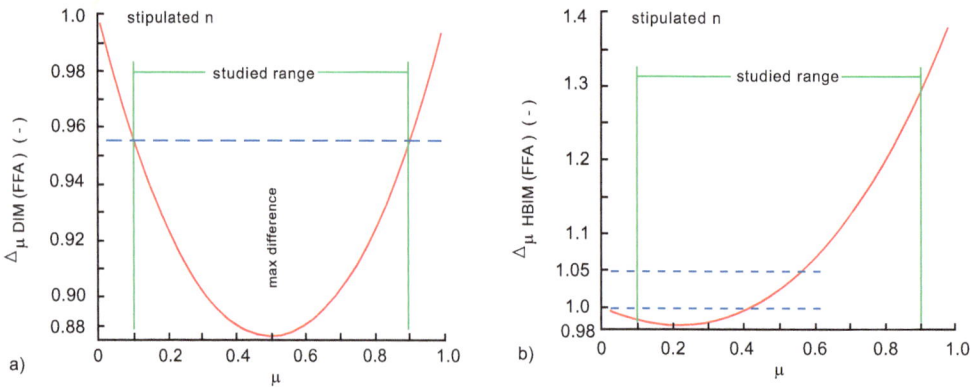

Fig. (3.1). Relative ratios of the penetration depths calculated with the FFA concept and the correct HBIM, and DIM technologies. a) Case of FFA concept and DIM; b) Case of FFA concept and HBIM. The horizontal dashed lines mark two levels of discrepancy between the penetration depths: exact matching at 1.0 and a difference of 0.05.

3.4.2.4. Matching the Penetration Depths of DIM and HBIM

Now, we turn again to the penetration depths developed by the correct approaches of HBIM and DIM solutions. From simple physical reasons, the penetration depths should be one and the same, that is $\delta_{\mu(HBIM)} = \delta_{\mu(DIM)}$, irrespective of the integration technique applied. Therefore, we should have

$$\sqrt{D_{\mu 0}t^\mu}\sqrt{\frac{n(n+1)}{\Gamma(1+\mu)}} = \sqrt{D_{\mu 0}t^\mu}\sqrt{\frac{(n+1)(n+2)}{(m+1)\Gamma(1+\mu)}} \qquad (3.47)$$

The straightforward solution of (3.47) is $n = 2/m$ and therefore, the exponent of the approximate parabolic profile is reciprocal to the parameter of the nonlinearity m. For $m > 1$ (slow diffusion) it follows that that we should have $n < 1$, while for $0 < m < 1$ we get $n > 1$. Further, this relationship excludes the linear case with $m = 0$ but the values of n established by matching the penetration depths will be considered as ones close to the upper limits (see the comments in the next section) of the exponent for given μ and n when the approximate profile optimization is applied.

3.5. APPROXIMATE PROFILE, SIMILARITY VARIABLE AND RELATED ISSUES

Now, we continue with the DIM solution and following the results (3.26) and (3.30) about $\delta_\mu(t)$ the approximate solution of (3.2) can be expressed as

$$u_a(x,t) = \left(1 - \frac{x}{\sqrt{\sqrt{D_{\mu 0}}t}\sqrt{\frac{(n+1)(n+2)}{(m+1)\Gamma 1+\mu}}}\right)^n \qquad (3.48)$$

The general form of the parabolic profile (3.48) is a function of the dimensionless ratio $\xi = x/\delta$. After the development of expression of the penetration depth $\delta(t)$ the ratio x/δ defines a new similarity variable $\eta_\mu = x/\sqrt{D_{\mu 0}t^\mu}$. For $\mu = 1$ we get the classical Boltzmann similarity variable $\eta_{\mu=1} = x/\sqrt{D_0 t}$.

Alternatively, the parabolic profile (3.48) defines an *effective similarity variable* $\eta_{eff(\mu,m)} = x/\sqrt{D_{eff(\mu,m)}t^\mu}$, where the effective diffusivity is $D_{eff(\mu,m)} = D_{\mu 0}/((m + 1)\Gamma(1 + \mu))$. Otherwise, we may present the effective diffusivity as $D_{eff(\mu,m)} = D_{\mu 0}p(\mu, m)$ with a retardation factor $p(\mu, m) =$

$1/((m+1)\Gamma(1+\mu)) < 1$ which incorporates the effects of the subdiffusion and the nonlinearity of the diffusion coefficient; the effect of the parameter is stronger taking into account that $\Gamma(1+\mu)$ for $0 < \mu < 1$ has an order of magnitude of unity.

In terms of the effective diffusion coefficient and the effective similarity variable the approximate profile reads

$$u_a(x,t) = \left(1 - \frac{x}{\sqrt{D_{eff(\mu,m}t^\mu}\sqrt{(n+1)(n+2)}}\right)^n = \left(1 - \frac{\eta_{(eff(\mu,m)}}{\sqrt{(n+1)(n+2)}}\right)^n \quad \textbf{(3.49)}$$

The second version of (3.49) resembles formally the classical integer-order diffusion solution [48, 50] when $\mu = 1$ but the only difference is the use of the effective diffusivity similarity variable $\eta_{eff(\mu,m)}$ incorporating the retardation effects of both fractional order μ and the *slow diffusion* represented by the parameter m.

The main problem arising in the integral-balance solutions with the profile (3.17) comes from the fact that the Goodman boundary conditions (3.3) are satisfied for any value of n [48, 50]. Precisely, the profile (3.17) can be presented in a general form as $u_a = a + b(1 - cx)^n$ with 4 parameters (a,b,c and n) that should be defined through the boundary conditions. Actually, the Goodman boundary conditions (3.3) are only 3 and they define 3 parameters (a,b and $c = 1/\delta$) but to define the exponent n we need 4 conditions. This problem has been analyzed in [48, 50].

Therefore, additional constrains should be applied in order to define the value of n corresponding to a solution with a minimal error of approximation. The main inherent problem of the integral-balance solutions is that the approximate profile satisfies the integral balance equation (eq. (3.10) or eq. (3.14)), for example) but not the original diffusion equation (3.2). The approach to establish the exponent n by matching $\delta\mu(DIM)$ and $\delta\mu(HBIM)$ is physically based but it does not communicate to the condition the approximate profile to satisfy the original diffusion model. However, the matching approach gives some upper limits of values of n that will be used in the next optimizations of the approximate profile; recall the right-hand side of the model (3.2) corresponds to *degenerate diffusion behaviour*. The determination of the error measure and the optimal exponent are at issue in the next sections

3.5.1. Error Measure of Approximate Solution

Therefore, we take the liberty to estimate the exponent n through minimization of the residuals when the approximate profile (3.17) should satisfy directly the original diffusion equation (53). This point is of primary importance because the integral balance method [38, 50] is a zero-moment method [53]. In this way we may define the residual function $R(x, t)$ as

$$R(x,t) = \frac{\partial^{\mu} u_a}{\partial t^{\mu}} - \frac{\partial}{\partial x}\left(D_{\mu 0} u_a^m \frac{\partial u_a}{\partial x} \right) \qquad (3.50)$$

The residual function should be zero if u_a matches the exact solution; otherwise it should attain a minimum for a certain value of the exponent n. To avoid any problems in evaluation of the fractional derivative of u_a we will transform the moving boundary domain $0 \le x \le \delta(t)$ into a fixed one $0 \le \xi \le 1$ by introduction of the variable (the Zener coordinate) $\xi = x/\delta$ [54]. In the ζ-space the approximate profile can be expressed as

$$u_a = (1 - x/\delta)^n \Rightarrow V(\xi, t) = (1 - \xi)^n \qquad (3.51)$$

In this context, the simple heat-balance integral (see eq.(3.5)) is $\int_0^1 R(\xi, t)d\xi$, while the double integral (see eq. (3.1)) is $\int_1^0 \int_z^1 R(\xi, t)d\xi dz$, where z is a dummy variable. Moreover, in ξ-space the minimization procedures focuses on minimization of the squared-error function $E = \int_0^1 (R(\xi, t, n, \mu))^2 d\xi$.

3.5.2. Approximate Fractional Derivative

Therefore, in ξ-space we have

$$\frac{\partial^{\mu} u_a}{\partial t^{\mu}} \rightarrow \frac{\partial^{\mu} V(\xi, t)}{\partial t^{\mu}} = \frac{\partial^{\mu}}{\partial t^{\mu}}(1 - \xi)^n \qquad (3.52)$$

and the approximation technique is explained next.

The approximation technique was developed in [26] by expression of the assumed profile $V(\xi) - (1 - \xi)^n$ as a converging series $V_a(\xi) \approx \sum_{k=0}^{N} b_k \xi^k, 0 \le \xi \le 1$. This series converges rapidly that easily can be proved by the ratio test [55] and as an example only we have (up to $N = 4$)

$$V_a(\xi) \approx 1 - n\xi + \frac{n(n-1)}{2}\xi^2 - \frac{n(n-1)(n-2)}{6}\xi^3 + \frac{n(n-1)(n-2)(n-3)}{24}\xi^4 + O(\xi^5) \quad (3.53)$$

Now, recall that with $(\xi = (x/\sqrt{D_{\mu 0}})t^{\frac{\mu}{2}})/F_{n,m}$,where $F_{n,m}$ is the numerical factor in the expression of $\delta(t)$, the series (3.53) can be presented as (we set $N = 3$, as example)

$$V_a(\xi) \approx 1 - n\frac{x}{\xi}t^{-\frac{\mu}{2}} + \frac{1}{2}n(n-1)\xi^2 t^{-\mu} - \frac{1}{6}n(n-1)(n-2)\xi^3 t^{-\frac{3}{2}\mu} \quad (3.54)$$

Since all terms are power-law functions of time, we may apply easily to each of them the rule $_{RL}D_t(b_k t^{\lambda}) = b_k(\Gamma(1+\lambda/\Gamma(1+\lambda-\mu))t^{\lambda-\mu}$ namely:

$$_{RL}D_t^{\mu}V(\xi) \approx \sum_{k=0}^{N} c_k t^{k-\mu}, c_k = b_k(\Gamma(1+k/\Gamma(1+k-\mu)) \quad (3.55)$$

Now, we may rearrange (3.5.2) in a form where each term contains a power of ξ, and to expressed as a product of a term dependent on ξ and $t^{-\mu}$, namely [26]

$$_{RL}D_t^{\mu}V(\xi) \approx_{RL} \Phi(n,\xi,\mu)t^{-\mu} \quad (3.56)$$

Therefore, the residual function can be approximated as

$$R_{RL}(\xi,t) = \left(_{RL}D_t^{\mu} \sum_{k=0}^{N} b_k \xi^k - D_{\mu 0}\left(\frac{m}{2}\right)V^{m-1}\left(\frac{\partial V}{\partial x}\right)^2 \right) \quad (3.57)$$

With Caputo derivative of the series expansion $V_a(\xi) \approx \sum_{k=0}^{N} b_k \xi^k$ and avoiding cumbersome expressions we get $_c D_t^{\mu}V(\xi) \approx_{RL} \Phi(n,\xi,\mu)t^{-\mu}$ [26]. It is easy to check that $(_{RL}\Phi(n,\xi,\mu)t^{-\mu} -_{RL}\Phi(n,\xi,\mu)t^{-\mu}) = 1/\Gamma(1-\mu)$ which is independent of n. The residual function has the same construction as that presented by (3.57) [26].

3.5.3. The Exponent n at the Boundaries of the Diffusion Layer

In this context, at the beginning of this analysis we should estimate how the residual function behaves at the boundaries of the diffusion layer: *i.e.* at $x = 0$ and $x = \delta$.Then, for example at $x = 0$, and with δ from the DIM solution eq. (3.28) and (3.57) we have

$$R_{RL}(0,t) = \left(\Phi(n,0,\mu)t^{\mu} - D_{\mu 0}\left(\frac{m}{\delta^2}\right)n^2 \right) = \frac{\bar{R}_{RL}(0,t)}{t^{\mu}} \quad (3.58)$$

$$\bar{R}_{RL}(0,t) = \left(\frac{1}{\Gamma(1-\mu)} - m(m+1)\Gamma(1+\mu)\frac{n^2}{(n+1)(n+2)} \right) \quad (3.59)$$

where $\Phi(n, 0, \mu) = c_k = b_k[\Gamma(1 + k)/\Gamma(1 + k - \mu)]$ for $k = 0$ since all terms in the series approximations of $_{RL}D_t V_a(\xi) \approx \Phi(n, \xi, \mu)t^\mu$ for $k \geq 1$ are dependent on x (*i.e.* on ξ). The solution of $\bar{R}_{RL}(0, t) = 0$ with respect to n for various μ and m are summarized in Table **3.1**. The plot in Fig. (**3.2**) reveals a decaying behavior of n with increase in both μ and m and in general we get $n < 1$ for large m. It is worth remarking that with $n < 1$ the approximate profile (3.17) generates convex plots as solutions [45], while for $n > 1$ the profiles are concave in shape. The definition of the exponent n at $x = 0$ can be considered as *a local calibration of the parabolic profile* (3.48).

Table 3.1. Optimal exponents of the approximate profile at $x = 0$.

$\mu \downarrow, m \rightarrow$	0	1	1.5	2	3	4	5
0.1	3.906	3.462	1.242	0.989	0.577	0.408	0.316
0.2	3.477	3.187	1.153	0.945	0.557	0.395	0.306
0.3	3.529	2.694	1.065	0.880	0.524	0.374	0.290
0.4	3.150	2.345	0.978	0.795	0.481	0.345	0.269
0.5	3.347	1.894	0.890	0.696	0.429	0.310	0.243
0.6	2.596	1.470	0.802	0.587	0.369	0.269	0.212
0.7	2.022	1.089	0.712	0.472	0.303	0.223	0.177
0.8	2.643	0.750	0.619	0.352	0.231	0.172	0.137
0.9	2.310	0.437	0.519	0.222	0.150	0.113	0.091
0.99	2.021	0.108	0.902	0.060	0.042	0.032	0.026

Further, in order to satisfy the Goodman boundary conditions $u_a(\delta, t) = 0$ and $\partial u_a(, t)/\partial x = 0$ we set $x \rightarrow 1$ and consequently $\Phi(n, 1, \mu) \rightarrow 1$. However, the behaviour of the exponent should be determined by the behaviour of the second term of (3.57), namely

$$\lim_{x \to \delta} R(\delta, t) = 1 - D_{\mu 0}\left(\frac{m}{\delta^2}\right) V^{m-1}\left(\frac{\partial V}{\partial \xi}\right)^2 \equiv \frac{m}{\delta^2}(1 - \xi)^{n(m-1)}n^2(1 - \xi)^{2(n-1)} \quad (3.60)$$

It follows from (3.60) that equation (3.2) is satisfied at $x = \delta$ by the approximate profile when the exponent is positive (as it required by the classic postulation of Goodman the profile to be a positive decaying function describing a relaxation process). This condition needs

$$n(m - 1) + 2(n - 1) > 0 \Rightarrow n > \frac{2}{m+1} \tag{3.61}$$

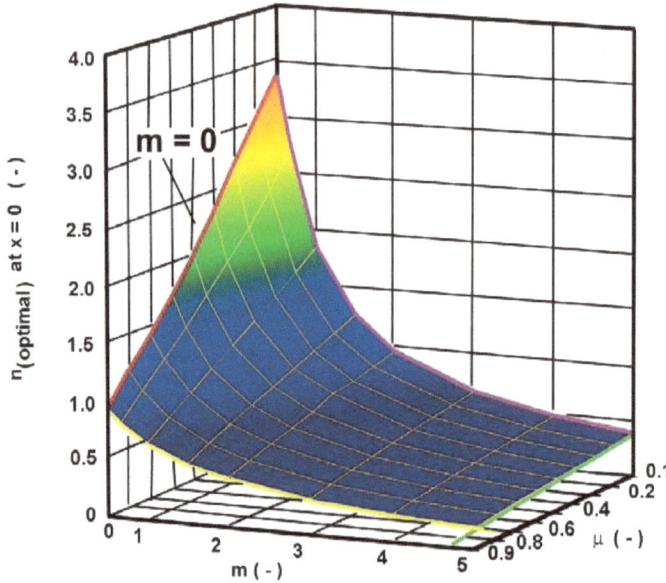

Fig. (3.2). The exponent n defined at $x = 0$ as a function of μ and m

The condition (3.61) confirms the results established in [45] where it was established that with $\mu = 1$ the exponents of the parabolic profile follow the rule $n \approx 1/(m + 1)$ when the Dirichlet problem is at issue. For $\mu = 1$ and $m = 0$ we get the well-known condition $n > 2$ [5, 52]. The estimations about the behaviour of n when $\xi \to 1$ is practically the same as that observed when $\xi \to 0$, that is n decreases with increase in m. The effect of μ on n is almost the same but the exponent is more strongly affected by m rather than μ.

3.5.4. Minimization of the Global Error of Approximation

The error measure is defined as $E(n, \mu, m, t) = \int_0^\delta [R(x, t)]^2 dx$ is, in fact, the Langford criterion [56] defined for integer order integral-balance solutions. The approach follows the idea explored in the integer-order problems [45, 46, 48, 52] and already used in solutions of subdiffusion equations by HBIM [24, 25] and DIM [26]. In ξ-space this means

$$E(n,\mu,m,t) = \int_0^1 \left[\Phi(n,\xi,\mu)t^{-\mu}d\xi - D_{\mu 0}\left(\frac{m}{\delta^2}\right)n^2 \right]^2 d\xi = \left[\frac{e(n,m,\mu)}{\delta^2} \right] \quad (3.62)$$

The last form of (3.62) indicates that E decreases in time, *i.e.* $E \equiv 1/\delta^2 = e(n,m,\mu)/\delta^2 \equiv 1/t^{2\mu}$ (see similar cases in [26, 45, 46], while the nominator $e(n,m,\mu) = \int_0^1 [r(\xi,t)]^2 d\xi$ is time-independent. In the development of the expression (3.62) we use the fact that the ratio $\delta^2/t^\mu D_{\mu 0}$ is time-independent.

The solution of (3.62), that is the minimization with respect to n, for various $0 < \mu < 1$ and $m > 0$, provides the optimal values of the exponent n_{opt} summarized in Table **3.2**. The numerical results in Table **3.2** reveal two main tendencies:

a)   The optimal exponent decreases with increase in the fractional order μ when the parameter of the non-linearity m is fixed.

b)   The optimal exponent decreases with increase in the parameter of the nonlinearity m when the fractional order μ is fixed . The same behaviour was drawn by the estimations about the exponent n at $x = \xi = 0$. In general, the maxima of the mean-squared error of approximation do not exceed 10^{-5} in some particular cases, but dominantly it is of order of 10^{-7}.

These are reasonable physical results because both the fractional order μ and the nonlinearity parameter m are related to retardation in time of the front $\delta(t)$. Precisely, the decrease in μ shifts *to the right* the distributions (with $\eta_\mu = x/\sqrt{D_{\mu 0}t^\mu}$ as independent variable) of the diffusant with respect to the normal diffusion ($\mu = 1, m = 0$), which is a well-known fact [24, 25, 26] as it shown in Fig. (**3.3a**). When the diffusion is performed with $\mu = 1$ but with $m > 0$ (classical slow diffusion), the increase in m shifts the distributions *to the left* (with $\eta_\mu = x/\sqrt{D_{\mu 0}t^\mu}$ as independent variable) of that corresponding to $m = 0$; the increase in reduces the length of $\delta(t)$. Moreover, the profiles of the distribution become convex and steep and cross the abscissa almost vertically, a fact well known from the classical similarity solutions of degenerate equations [27-29] as well as from their approximate solutions [30, 45] as it is shown in Fig. (**3.3b**).

Table 3.2. Optimal exponents of the approximate profile.

$\mu \downarrow, m \rightarrow$	0	0.5	2/3	1	1.5	2	3	4	5
.1	3.870	2.283	2.098	2.021	0.191	0.162	0.124	0.104	0.085
$E \times 10^7$	3.178	3.427	3.928	3.875	1.908	1.185	4.250	1.842	1.751
0.2	3.411	2.162	2.034	2.102	0.183	0.155	0.118	0.096	0.0812
$E \times 10^7$	2.271	2.602	5.120	1.207	4.409	4.293	5.236	5.467	4.402
0.3	3.041	2.074	2.007	2.305	1.721	1.460	1.110	0.090	0.076
$E \times 10^7$	4.812	1.905	2.564	2.931	3.801	12.010	27.631	169.912	1.127
0.4	2.750	2.027	2.036	2.830	0.159	0.135	0.103	0.083	0.070
$E \times 10^7$	0.420	0.404	3.307	2.264	11.372	15.593	2.362	2.443	4.167
0.5	2.538	2.043	2.171	0.541	0.144	0.122	0.093	0.075	0.063
$E \times 10^7$	3.667	0.360	2.221	28.011	19.743	23.415	201.182	22.351	23.316
0.6	2.412	2.185	2.574	0.517	0.125	0.105	0.081	0.065	0.055
$E \times 10^7$	4.071	2.569	0.670	21.062	5.245	15.743	57.521	140.001	0.237
0.7	1.214	0.771	0.613	0.493	0.101	0.086	0.066	0.053	0.044
$E \times 10^7$	88.77	97.542	0.119	4.863	7.556	2.389	2.803	86.901	7.912
0.8	1.132	0.644	0.582	0.468	0.073	0.062	0.047	0.038	0.032
$E \times 10^7$	18.883	14.112	8.138	7.839	2.449	2.787	34.175	4.209	413.201
0.9	1.052	0.767	0.541	0.437	0.040	0.033	0.025	0.020	0.017
$E \times 10^7$	0.329	44.507	8.880	0.903	0.834	5.372	6.106	1.598	1.568

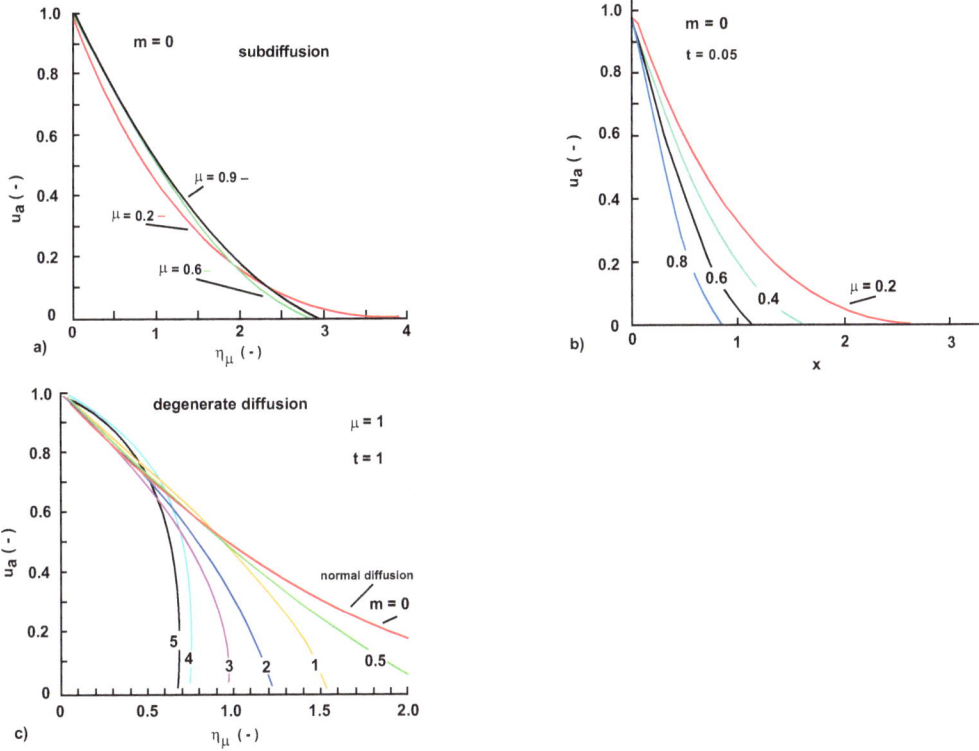

Fig. (3.3). Typical approximate solutions of the integral-balance method (double integration): a) Typical subdiffusion with a constant diffusion coefficient, *i.e.* for $m = 0$. Similarity variable η_μ as independent variable; b) Typical subdiffusion with a constant diffusion coefficient, *i.e.* for for $m = 0$. Presentation with the space coordinate x as independent variable at a fixed time t=0.05; c) Typical degenerate diffusion profiles for $\mu = 1, t = 1$ and various m. Similarity variable η_μ is the independent variable.

3.6. NUMERICAL EXPERIMENTS AND DATA ANALYSIS

The numerical experiments reported in this section visualize the developed analytical solutions and evaluate the results developed in several directions, among them:

1) Profile expressions by the similarity variable $\eta_\mu = x/\sqrt{D_{0\mu} t^\mu}$

2) Discrimination of the profile behaviour as a result of the simultaneous effect of the subdiffusion by the fractional order μ and the degenerate diffusion *via* the parameter m

3)   Profile expressions by the effective similarity variable $\eta_{eff(\mu,m)} = x/\sqrt{D_{eff(\mu,m)}t^{\mu}}$

4)   Expression of the approximate profiles as explicit functions in terms of the physical coordinate x and the time t taking into account the real order of magnitude of the diffusivity in subdiffusion systems.

3.6.1. Profiles Expression through the Similarity Variable $\eta_{\mu} = x/\sqrt{D_{0\mu}t^{\mu}}$

The profiles expressed as functions of the similarity variable η_{μ} exhibits two distinct behaviours. First, for low values of the parameter m in the range $0 < m < 1$ the approximate profiles demonstrate *classical subdiffusion behaviour* with weak effects of the *degeneracy of the diffusion* (the nonlinearity in the diffusion coefficient). The plots in Fig. (**3.4**) reveal that the effect of the degenerate diffusion is stronger as the fractional order μ is closer to 1 (see Fig. **3.4 b,c,d**) . Precisely, with increase in the value of m the effect of the degenerate diffusion becomes more pronounced and easily detected *by changes in the curve behaviours from concave to convex*. Irrespective of the behaviour of the approximate solutions, either convex or concave, the slopes of the profiles decrease with increase in the fractional order μ. These results confirm the observations of Das *et al.* [19] because the slope of the profile (generally negative) is

$$\frac{du_a}{d\eta_\mu} = -\frac{n}{(n+1)(n+2)}\sqrt{(m+1)\Gamma(1+\mu)}\left(1 - \eta_\mu\sqrt{\frac{(m+1)\Gamma(1+\mu)}{(n+1)(n+2)}}\right)^{n-1} \quad (3.63)$$

For example, for $\eta_\mu = 0$ corresponding to $x = 0$ we get $du_a/d\eta_\mu = -(1/p_{(\mu,m)})(n/\sqrt{(n+1)(n+2)})$ where $p_{(\mu,m)} = 1/\sqrt{(m+1)\Gamma(1+\mu)}$.

To precise these comments, let us focus the attention on the numerical factor $p_{(\mu,m)}$ which is able to make the penetration depth shorter or larger with respect to the case $\delta(\mu = 1, m = 0)$. In the factor $p_{(\mu,m)}$ the term $\Gamma(1+\mu)$ corresponding to $0 < \mu < 1$ is of order of magnitude of unity because $\Gamma(1) = \Gamma(2 = 1$. However, for $0 < \mu < 0.5$, the term $\Gamma(1+\mu)$ increases as μ decreases, while for $0.5 < \mu < 1.0$ the behaviour of is just the opposite (see Fig. **3.5a**). Since by definition $m > 0$,then the variation of in the range $0 < m < 1$ results in $0.5 < p_{(\mu,m)} < 1.05$ (see Fig. **3.5b**), while for $1 < m < 5$, the range of variations of $p_{(\mu,m)}$ is $0.15 < p_{(\mu,m)} < 0.55$ (see Fig. **3.5c**). These plots (see Fig. **3.5d** for the entire behaviour of $p_{(\mu,m)}$) indicate that the effect of the non-linear parameter m is stronger than the effect fractional order μ.

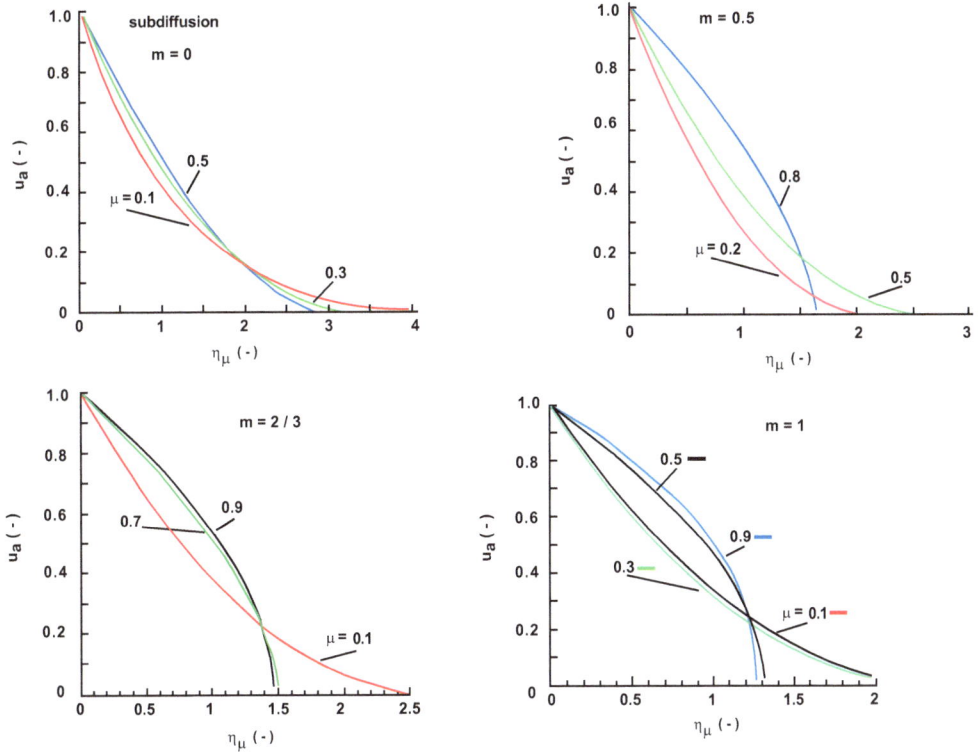

Fig. (3.4). Approximate profiles developed by the double-integration technique expressed as functions of the similarity variable η_μ. Weak effects in the nonlinearity of the diffusion coefficient (low m) and a strong subdiffusion performance. a)Typical strong subdiffusion ($m = 0$) for $0 < \mu < 1$. b) Approximate profiles for $m = 0.5$. Alternations in the profile shape with variations in the fractional order μ are easily detectable. c) Approximate profiles for $\mu = 2/3$. d) Approximate profiles for $m = 1$ where it is possible easily to detect that the subdiffusion dominates only for low μ.

The second group of approximate profiles exhibit *strong generate diffusion behaviour* corresponding to cases with $m > 1$, in the present study for $m > 1.5$, albeit for $m = 1$ this behaviour is observed when $\mu > 0.7$. The common convex shape behaviour of the approximate profiles encompasses the entire range $0 < \mu < 1$ (see Fig. **3.6**) and its is well demonstrated for $\mu > 0.7$. Additionally, the plots in Fig. (**3.7**) show approximate profiles for various values of the parameter m in three distinct points of the subdiffusion range. In general, with increase in m the penetration depth becomes shorter, as mentioned earlier, and the steep branches of the profiles cross the abscissa at shorter distances, too. The latter is a well-known fact from the integer-order degenerate diffusion equations [28-30, 45].

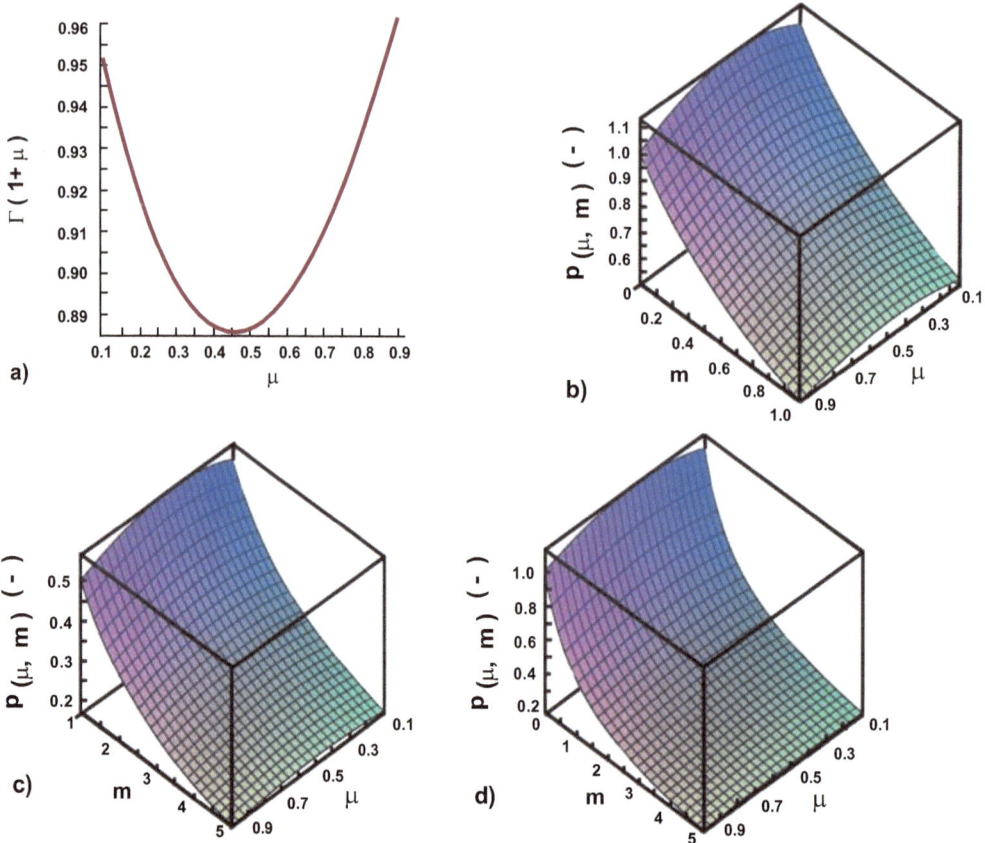

Fig. (3.5). Variations of the factor $p_{(\mu,m)}$ with m and μ. a) Variations of $\Gamma(1+\mu)$ for $0 < \mu < 1$. b) Variations of $p_{(\mu,m)}$ for $0 < m < 1$. c) Variations of $p_{(\mu,m)}$ for $1 < m < 5$. d) Variations of $p_{(\mu,m)}$ for $0 < m < 5$.

3.6.2. Competition Between the Subdiffusion and the Degenerate Diffusion

The approximate profile developed by the double-integration technique and the consequent optimization procedure exhibits two distinct behaviours (shapes) depending on the value of the exponent n. Based on the numerical data collected and the plots discussed to this point, it is obvious that the threshold value to change the profile shape is $n = 1$. If $n < 1$ we get convex profiles, otherwise for $n > 1$ we obtain concave profiles.

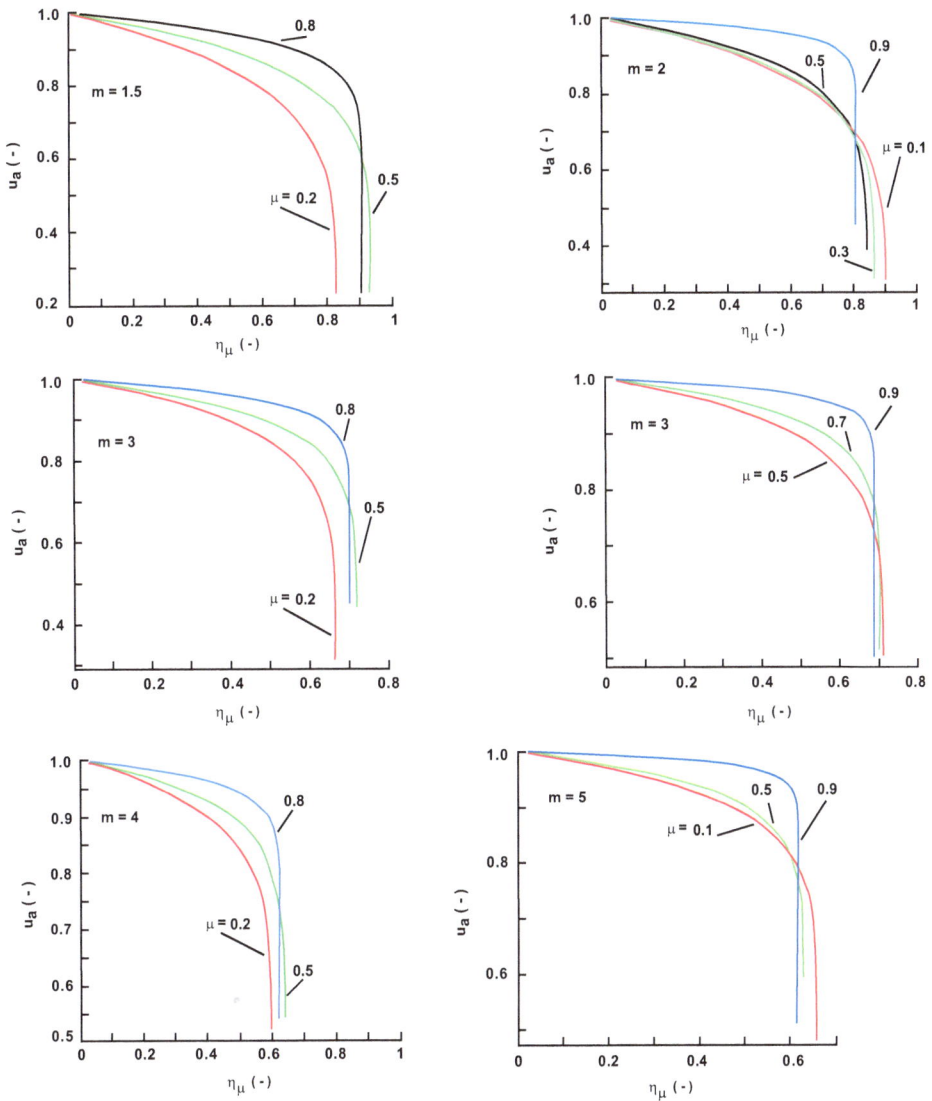

Fig. (3.6). Approximate profiles developed by the double-integration technique expressed as functions of the similarity variable η_μ. Strong effects of the nonlinearity of the diffusion coefficient ($m > 1$) and strong subdiffusion performance. a) Approximate profiles for $m = 1.5$; b) Approximate profiles for $m = 2$ c) Approximate profiles for $m = 3$. Cases with $\mu = 0.2$, $\mu = 0.5$ and $\mu = 0.8$. d) Approximate profiles for $m = 3$. Cases with $\mu > 0.5$ (weak effects of the subdiffusion). The fronts of the profiles are steeper that those in Fig. **3.6c**. e) Approximate profiles for $m = 4$. The strong effect of m suppresses all effects of the subdiffusion over the entire range $0 < \mu < 1$. f) Approximate profiles for $m = 5$. The effect of m is strong and results in convex profiles with steep fronts.

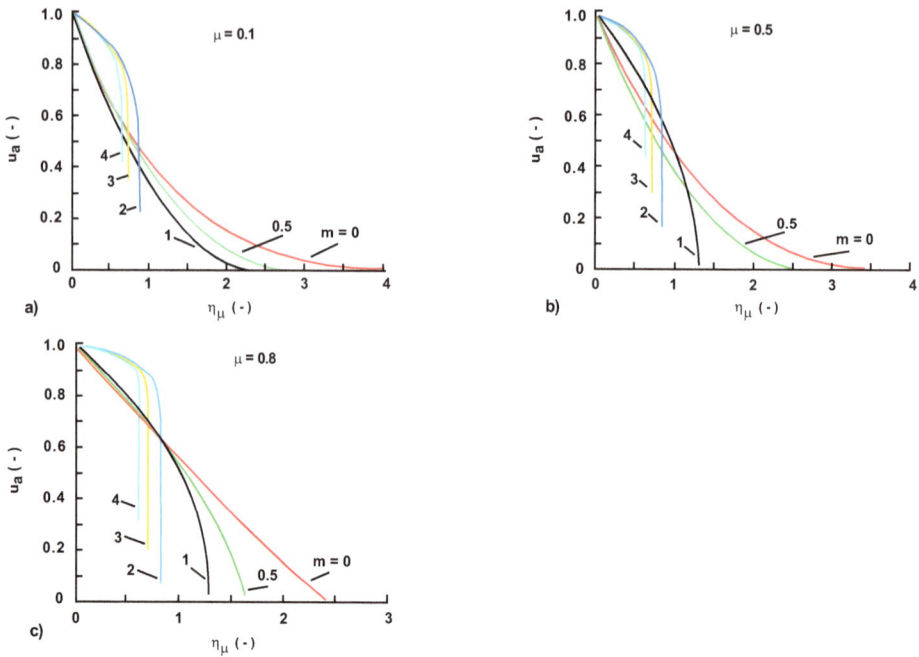

Fig. (3.7). Approximate profiles as functions of the similarity variable η_μ demonstrating the effect of the nonlinearity of the diffusion coefficient (*via m*) in 3 distinct points of the subdiffusion range. a) Solutions for $\mu = 0.1$ and various m; b) Solutions for $\mu = 0.5$ and various m; c) Solutions for $\mu = 0.8$ and various m .

These are direct results of the numerical simulations, but we may relate them, to same extent, to the mean square displacement $\langle x^2 \rangle = \int_0^\delta u(x,t)x^2 dx$. With the parabolic profile (3.17) we get

$$\langle x^2 \rangle = \frac{\delta^2}{(n+1)(n+2)} = 2(D_{\mu 0}t^\mu)(m+1)\Gamma(1+\mu) = 2(D_{\mu 0}t^\mu)\left(1/p_{(\mu,m)}\right) \text{(3.64)}$$

Therefore, the subdiffusion behaviour remains but the slopes of the plots $\langle x^2 \rangle \equiv t^\mu$ depend on the factor $\left(2/p_{(\mu,m)}\right) = 2(m+1)\Gamma(1+\mu)$ (see the definition (3.63). For the threshold value $m = 1$ we get $\left(2/p_{(\mu,m)}\right) = 4\Gamma(1+\mu)$.

3.6.3. Profile Expressions by the Effective Similarity Variable $\eta_{eff} = x/\sqrt{D_{eff(\mu,m)}t^\mu}$

It seems attractive to present the approximate profile by means of the effective similarity variable η_{eff} because the expression is simpler than that where η_μ is used. Further, all non-linear effects are incorporated in the effective diffusivity $D_{eff(\mu,m)}$ and the solution resembles formally that of the linear integer-order problem ($m = 0$ and $\mu = 1$). However, the principle question immediately arising is: Do the effects of the fractional order μ and the degenerate behaviour *via* m on the behaviours of the approximate profiles could be clearly distinguished by this approach? The answer to this question is presented in Figs. (**3.8a, b, c**). In general, these plots resemble Fig. (**3.7a,b,c**), but unfortunately the effects of the subdiffusion order μ and the nonlinearity in the diffusivity *via* m are not easily distinguishable. In this context, as a support of the previous standpoint, we can see from Fig. (**3.7d**), that the effect of the fractional order μ on the correction factor $\left(1/p_{(\mu,m)}\right)$ is weaker than that of the parameter m; in addition, the effects of μ are more distinct for low m (that is for $m \leq 1$) and are suppressed with increase in m (see also Fig. **3.7c**).

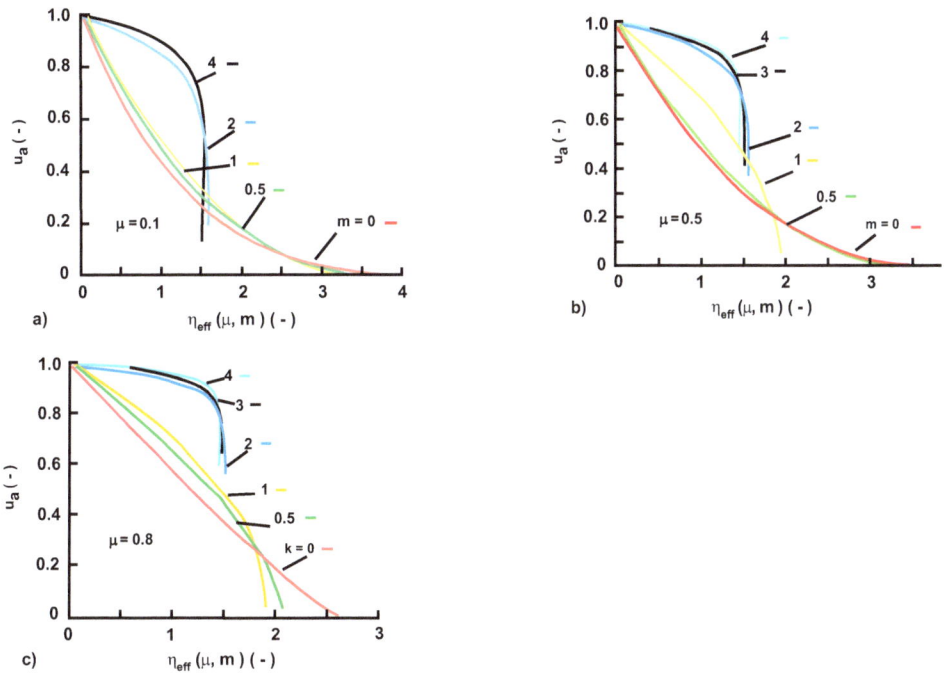

Fig. (3.8). Approximate profiles presented as functions of the effective similarity variable $\eta_{eff} = x/\sqrt{D_{eff(\mu,m)}t^{\mu}}$. a) Strong subdiffusion effects ($\mu = 0.1$) and variations of the nonlinearity of the diffusion coefficient *via* m. b) Moderate subdiffusion ($\mu = 0.5$) and various m. c) Weak subdiffusion ($\mu = 0.8$) and various m .

3.6.4. Direct Expression of the Profiles by Physical Variables x and t

Direct expression of the approximate solution in terms of the spatial coordinate x and the time t are common for the modern method such Adomian decomposition method [19, 34], variational iteration method [19], homotopy perturbation method [35], *etc.* In the context of the problem at issue, in contrast to the cases where similarity variables are used [37] (and the integral balance method used in this study) the approximate solutions of Das [19, 21, 32] have been developed as a series of terms containing explicitly powers of x and t. Moreover, there is a strong feature in these approximate solutions [19, 21, 32] specifically, the diffusion coefficient $D_{\mu 0}$ is accepted equal to 1 . For the sake of simplicity of calculations, this common mathematical approach is convenient, but quite unphysical. Precisely, even for diffusion described by the Fick law the diffusivity is order of magnitude $O(10^{-5} - O(10^{-6}))$ while for subdiffusion systems [57, 58, 59, 60] its is of order $O(10^{-9} - O(10^{-10}))$. When the diffusivity $D_{\mu 0}$ is incorporated in the similarity variable η_μ its order of magnitude does not affect the solution as it was demonstrated in [39]. Besides, the order of magnitude of $D_{\mu 0}$ does not affect the determination of the optimal exponent of the integral-balance solutions (see [39] and the optimization procedure in this work). However, the use of the proper order of magnitude in direct simulations when explicit terms contain powers of x and t is mandatory because this allows evaluating the real contributions of the terms of these series and their order of magnitudes. As a direct support of this preamble we present simulations of the developed approximate solutions with two values of the diffusion coefficient: $D_{\mu 0} = 1.10^{-9} m^2/s^\mu$ and $_{D\mu 0} = 1.10^{-9} m^2/s^\mu$. The values $x = 1.10^{-5}$ and $t = 1000s$ in the following examples are chosen for reasons of easy calculations only.

The plots in Fig. (**3.9**) a, b show how the penetration depth $\delta(t)$, the key elements of the integral-balance solution propagates in time and the effect on its length when the order of magnitudes of $D_{\mu 0}$ is changed, even in the simplest case for $m = 0$. Because $\delta \equiv \sqrt{D_{0\mu}}$, the shift of the order of from 10^{-9} to 10^{-10} results is a retardation factor (*i.e.* δ becomes shorter) of about 0.316 only as a contribution of the diffusivity change.

Further, the time evolutions of the profiles at a fixed distance from the surface clearly demonstrate the effect of the order of magnitude of the diffusion coefficient: The profiles with $D_{\mu 0} = 1.10^{-9} m^2/s^\mu$ grow faster than the ones with $D_{\mu 0} = 1.10^{-10} m^2/s^\mu$ as it demonstrated in Fig. (**3.10a, b**); the time to reach one an the same level ($u_a \approx 0.75$, for instance) is about 50 times shorter for the case with the larger diffusivity. Otherwise, if we need profiles along the space coordinate for a fixed time, then the plots in Fig. (**3.11a, b**) clearly demonstrate the simultaneous effect of the fractional order μ and the difference in the order of magnitudes of the diffusion coefficient in case of weak non-linearity ($m = 0.5$). The character of these profiles is the same as that demonstrated by expressions through the similarity variable η_μ. The case with $m = 1$, where we can observe alternation in the shape of the profile by variations in the fractional order μ is shown in Fig. (**3.12a, b**). Further, the strong non-linearity of the diffusion coefficient for $m = 3$ and the effect the order of magnitudes of the diffusion coefficient are shown in Fig. (**3.13a, b**).

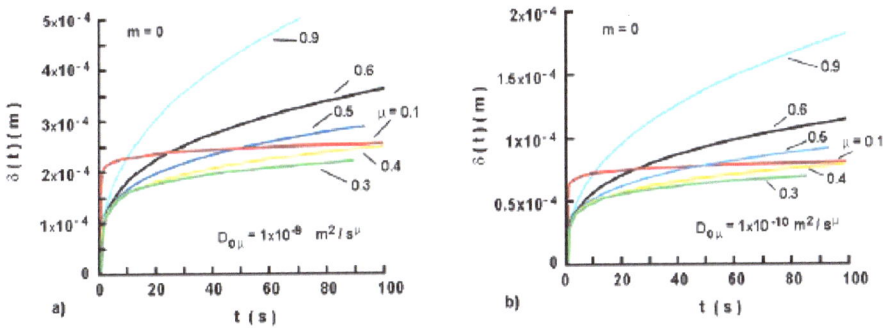

Fig. (3.9). Penetration depths $\delta(t)$ as functions of the time for two different values of the diffusion coefficient $D_{\mu 0}$. a) The case for $D_{0\mu} = 1.10^{-9} m^2/s^\mu$. b) The case for $D_{0\mu} = 1.10^{-10} m^2/s^\mu$.

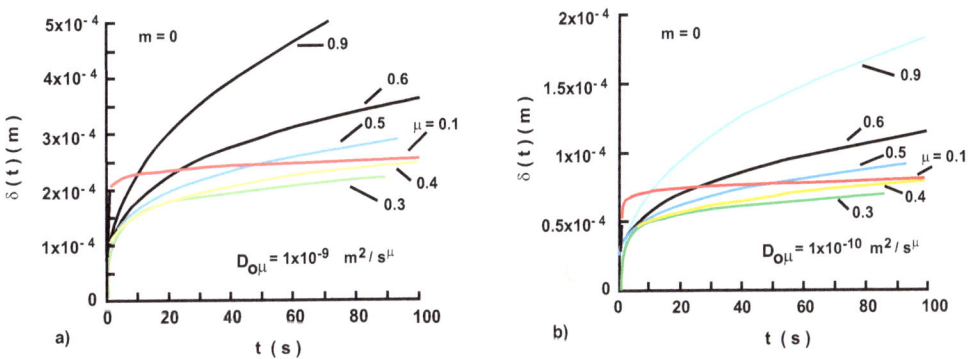

Fig. (3.10). Time evolution of the approximate profiles at $x = 1.10^{-5} m$. Effect of the order of magnitude of the diffusion coefficient. a) The case for $D_{0\mu} = 1.10^{-9} m^2/s^\mu$. b) The case for $D_{0\mu} = 1.10^{-10} m^2/s^\mu$.

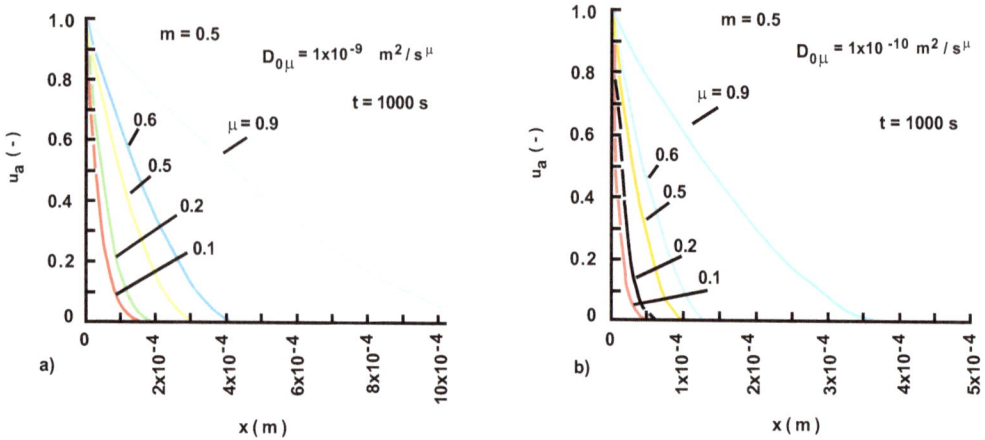

Fig. (3.11). Approximate solutions expressed against the space variable x at fixed time $t = 1000s$. Weak nonlinearity of the diffusion coefficient (represented by m). Effect of the order of magnitude of the diffusion coefficient $D_{\mu 0}$. a) The case for $D_{0\mu} = 1.10^{-9}m^2/s^\mu$. b) The case for $D_{0\mu} = 1.10^{-10}m^2/s^\mu$.

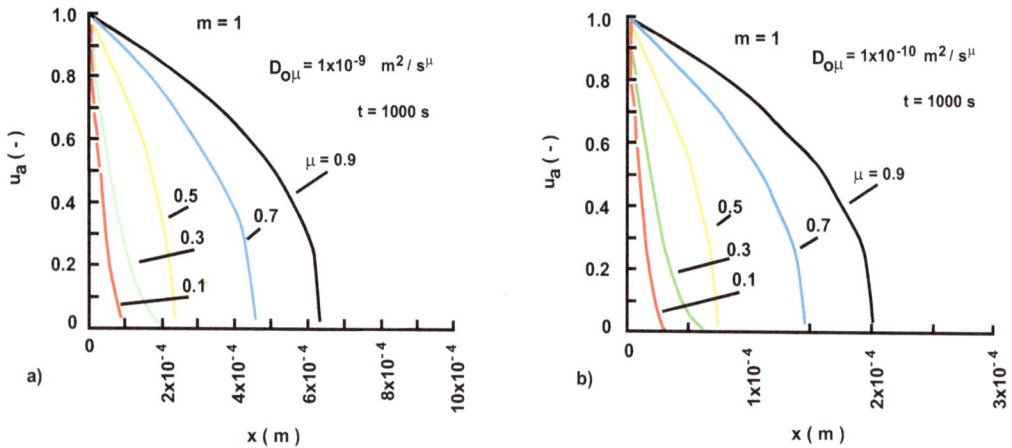

Fig. (3.12). Approximate solutions expressed against the space variable x and fixed time $t = 1000s$. Moderate nonlinearity of the diffusion coefficient ($m = 1$). Effect of the order of magnitude of the diffusion coefficient $D_{\mu 0}$. The alternation in the shape of the profile depends on the value of μ. a) The case for $D_{0\mu} = 10.10^{-9}m^2/s^\mu$. b) The case for $D_{0\mu} = 10.10^{-10}m^2/s^\mu$.

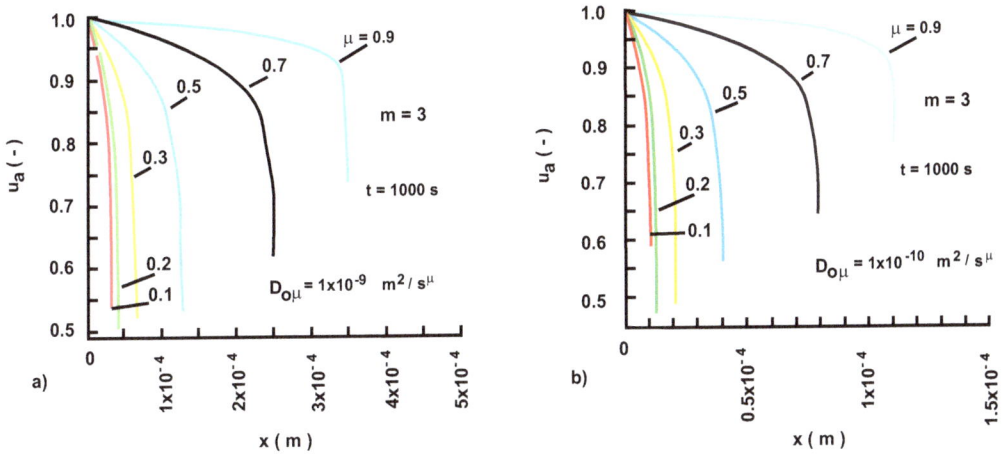

Fig. (3.13). Approximate solutions expressed against the space variable x and fixed time $t = 1000s$. Strong nonlinearity of the diffusion coefficient ($m = 3$) resulting in strong convex profiles with steep fronts over the entire range $0 < \mu < 1$. Effect of the order of magnitude of the diffusion coefficient $D_{\mu 0}$ a) The case for $D_{0\mu} = 1.10^{-9} m^2/s^{\mu}$. b) The case for $D_{0\mu} = 1.10^{-10} m^2/s^{\mu}$.

3.6.5. Comparative Analysis

In the context of a comparative analysis, the graphical results of Das *et al.* [32] obtained with $D_{0\mu} = 1$ and $m = 0.5$, $m = 2/3$, $m = 1.5$ and $m = 2$ for $x = 1$, also reveal slight changes in the profile shape from concave to convex (Fig. **3.2** in ref.[32]). However, an analysis of the solution behaviour like that developed in the present study has not been performed in [32]. In the context of the present point, we have to refer to the work of Sun *et al.* [36] where the profiles (solutions) of the fractal Richards equation with a power-law diffusivity expressed through x and t were classified by means of the non-Boltzmann scaling x/t^{μ} as a parameter. This is equivalent to the approach used in this work by discrimination of the profile with respect to the fractional order μ.

The approximate numerical solution of Sun *et al.* [36] with $m = 2$ ($D_{0\mu} = 10^{-4}$ and $t = 100s$) based on the experimental work [61] resulted in convex profiles for $\mu/2 = 0.4, \mu/2 = 0.5$, and $\mu/2 = 0.6$ (see Fig. **3.2b** in ref. [36]). Even though only the former fractional order was used in the present analysis this is in agreement with the developed results. In view to establish what is the real value of the non-linear parameter m Sun *et al.* [36] did a regression analysis of the experimental data of El-Ghany *et al.* [61] about water propagation in a fired-clay brick. They obtained $D_{0\mu} \approx 0.075$ and $m \approx 1.75$ represented by profiles (smooth convex shapes) similar to ones in the present work obtained for $\mu < 0.5$ and $1 < m < 2$. Further, for water penetration in a siliceous brick material [61], the corresponding values are $D_{0\mu} \approx 0.0.98$ and $m \approx 8.2$; the profiles (convex shapes with steep front sections) correspond to those obtained with any μ in the subdiffusion range and $m > 2$.

These are only examples showing particular cases and the possibility of the developed solutions to model certain practical situations. The real generalization of the results developed by the integral balance solutions, however, is the approach allowing establishing interrelations between the variables and parameters controlling the process of diffusion, as well as their physical relevance through the similarity variable. Besides, the comparison with the results of Sun *et al.* [36] (developed numerically) and the experimental results of Ghany *et al.* [61] confirms the physical relevance of the developed approximate integral-balance solutions.

3.7. CONCLUSION

The chapter presents an efficient approximate method applying the double integral-balance techniques to a non-linear diffusion equation involving time-fractional derivatives. The solution performed by a general parabolic profile with unspecified exponent gives more freedom in the tuning of the approximate solution (by definition of optimal exponents) in contrast to the cases when polynomial profiles (quadratic or cubic) are used.

The important feature of the finite penetration depth concept and the double-integration method is the possibility to crate closed-form solutions allowing post-solution analyzes, which for instance are impossible with the exact solutions expressed as infinite series. The double integration method provides straightforward analytical solutions with both the Riemann-Liouville and the Caputo derivatives. The method was exemplified by the Dirichlet problem with zero initial conditions for the sake of simplicity in the demonstration of the technology of the solutions.

Principle point in the developed approximate solutions is the determination of the optimal exponents. In general, the values of the optimal exponent decrease with increase in the fractional order μ and the nonlinear parameter m of the diffusion coefficient.

The solutions developed allow clearly to demonstrate the competition between the nonlinearities imposed by the fractional order μ and the parameter m and how it changes the shape of the approximate profiles from concave (weak effect of m) to convex (strong effect of m) and *vice versa*.

CONSENT FOR PUBLICATION

Not applicable.

CONFLICT OF INTEREST

The authors declare no conflict of interest, financial or otherwise.

ACKNOWLEDGEMENTS

Declared none

REFERENCES

[1]   N. Korabel, R. Klages, A. V. Chechkin, I. M. Sokolov, and V. Y. Gonchar, "Fractal properties of anomalous diffusion in intermittent maps," *Phys. Rev. E*, vol. 75, p. 036213, Mar. 2007.

[2]   T. Kosztołowicz, "Subdiffusion in a system with a thick membrane," *J. Membrane Science*, vol. 320, pp. 492–499, Jul. 2008.

[3]   A. Dokoumetzidis and P. Macheras, "Fractional kinetics in drug absorption and disposition processes," *J. pharmacokinetics and pharmacodynamics*, vol. 36, pp. 165–178, Apr. 2009.

[4]   R. Nigmatullin, "The realization of the generalized transfer equation in a medium with fractal geometry," *Physica Status Solidi (B)*, vol. 133, pp. 425–430, Jan. 1986.

[5]   Y. Li, G. Farrher, and R. Kimmich, "Sub-and superdiffusive molecular displacement laws in disordered porous media probed by nuclear magnetic resonance," *Phys. Rev. E*, vol. 74, p. 066309, Dec. 2006.

[6]   B. Berkowitz, H. Scher, and S. E. Silliman, "Anomalous transport in laboratory-scale, heterogeneous porous media," *Water Resources Research*, vol. 36, pp. 149–158, Jan. 2000.

[7]   Y. Povstenko, "Time-fractional radial diffusion in a sphere," *Nonlinear Dyn.*, vol. 53, pp. 55-65, Jul. 2008.

[8]   R. Metzler and J. Klafter, "The random walk's guide to anomalous diffusion: a fractional dynamics approach," *Physics reports*, vol. 339, pp. 1–77, Dec. 2000.

[9]   R. Metzler and J. Klafter, "The restaurant at the end of the random walk: recent developments in the description of anomalous transport by fractional dynamics," *J. Phys. A*, vol. 37, p. R161, Jul. 2004.

[10]   J. Trujillo, "Fractional models: Sub and super-diffusive, and undifferentiable solutions," *Innovation in Engineering Computational Technology*, pp. 371–402, 2006.

[11]   H. Sun, W. Chen, and Y. Chen, "Variable-order fractional differential operators in anomalous diffusion modeling," *Phys. A*, vol. 388, pp. 4586–4592, Nov. 2009.

[12]   W. Chen, H. Sun, X. Zhang, and D. Korosak, "Anomalous diffusion modeling by fractal and fractional derivatives," *Comput. Math. Appl.*, vol. 59, pp. 1754–1758, Mar. 2010.

[13]   W. Chen, Y. Liang, and X. Hei, "Structural derivative based on inverse mittag-leffler function for modeling ultraslow diffusion," *Fract. Calc. Appl. Anal.*, vol. 19, pp. 1250–1261, Oct. 2016.

[14]   J. Nakagawa, K. Sakamoto, and M. Yamamoto, "Overview to mathematical analysis for fractional diffusion equationsnew mathematical aspects motivated by industrial collaboration," *J. Math-for-*

Industry, vol. 2, no. 10, pp. 99–108, 2010.

[15] B. N. Achar and J. W. Hanneken, "Fractional radial diffusion in a cylinder," *J. Molecular Liquids*, vol. 114, pp. 147–151, Sep. 2004.

[16] V. D. Djordjevic and T. M. Atanackovic, "Similarity solutions to nonlinear heat conduction and burgers/korteweg–devries fractional equations," *J. Comput. Appl. Math.*, vol. 222, pp. 701–714, Dec. 2008.

[17] T. Langlands, "Solution of a modified fractional diffusion equation," *Phys. A*, vol. 367, pp. 136–144, Jul. 2006.

[18] Y. F. Luchko and H. Srivastava, "The exact solution of certain differential equations of fractional order by using operational calculus," *Comput. Math. Appl.*, vol. 29, pp. 73–85, Apr. 1995.

[19] S. Das, P. K. Gupta, and P. Ghosh, "An approximate analytical solution of nonlinear fractional diffusion equation," *Int. J. Nonlinear Science*, vol. 12, pp. 339–346, Dec. 2011.

[20] K. Diethelm, N. J. Ford, A. D. Freed, and Y. Luchko, "Algorithms for the fractional calculus: a selection of numerical methods," *Comput.Methods in Appl.Mech.Eng.*, vol. 194, pp. 743-773, Feb. 2005.

[21] S. Das, "A note on fractional diffusion equations," *Chaos Solitons Fractals*, vol. 42, pp. 2074–2079, Nov. 2009.

[22] A. Al-rabtah, V. S. Ert"urk, and S. Momani, "Solutions of a fractional oscillator by using differential transform method," *Comput. Math. Appl.*, vol. 59, pp. 1356–1362, Feb. 2010.

[23] J. Hristov, "Heat-balance integral to fractional (half-time) heat diffusion sub-model," *Thermal Science*, vol. 14, pp. 291–316, Mar. 2010.

[24] J. Hristov, "Starting radial subdiffusion from a central point through a diverging medium (a sphere): heat-balance integral method," *Thermal Science*, vol. 15, pp. 5–20, Jan. 2011.

[25] J. Hristov, "Approximate solutions to fractional subdiffusion equations," *Eur. Phys. J.Spec. Top.*, vol. 193, pp. 229–243, Mar. 2011.

[26] J. Hristov, "Double integral-balance method to the fractional subdiffusion equation: approximate solutions, optimization problems to be resolved and numerical simulations," *J. Vib. Control*, vol. 23, pp. 2795–2818, Oct. 2017.

[27] J. Hill, "Similarity solutions for nonlinear diffusiona new integration procedure," *J. Engrg. Math.*, vol. 23, pp. 141–155, Jun. 1989.

[28] N. Smyth and J. Hill, "High-order nonlinear diffusion," *IMA J. Appl. Math.*, vol. 40, pp. 73-86, Jan. 1988.

[29] R. Pattle, "Diffusion from an instantaneous point source with a concentration-dependent coefficient," *Quart. J. Mech. Appl. Math.*, vol. 12, pp. 407–409, Jan. 1959.

[30] S. Prasad and J. Salomon, "A new method for the analytical solution of a degenerate diffusion equation," *Adv. Water. Resour.*, vol. 28, pp. 1091–1101, Oct. 2005.

[31] M. L. Gandarias, "New symmetries for a model of fast diffusion," *Phys. Lett. A*, vol. 286, pp. 153–160, Jul. 2001.

[32] S. Das, K. Vishal, and P. Gupta, "Solution of the nonlinear fractional diffusion equation with absorbent term and external force," *Appl. Math. Model.*, vol. 35, pp. 3970–3979, Aug. 2011.

[33] J.-S. Guo and Y.-J. L. Guo, "On a fast diffusion equation with source," *Tohoku Math. J. (2)*, vol. 53, no. 4, pp. 571–579, 2001.

[34] G. Adomian, *Solving frontier problems of physics: the decomposition method*, vol. 60. Springer: New York, 2013.

[35] J.-H. He, "Homotopy perturbation technique," *Comput. Methods Appl. Mech. Engrg.*, vol. 178, pp. 257–262, Aug. 1999.

[36] H. Sun, M. M. Meerschaert, Y. Zhang, J. Zhu, and W. Chen, "A fractal richards equation to capture the non-boltzmann scaling of water transport in unsaturated media," *Adv. Water. Resour.*, vol. 52, pp. 292–295, Feb. 2013.

[37] Ł. Płociniczak and H. Okrasi'nska, "Approximate self-similar solutions to a nonlinear diffusion equation with time-fractional derivative," *Phys. D*, vol. 261, pp. 85–91, Oct. 2013.

[38] T. R. Goodman, "Application of integral methods to transient nonlinear heat transfer," *Adv. Heat Transfer*, vol. 1, pp. 51–122, Dec. 1964.

[39]   J. Hristov, "A short-distance integral-balance solution to a strong subdiffusion equation: a weak power-law profile," *Int. Rev. Chem. Eng.*, vol. 2, pp. 555–563, Sep. 2010.

[40]   J. Hristov, "Transient flow of a generalized second grade fluid due to a constant surface shear stress: an approximate integral-balance solution," *Int. Rev. Chem. Eng.*, vol. 3, pp. 802–809, Nov. 2011.

[41]   J. Hristov, "Integral-balance solution to the stokes first problem of a viscoelastic generalized second grade fluid," *Thermal Science*, vol. 16, no. 2, pp. 395–410, 2012.

[42]   J. Hristov, "A note on the integral approach to non-linear heat conduction with Jeffrey's fading memory," *Thermal Science*, vol. 17, no. 3, pp. 733–737, 2013.

[43]   J. Hristov, "Approximate solutions to time-fractional models by integral balance approach,"

[44]   J. Hristov, "Diffusion models with weakly singular kernels in the fading memories: How the integral-balance method can be applied?," *Thermal Science*, vol. 19, pp. 947–957, May 2015.

[45]   J. Hristov, "Integral solutions to transient nonlinear heat (mass) diffusion with a power-law diffusivity: a semi-infinite medium with fixed boundary conditions," *Heat Mass Transfer*, vol. 52, pp. 635–655, Mar. 2016.

[46]   A. Fabre and J. Hristov, "On the integral-balance approach to the transient heat conduction with linearly temperature-dependent thermal diffusivity," *Heat Mass Transfer*, vol. 53, pp. 177–204, Jan 2017.

[47]   I. Podlubny, *Fractional Differential Equations*, vol. 198. Academic Press: San Diego, 1998.

[48]   T. Myers, "Optimizing the exponent in the heat balance and refined integral methods," *Int. Comm. Heat Mass Transfer*, vol. 36, pp. 143–147, Feb 2008.

[49]   N. Sadoun, E.-K. Si-Ahmed, and P. Colinet, "On the refined integral method for the one-phase stefan problem with time-dependent boundary conditions," *Appl. Math. Model.*, vol. 30, pp. 531–544, Jun.

[50]   J. Hristov, "The heat-balance integral method by a parabolic profile with unspecified exponent: Analysis and benchmark exercises," *Thermal Science*, vol. 13, no. 2, pp. 27–48, 2009.

[51]   J. Hristov, "Transient flow of a generalized second grade fluid due to a constant surface shear stress: an approximate integral-balance solution," *Int. Rev. Chem. Eng.*, vol. 3, pp. 802–809, Nov 2010.

[52]   S. L. Mitchell and T. Myers, "Application of standard and refined heat balance integral methods to one-dimensional stefan problems," *SIAM review*, vol. 52, pp. 57–86, Feb.

[53]   W. F. Ames, *Nonlinear partial differential equations in engineering*, vol. 18. Academic Press: New York, 1965.

[54]   C. Zener, "Theory of growth of spherical precipitates from solid solution," *J. Appl. Phys.*, vol. 20, pp. 950–953, Oct 1949.

[55]   K. Oldham and J. Spanier, *The Fractional Calculus*, vol. 111. Academic Press: New York, 1974.

[56]   D. Langford, "The heat balance integral method," *Int. J. Heat Mass Transfer*, vol. 16, pp. 2424–2428, Dec.

[57]   T. Kosztołowicz, K. Dworecki, *et al.*, "How to measure subdiffusion parameters," *Phys. Rev. Lett.*, vol. 94, p. 170602, May 2005.

[58]   T. Kosztolowicz, "Transport in diffusive–subdiffusive system," *Acta Phys. Polon. B*, vol. 36, p. 1635, May 2005.

[59]   T. Kosztołowicz, K. Dworecki, *et al.*, "Measuring subdiffusion parameters," *Phys. Rev. E*, vol. 71, p. 041105, Apr 2005.

[60]   K. Dworecki, A. ´ Slzak, B. Ornal-Wasik, and S. Wasik, "Evolution of concentration field in a membrane system," *J Biochem. Biophys. Methods*, vol. 62, pp. 153-162, Feb 2005.

[61]   A. E.-G. El Abd and J. J. Milczarek, "Neutron radiography study of water absorption in porous building materials: anomalous diffusion analysis," *J. Phys D: Appl. Phys*, vol. 37, p. 2305, Jul 2004.

Analysis of Solution and System Identification of Coupled Fractional Delay Differential Equation by Shifted Jacobi Polynomials

B. Ganesh Priya, P. Muthukumar and P. Balasubramaniam[*]

Department of Mathematics, The Gandhigram Rural Institute-Deemed University, Gandhigram 624 302, India

Abstract: This chapter is concerned with a class of linear delay differential equation with two fractional orders. Based on the shifted Jacobi operational matrix method, the numerical approximate solution as well as the parameters estimation for the proposed system are obtained. The maximum error bounds of approximate solution and operational matrix for fractional integration are analyzed. The main results are illustrated with suitable examples.

Keywords: Fractional delay differential equation, Linear estimation, Operational matrix, Shifted Jacobi polynomial.

AMS Subject Classification: 33C45, 34A08, 34K28.

4.1. INTRODUCTION

In literature, analytic results on the existence and uniqueness of solutions for delay differential equations of integer order as well as fractional order have been investigated by many authors (see [1, 2] and references therein). Also, in recent years, fractional calculus have found wide applications in the field of engineering and physics [3, 4, 5] and it may be considered a new topic as well. In general, most of the fractional differential equations (FDEs) and fractional delay differential equations (FDDEs) are not solvable towards exact solutions. Therefore, there has been significant interest in developing numerical methods for solving FDEs [6] and FDDEs [7]. During the last decades, several methods have been used to solve FDDEs (see [8, 9, 10, 11] and references therein). In particular, orthogonal polynomials have received considerable attention for solving various

---
[*]**Corresponding author P. Balasubramaniam:** Department of Mathematics, The Gandhigram Rural Institute-Deemed University, Gandhigram 624 302, India; Tel: 91-451-2452371; Fax: 91-451-2454466; E-mail: balugru@gmail.com

types of problems in fractional calculus [12]. The main characteristic behind this approach is the dynamical problem reduces to a system of algebraic equations. Most importantly, the shifted Jacobi polynomial is used for solving the FDEs [13] and FDDEs [14].

In general, it is important to analyze and identify the parameters of many dynamical systems. Particularly, orthogonal polynomials are predominantly used by many authors (see [15-18]) to deal the system identification and parameter estimation of integer order linear time delay system.

In this chapter, we extend the shifted Jacobi polynomial to find approximate solutions and estimation of unknown parameters for a coupled delay differential equations with two different fractional orders.

4.2. PRELIMINARIES

In this section, we recall some basic definitions of fractional calculus and necessary properties of shifted Jacobi polynomials, operational matrices for fractional integration and Jacobi delay function (see [14] for more details).

4.2.1. Fractional Calculus

Definition 4.1. *The Riemann-Liouville(R-L) fractional integral of order $\alpha \in \mathbb{R}_+$ of a function $x(t) \in L^1([t_0, t_f], \mathbb{R})$ is defined by*

$$I_{t_0^+}^\alpha x(t) = \frac{1}{\Gamma(\alpha)} \int_{t_0^+}^t \frac{x(s)}{(t-s)^{1-\alpha}} ds,$$

provided that the integral on right hand side exists, where $\Gamma(\cdot)$ is the gamma function.

Definition 4.2. *The Caputo derivative of order α with the lower limit t_0^+ for a function $x(t) \in L^1([t_0, t_f], \mathbb{R})$ can be written as*

$$^c D_{t_0^+}^\alpha x(t) = \frac{1}{\Gamma(n-\alpha)} \int_{t_0^+}^t \frac{x^{(n)}(s)}{(t-s)^{\alpha+1-n}} ds = I^{n-\alpha} x^{(n)}(t), 0 \le n-1 < \alpha < n,$$

provided that the right hand side is point-wise defined on $[t_0, \infty)$ where $n = [\alpha] + 1, [\alpha]$ denotes the integral value of α.

From the Definition 4.1 and Definition 4.2, the following results hold:

$$^cD_{0+}^{\alpha}\theta = 0, \text{ for a constant } \theta,$$

$$I_t^{\alpha}\left(^cD_{t_0^+}^{\alpha}x(t)\right) = x(t) - \sum_{k=0}^{n-1} x^{(k)}(t_0^+)\frac{(t-t_0)^k}{k!}, 0 \leq n-1 < \alpha < n. \quad (4.1)$$

$$^cD_{0+}^{\alpha}x^{\beta} = \begin{cases} 0, & \text{for } \beta < \lceil\alpha\rceil, \\ \dfrac{\Gamma(1+\beta)}{\Gamma(1+\beta-\alpha)}x^{\beta-\alpha}, & \text{for } \beta \geq \lceil\alpha\rceil, \end{cases}$$

$$I_{0+}^{\alpha}x^{\beta} = \frac{\Gamma(1+\beta)}{\Gamma(1+\beta+\alpha)}x^{\alpha+\beta}.$$

Here the ceiling function $\lceil\alpha\rceil$ is denoted as smallest integer greater than or equal to α.

Definition 4.3. *The beta integral formula for any finite interval* $[a,b] \in \mathbb{R}$ *is defined as*

$$\mathcal{B}(u,v) = \int_a^b (s-a)^{u-1}(b-s)^{v-1}ds = (b-a)^{u+v-1}\frac{\Gamma(u)\Gamma(v)}{\Gamma(u+v)},$$

where $u, v > 0$.

4.2.2. Shifted Jacobi Polynomials

In this section, we study some necessary concepts about the shifted Jacobi polynomials, which are useful for the proof of main results.

4.2.2.1. Properties of Shifted Jacobi Polynomials

The Jacobi polynomial $P_i^{(a,b)}(z)$ are orthogonal with Jacobi weight function $w(z) = (1-z)^a(z-1)^b$ over the interval $-1 \leq z \leq 1$, it means that

$$\int_{-1}^1 P_i^{(a,b)}(z)P_j^{(a,b)}(z)w(z)dz = \gamma_i^{(a,b)}\delta_{ij},$$

where δ_{ij} is the Kronecker function and

$$\gamma_i^{(a,b)} = \frac{2^{a+b+1}\Gamma(i+a+1)\Gamma(i+b+1)}{(2i+a+b+1)\Gamma(i+1)\Gamma(i+a+b+1)}.$$

For any $a, b > -1$, the weight function $w(z) \in L^1([-1,1], \mathbb{R})$. The Jacobi polynomials are generated from the following recurrence relation

$$P_0^{(a,b)}(z) = 1, P_1^{(a,b)}(z) = \frac{1}{2}[(a + b + 2)z + a - b],$$

$$\vdots$$

$$2i(i + a + b)(2i + a + b - 2)P_i^{(a,b)}(z)$$

$$= (2i + a + b - 1)[(2i + a + b)(2i + a + b - 2)z + a^2 - b^2]$$

$$\times P_{i-1}^{(a,b)}(z) - 2(i + a - 1)(i + b - 1)(2i + a + b)P_{i-2}^{(a,b)}(z).$$

In general, the ith degree Jacobi polynomial is given by the following relation

$$P_i^{(a,b)}(z) = (-1)^i \frac{(b+1)_i}{\Gamma(i+1)} \sum_{k=0}^{i} \frac{(-i)_k(i+a+b+1)_k}{(b+1)_k\Gamma(k+1)} (\frac{1+z}{2})^k,$$

where,

$$(b + 1)_0 = 1, (b + 1)_i = (b + 1)(b + 2) \cdots (b + i) = \frac{\Gamma(b+1+i)}{\Gamma(b+1)}.$$

In order to study the system in an arbitrary interval $[t_0, t_f]$, we change an independent variable $z \in [-1,1]$ into time variable t between t_0 and t_f by using the relation $z = \frac{2t-t_0-t_f}{t_f-t_0}$. Let the shifted Jacobi polynomial $P_i^{(a,b)}\left(\frac{2t-t_0-t_f}{t_f-t_0}\right)$ be denoted by $J_i^{(a,b)}(t)$ and obtained as

$$J_0^{(a,b)}(t) = 1,$$

$$J_1^{(a,b)}(t) = -[\frac{(a+1)t_0+(b+1)t_f}{t_f-t_0}] + [\frac{(a+b+2)t}{t_f-t_0}],$$

$$\vdots$$

$$J_i^{(a,b)}(t) = (-1)^i \frac{(b+1)_i}{\Gamma(i+1)} \sum_{k=0}^{i} \frac{(-i)_k(i+a+b+1)_k}{(b+1)_k\Gamma(k+1)} [\frac{t-t_0}{t_f-t_0}]^k.$$

Now we write the recurrence relation of the above shifted Jacobi polynomial is given as follows:

$$2i(i + a + b)(2i + a + b - 2)J_i^{(a,b)}(t)$$

$$= (2i + a + b - 1)[(2i + a + b)(2i + a + b - 2)(\frac{2t - t_0 - t_f}{t_f - t_0})$$

$$+a^2 - b^2]J_{i-1}^{(a,b)}(t) - 2(i + a - 1)(i + b - 1)(2i + a + b)J_{i-2}^{(a,b)}(t).$$

In general, the ith order shifted Jacobi polynomial can be represented as

$$J_i^{(a,b)}(t) = \sum_{k=0}^i \frac{(-1)^{i-k}\Gamma(i+b+1)\Gamma(i+k+a+b+1)}{\Gamma(k+b+1)\Gamma(i+a+b+1)\Gamma(i-k+1)\Gamma(k+1)} [\frac{t-t_0}{t_f-t_0}]^k. \quad \textbf{(4.2)}$$

The shifted Jacobi polynomial (4.2) with respect to the weight function $W^{(a,b)}(t) = (t - t_0)^b(t_f - t)^a$ is defined in the interval $[t_0, t_f]$ and it satisfies the following orthogonality property:

$$\int_{t_0}^{t_f} W^{(a,b)}(t)J_i^{(a,b)}(t)J_j^{(a,b)}(t)dt = \begin{cases} 0, & i \neq j, \\ h_{ij}^{(a,b)}, & i = j, \end{cases}$$

where,

$$h_{ij}^{(a,b)} = \frac{\Gamma(i+a+1)\Gamma(i+b+1)(t_f-t_0)^{a+b+1}}{\Gamma(i+a+b+1)\Gamma(i+1)(2i+a+b+1)}.$$

Remark. From the shifted Jacobi polynomials (4.2), the following results can be easily obtained

1. $J_i^{(a,b)}(t_0) = (-1)^i \binom{i + b}{i}$.

2. $J_i^{(a,b)}(t_f) = \binom{i + a - 1}{i}$.

3. $\frac{d^j}{dt^j}J_i^{(a,b)}(t) = \frac{\Gamma(i+a+b+j+1)}{\Gamma(i+a+b+1)}J_{i-j}^{(a+j,b+j)}(t)$.

Lemma 4.1. *The shifted Jacobi polynomial $J_i^{(a,b)}(t)$ defined in (4.2) which also satisfies*

$$J_i^{(a,b)}(t) = \sum_{k=0}^i J_k^{(i)} \left(\frac{t-t_0}{t_f-t_0}\right)^k,$$

where $J_k^{(i)} = (-1)^{(i-k)} \binom{i+a+b+k}{k} \binom{i+b}{i-k}, k = 0,1,\ldots,m.$

Proof. The proof of this lemma is similar to the proof of Lemma 1 in [6] by applying the equation

$$J_k^{(i)} = \frac{1}{k!} \frac{d^k}{dt^k} J_i^{(a,b)}(t) \Big|_{t=t_0}.$$

4.2.2.2. Function Approximation Using the Shifted Jacobi Polynomials

Definition 4.4. *[19] Let $x(t) \in L^2([t_0, t_f], \mathbb{R})$ can be approximated by using the shifted Jacobi polynomials as*

$$x(t) \simeq \sum_{i=0}^m x_i J_i^{(a,b)}(t) = X^T J^{(a,b)}(t), \tag{4.3}$$

where $X^T = [x_0 \quad x_1 \quad x_2 \quad \cdots \quad x_m], J^{(a,b)}(t) = \left[J_0^{(a,b)}(t) \quad J_1^{(a,b)}(t) \quad \cdots \quad J_m^{(a,b)}(t) \right]^T.$

Here the coefficients x_i satisfies

$$x_i = \frac{1}{h_{ij}^{(a,b)}} \int_{t_0}^{t_f} x(t) J_i^{(a,b)}(t) W^{(a,b)}(t) dt, i = 0,1,2,\ldots,m.$$

Remark. In particular, any constant function $x(t) = \theta, t \in [t_0, t_f]$, can be expressed as:

$$x(t) \simeq X^T J^{(a,b)}(t), \text{ where } X^T = [\theta \quad 0 \quad \cdots \quad 0]_{1 \times m+1}. \tag{4.4}$$

For example $x(t) = 5, 0 \le t \le 1, a = 0, b = 0$ and $m = 3$ then the approximation (4.4) is written as

$$x(t) \simeq [5 \quad 0 \quad 0 \quad 0][1 \quad 2t-1 \quad 6t^2-6t+1 \quad 20t^3-30t^2+12t-1]^T = 5.$$

From the Definition 4.4, the approximation of any delay function $x(t-\tau) \in L^2([t_0 - \tau, t_f], \mathbb{R})$ can be written as (see [14])

$$x(t-\tau) \simeq X^T J^{(a,b)}(t-\tau) - X^T H_{\iota} J^{(a,b)}(t). \tag{4.5}$$

Here

$$H_\tau = \begin{pmatrix} h_{0,0} & 0 & 0 & \cdots & 0 \\ h_{1,0} & h_{1,1} & 0 & \cdots & 0 \\ \vdots & \cdots & \ddots & \cdots & \vdots \\ \vdots & \cdots & \cdots & \ddots & \vdots \\ h_{m,0} & h_{m,1} & h_{m,2} & \cdots & h_{m,m} \end{pmatrix}_{(m+1)\times(m+1)}$$

is the operational matrix of shifted Jacobi delay function. The elements $h_{i,j}$ are obtained in the following recursion:

$$h_{0,0} = 1; h_{1,0} = -\tau(a+b+2)/(t_f - t_0);$$

$$h_{1,1} = 1; h_{i,j} = 0 \text{ for } j > i;$$

$$2i(i+a+b)(2i+a+b-2)h_{i,0} = (2i+a+b)(2i+a+b-1)(2i+a+b-2)$$

$$\times [\frac{2(a+1)(b+1)}{(3+a+b)(2+a+b)}h_{i-1,1} - \frac{a-b}{a+b+2}h_{i-1,0}]$$

$$+(2i+a+b-1)[a^2 - b^2 - \frac{2\tau}{t_f - t_0}(2i+a+b)(2i+a+b-2)]h_{i-1,0}$$

$$-2(i+a-1)(i+b-1)(2i+a+b)h_{i-2,0};$$

$$2i(i+a+b)(2i+a+b-2)h_{i,j} = (2i+a+b)(2i+a+b-1)(2i+a+b-2)$$

$$\times [\frac{2j(j+a+b)}{(2j+a+b)(2j+a+b-1)}h_{i-1,j-1}$$

$$-\frac{a^2 - b^2}{(2j+a+b)(2j+a+b+2)}h_{i-1,j}$$

$$+\frac{2(j+a+1)(j+b+1)}{(2j+a+b+3)(2j+a+b+2)}h_{i-1,j+1}]$$

$$+(2i+a+b-1)[a^2 - b^2 - \frac{2\tau}{t_f - t_0}(2i+a+b)(2i+a+b-2)]h_{i-1,j}$$

$$-2(i+a-1)(i+b-1)(2i+a+b)h_{i-2,j}, \text{ for } j \geq 1.$$

In order to find the approximate value of fractional order integration, the following lemma is used in the main result.

Lemma 4.2. (Operational matrix for fractional integration). *Let $J^{(a,b)}(t)$ be the shifted Jacobi vector and $\eta > 0$ then*

$$I_{t_0+}^{\eta} J^{(a,b)}(t) \simeq P_{t_0+}^{\eta} J^{(a,b)}(t), \tag{4.6}$$

where $P_{t_0+}^{\eta}$ is the $(m+1) \times (m+1)$ operational matrix of fractional integration of order η in the R-L sense and is defined as follows:

$$P_{t_0+}^{\eta} = \begin{pmatrix} \Omega_{\eta}^{(a,b)}(0,0) & \Omega_{\eta}^{(a,b)}(0,1) & \cdots & \Omega_{\eta}^{(a,b)}(0,m) \\ \Omega_{\eta}^{(a,b)}(1,0) & \Omega_{\eta}^{(a,b)}(1,1) & \cdots & \Omega_{\eta}^{(a,b)}(1,m) \\ \vdots & \cdots & \cdots & \cdots \\ \Omega_{\eta}^{(a,b)}(i,0) & \Omega_{\eta}^{(a,b)}(i,1) & \cdots & \Omega_{\eta}^{(a,b)}(i,m) \\ \vdots & \cdots & \cdots & \cdots \\ \Omega_{\eta}^{(a,b)}(m,0) & \Omega_{\eta}^{(a,b)}(m,1) & \cdots & \Omega_{\eta}^{(a,b)}(m,m) \end{pmatrix},$$

where,

$$\Omega_{\eta}^{(a,b)}(i,j) = (t_f - t_0)^{\eta} \sum_{k=0}^{i} \frac{(-1)^{i-k}\Gamma(i+b+1)\Gamma(i+k+a+b+1)}{\Gamma(k+b+1)\Gamma(i+a+b+1)\Gamma(i-k+1)\Gamma(k+\eta+1)}$$

$$\times \sum_{f=0}^{j} \frac{(-1)^{j-f}\Gamma(j+f+a+b+1)\Gamma(a+1)\Gamma(f+k+\eta+b+1)\Gamma(j+1)(2j+a+b+1)}{\Gamma(j+a+1)\Gamma(f+b+1)\Gamma(j-f+1)\Gamma(f+1)\Gamma(f+k+a+b+\eta+2)}.$$

Proof. The proof is similar to the proof of the Theorem 3.2 in [14]. Hence it is omitted.

4.3. ESTIMATION OF UPPER BOUNDS FOR ERROR

In this section, we provide an analytical expression for the error of approximate solution using the shifted Jacobi polynomials.

We know that the shifted Jacobi polynomials are orthogonal, then one can write:

$$J^{(a,b)} = Span\{J_0^{(a,b)}(t), J_1^{(a,b)}(t), \dots J_m^{(a,b)}(t)\},$$

is a finite dimensional vector space. The sufficiently smooth function $x(t) \in L^2([t_0, t_f], \mathbb{R})$ has the unique best approximation form $J^{(a,b)}$, say $\tilde{x}(t) \in J^{(a,b)}$ which satisfies

$$|x(t) - \tilde{x}(t)| \le |x(t) - \hat{x}(t)|, \forall \hat{x}(t) \in J^{(a,b)}.$$

In [14], Muthukumar *et al.* shows that the error bound of the approximation $\tilde{x}(t)$ with $x(t)$ as

$$|x(t) - \tilde{x}(t)| \le C[\tfrac{t_f - t_0}{m+1}]^{m+2}\sqrt{t_f - t_0}, \tag{4.7}$$

where,

$$C = \frac{1}{4} \max_{t \in [t_0, t_f]} \left| \frac{d^{m+1}x(t)}{dt^{m+1}} \right|.$$

Suppose the function $x(t)$ is $m+1$ times continuously differentiable, the following lemma gives an upper bound for the error estimate.

Lemma 4.3. *Let the function $x(t) \in L^2([t_0, t_f], \mathbb{R})$ is $m+1$ times continuously differentiable for $t_0 > 0$, denoted $x \in C^{m+1}([t_0, t_f], \mathbb{R})$. If $\tilde{x}(t) = X^T J^{(a,b)}(t)$ is the best approximation to x from $J^{(a,b)}$ then the error bound is obtained as follows:*

$$|x(t) - \tilde{x}(t)| \le \frac{\lambda}{(m+1)!}(t_f - t_0)^{m+a+b+2}B(a+1, m+b+2), \tag{4.8}$$

where $\lambda = max_{t \in [t_0, t_f]}|x^{(m+1)}(t)|$, $x^{(m+1)}$ is the $(m+1)^{th}$ derivative of x with respect to t.

Proof. Consider the mth degree Taylor polynomial of $x(t) \in L^2([t_0, t_f], \mathbb{R})$, with respect to t_0 is given as

$$\hat{x} = x(t_0) + x'(t_0)(t - t_0) + \cdots + x^{(m)}(t_0)\frac{(t - t_0)^m}{m!}, t \in [t_0, t_f],$$

which implies from remainder formula that the error bound satisfies

$$|x - \hat{x}| \le \left| x^{(m+1)}(\xi)\frac{(t-t_0)^{m+1}}{(m+1)!} \right|, \xi \in (t_0, t), t \in [t_0, t_f].$$

Since $\tilde{x}(t) = X^T J^{(a,b)}(t)$ is best approximation to x form $J^{(a,b)}$ and $\hat{x} \in J^{(a,b)}$ (see [6]), we have

$$|x(t) - \tilde{x}(t)| \le |x - \hat{x}| \le \frac{\lambda}{(m+1)!} \int_{t_0}^{t_f} (t - t_0)^{m+1} W^{(a,b)}(t) dt.$$

Since $W^{(a,b)} = (t - t_0)^b (t_f - t)^a$ and it is always positive for all $t \in [t_0, t_f]$, we have

$$|x(t) - \tilde{x}(t)| \le \frac{\lambda}{(m+1)!} \int_{t_0}^{t_f} (t - t_0)^{b+m+1}(t_f - t)^a dt,$$

$$= \frac{\lambda}{(m+1)!} (t_f - t_0)^{m+a+b+2} B(a + 1, m + b + 2).$$

As $m \to \infty$, the above inequality shows that the approximation $\tilde{x}(t)$ converges to the exact function $x(t)$.

Also, the following Theorem 4.1 gives an upper bound for estimating the error of operational matrix for fractional integration $P_{t_0+}^{\eta}$ defined in (4.6). In first, we define error vector E_η as:

$$E_\eta = I_{t_0+}^{\eta} J^{(a,b)}(t) - P_{t_0+}^{\eta} J^{(a,b)}(t) = [E_{0,\eta} \quad E_{1,\eta} \quad \cdots \quad E_{m,\eta}]^T,$$

where $E_{k,\eta} = I_{t_0+}^{\eta} J_k^{(a,b)}(t) - \sum_{j=0}^{m} \Omega_\eta^{(a,b)}(k,j) J_j^{(a,b)}(t), k = 0,1,\ldots,m$.

Theorem 4.1. *Let the error function of R-L integration for shifted Jacobi polynomials $E_{k,\eta}: [t_0, t_f] \to \mathbb{R}$ be $m + 1$ times continuously differentiable for $t_0 > 0, E_{k,\eta} \in C^{m+1}([t_0, t_f], \mathbb{R})$ and $\eta < m + 1$, then the error bound satisfies the following inequality*

$$|E_{k,\eta}| \le \frac{(t_f - t_0)^{\eta+a+b}}{(m+1)!|\Gamma(\eta-m)|} \binom{k+a-1}{k} B(a + 1, m + b + 2). \qquad (4.9)$$

Proof. Let D^{m+1} be the $(m + 1)^{th}$ derivative operator with respect to t. In order to find $D^{m+1}\left(I_{t_0+}^{\eta} J^{(a,b)}(t)\right)$, we need the following inequality:

$$D^{m+1} I_{t_0+}^{\eta} \left(\frac{t-t_0}{t_f-t_0}\right)^k = (t_f - t_0)^{-k} D^{m+1}\left[\frac{\Gamma(1+k)}{\Gamma(\eta+k+1)}(t - t_0)^{k+\eta}\right],$$

$$= (t_f - t_0)^{-k} \frac{\Gamma(1+k)}{\Gamma(\eta+k+1)} \frac{(t-t_0)^{k+\eta-m-1}}{k+\eta-m-1},$$

$$\leq \frac{(t_f-t_0)^{\eta-m-1}}{|\Gamma(\eta-m)|}.$$

Then by using Lemma 4.1, we have

$$D^{m+1} I^{\eta}_{t_0^+} J^{(a,b)}(t) = \sum_{k=0}^{i} J_k^{(i)} D^{m+1} I^{\eta}_{t_0^+} \left(\frac{t-t_0}{t_f-t_0}\right)^k,$$

$$\leq \frac{(t_f-t_0)^{\eta-M}}{|\Gamma(\eta-M+1)|} \sum_{k=0}^{i} J_k^{(i)},$$

$$= \frac{(t_f-t_0)^{\eta-m-1}}{|\Gamma(\eta-m)|} J_k^{(a,b)}(t_f),$$

$$= \frac{(t_f-t_0)^{\eta-m-1}}{|\Gamma(\eta-m)|} \binom{k + a - 1}{k}.$$

Hence $I^{\eta}_{t_0^+} J^{(a,b)}(t)$ is $m + 1$ times differentiable function. By using Lemma 4.3, the required error bound (4.9) can easily obtained.

4.4. COUPLED FDDES WITH TWO DIFFERENT ORDERS

In this section, we find the numerical approximate solution and parameter identification for the following initial value problem of linear time-delay system with fractional orders $\alpha, \beta \in (0,1]$:

$$^{c}D^{\alpha}_{t_0^+} x(t) = A_{11}x(t) + A_{12}y(t) + B_{11}x(t - \tau) + B_{12}y(t - \tau) + u(t),$$

$$^{c}D^{\beta}_{t_0^+} y(t) = A_{21}x(t) + A_{22}y(t) + B_{21}x(t - \tau) + B_{22}y(t - \tau) + v(t),$$

$$t \in [t_0, t_f], \tag{4.10}$$

with the initial conditions

$$x(t) = \phi(t), y(t) = \psi(t), t_0 - \tau \leq t \leq t_0, \tag{4.11}$$

where $^{c}D^{\alpha}_{t_0^+}$ and $^{c}D^{\beta}_{t_0^+}$ denotes the Caputo fractional derivatives. Here $A_{ij}, B_{ij} \in \mathbb{R}^{n_i \times n_j}, i, j = 1,2$ are defined as

$$A_{ij} = \begin{bmatrix} a_1^{ij} & a_2^{ij} & \cdots & a_{n_j}^{ij} \\ a_{n_j+1}^{ij} & a_{n_j+2}^{ij} & \cdots & a_{2n_j}^{ij} \\ \cdots & \cdots & \ddots & \cdots \\ a_{(n_i-1)n_j+1}^{ij} & a_{(n_i-1)n_j+2}^{ij} & \cdots & a_{n_i n_j}^{ij} \end{bmatrix},$$

$$B_{ij} = \begin{bmatrix} b_1^{ij} & b_2^{ij} & \cdots & b_{n_j}^{ij} \\ b_{n_j+1}^{ij} & b_{n_j+2}^{ij} & \cdots & b_{2n_j}^{ij} \\ \cdots & \cdots & \ddots & \cdots \\ b_{(n_i-1)n_j+1}^{ij} & b_{(n_i-1)n_j+2}^{ij} & \cdots & b_{n_i n_j}^{ij} \end{bmatrix}.$$

Let $x(t), u(t) \in \mathbb{R}^{n_1}, y(t), v(t) \in \mathbb{R}^{n_2}$ be real valued matrices and let $\tau > 0$ be a fixed constant delay. The associated initial functions $\phi(t), \psi(t)$ in (4.11) are assumed to be continuous on the interval $[t_0 - \tau, t_0]$.

4.4.1. Numerical Solution of Coupled FDDEs

From the approximation (4.3) and (4.5), the shifted Jacobi series approximation of vectors $x(t), y(t), u(t), v(t), x(t - \tau)$ and $y(t - \tau)$ of system (4.10) are defined as follows. For all $t \in [t_0, t_f]$,

$$x(t) = X^T J^{(a,b)}(t), y(t) = Y^T J^{(a,b)}(t), u(t) = U^T J^{(a,b)}(t), v(t) = V^T J^{(a,b)}(t).$$

For $t \in [t_0, t_0 + \tau]$,

$$x(t - \tau) = \Phi^T J^{(a,b)}(t), y(t - \tau) = \Psi^T J^{(a,b)}(t).$$

For all $t \in [t_0 + \tau, t_f]$,

$$x(t - \tau) = X^T H_\tau J^{(a,b)}(t), y(t - \tau) = Y^T H_\tau J^{(a,b)}(t),$$

Here $J^{(a,b)}(t)$ and H_τ are defined in (4.2) and (4.5). The remaining parameters are given as follows:

$$X^T = \begin{bmatrix} x_{10} & \cdots & x_{1m} \\ \cdots & \ddots & \cdots \\ x_{n_1 0} & \cdots & x_{n_1 m} \end{bmatrix}, Y^T = \begin{bmatrix} y_{10} & \cdots & y_{1m} \\ \cdots & \ddots & \cdots \\ y_{n_2 0} & \cdots & y_{n_2 m} \end{bmatrix},$$

$$U^T = \begin{bmatrix} u_{10} & \cdots & u_{1m} \\ \cdots & \ddots & \cdots \\ u_{n_1 0} & \cdots & u_{n_1 m} \end{bmatrix}, V^T = \begin{bmatrix} v_{10} & \cdots & v_{1m} \\ \cdots & \ddots & \cdots \\ v_{n_2 0} & \cdots & v_{n_2 m} \end{bmatrix},$$

$$\Phi^T = \begin{bmatrix} \phi_{10} & \cdots & \phi_{1m} \\ \cdots & \ddots & \cdots \\ \phi_{n_1 0} & \cdots & \phi_{n_1 m} \end{bmatrix}, \Psi^T = \begin{bmatrix} \psi_{10} & \cdots & \psi_{1m} \\ \cdots & \ddots & \cdots \\ \psi_{n_2 0} & \cdots & \psi_{n_2 m} \end{bmatrix}.$$

In particular, at $t = t_0$, the initial functions $\phi(t)$ and $\psi(t)$ are defined as

$$\phi(t_0) = \begin{bmatrix} \hat{\phi}_{10} \\ \hat{\phi}_{20} \\ \vdots \\ \hat{\phi}_{n_1 0} \end{bmatrix}, \psi(t_0) = \begin{bmatrix} \hat{\psi}_{10} \\ \hat{\psi}_{20} \\ \vdots \\ \hat{\psi}_{n_2 0} \end{bmatrix}$$

and its corresponding approximations are given as follows

$$\phi(t_0) = \Phi_0^T J^{(a,b)}(t), \psi(t_0) = \Psi_0^T J^{(a,b)}(t),$$

$$\text{where } \Phi_0^T = \begin{bmatrix} \hat{\phi}_{10} & 0 & \cdots & 0 \\ \hat{\phi}_{20} & 0 & \cdots & 0 \\ \cdots & \cdots & \ddots & \cdots \\ \hat{\phi}_{n_1 0} & 0 & \cdots & 0 \end{bmatrix}_{n_1 \times m}, \Psi_0^T = \begin{bmatrix} \hat{\phi}_{10} & 0 & \cdots & 0 \\ \hat{\phi}_{20} & 0 & \cdots & 0 \\ \cdots & \cdots & \ddots & \cdots \\ \hat{\phi}_{n_2 0} & 0 & \cdots & 0 \end{bmatrix}_{n_2 \times m}.$$

In order to find the approximate solution of the system (4.10), the following two cases are analyzed:

Case 1: For the interval $t_0 \le t \le t_0 + \tau$, applying the Definition 4.1 and (4.1) to the system (4.10), gives

$$I_t^\alpha \left({}^c D_{t_0^+}^\alpha x(t) \right) = A_{11} I_{t_0^+}^\alpha x(t) + A_{12} I_{t_0^+}^\alpha y(t) + B_{11} I_{t_0^+}^\alpha x(t - \tau) + B_{12} I_{t_0^+}^\alpha y(t - \tau) + I_{t_0^+}^\alpha u(t),$$

$$x(t) = \phi(t_0) + A_{11} I_{t_0^+}^\alpha x(t) + A_{12} I_{t_0^+}^\alpha y(t) + B_{11} I_{t_0^+}^\alpha \phi(t - \tau) + B_{12} I_{t_0^+}^\alpha \psi(t - \tau) + I_{t_0^+}^\alpha u(t).$$

Similarly,

$$y(t) = \psi(t_0) + A_{21} I_{t_0^+}^\beta x(t) + A_{22} I_{t_0^+}^\beta y(t) + B_{21} I_{t_0^+}^\beta \phi(t - \tau) + B_{22} I_{t_0^+}^\beta \psi(t - \tau) + I_{t_0^+}^\beta v(t).$$

The well-defined shifted Jacobi approximations (4.3) and operational matrix for fractional integration (4.6) are applied to the above equation in the interval $t_0 \leq t \leq t_0 + \tau$, we have

$$X^T J^{(a,b)}(t) = \Phi_0^T J^{(a,b)}(t) + A_{11} X^T P_{t_0^+}^\alpha J^{(a,b)}(t) + A_{12} Y^T P_{t_0^+}^\alpha J^{(a,b)}(t)$$

$$+ B_{11} \Phi^T P_{t_0^+}^\alpha J^{(a,b)}(t) + B_{12} \Psi^T P_{t_0^+}^\alpha J^{(a,b)}(t) + U^T P_{t_0^+}^\alpha J^{(a,b)}(t).$$

$$Y^T J^{(a,b)}(t) = \Psi_0^T J^{(a,b)}(t) + A_{21} X^T P_{t_0^+}^\beta J^{(a,b)}(t) + A_{22} Y^T P_{t_0^+}^\beta J^{(a,b)}(t)$$

$$+ B_{21} \Phi^T P_{t_0^+}^\beta J^{(a,b)}(t) + B_{22} \Psi^T P_{t_0^+}^\beta J^{(a,b)}(t) + V^T P_{t_0^+}^\beta J^{(a,b)}(t).$$

By the Kronecker product ($\otimes$), matrix form of the above coupled system is written as,

$$\begin{bmatrix} X^T J^{(a,b)}(t) \\ Y^T J^{(a,b)}(t) \end{bmatrix} - \begin{bmatrix} (P_{t_0}^\alpha)^T \otimes A_{11} & (P_{t_0}^\alpha)^T \otimes A_{12} \\ (P_{t_0}^\beta)^T \otimes A_{21} & (P_{t_0}^\beta)^T \otimes A_{22} \end{bmatrix} \begin{bmatrix} X^T J^{(a,b)}(t) \\ Y^T J^{(a,b)}(t) \end{bmatrix}$$

$$- \begin{bmatrix} (P_{t_0}^\alpha)^T \otimes B_{11} & (P_{t_0}^\alpha)^T \otimes B_{12} \\ (P_{t_0}^\beta)^T \otimes B_{21} & (P_{t_0}^\beta)^T \otimes B_{22} \end{bmatrix} \begin{bmatrix} \Phi^T J^{(a,b)}(t) \\ \Psi^T J^{(a,b)}(t) \end{bmatrix} - \begin{bmatrix} U^T P_{t_0}^\alpha J^{(a,b)}(t) \\ V^T P_{t_0}^\beta J^{(a,b)}(t) \end{bmatrix}$$

$$- \begin{bmatrix} \Phi_0^T J^{(a,b)}(t) \\ \Psi_0^T J^{(a,b)}(t) \end{bmatrix} = 0.$$

Taking the transpose of the above matrix equation, we get

$$[X^T J^{(a,b)}(t) \quad Y^T J^{(a,b)}(t)]$$

$$- [X^T J^{(a,b)}(t) \quad Y^T J^{(a,b)}(t)] \begin{bmatrix} (P_{t_0}^\alpha)^T \otimes A_{11} & (P_{t_0}^\alpha)^T \otimes A_{12} \\ (P_{t_0}^\beta)^T \otimes A_{21} & (P_{t_0}^\beta)^T \otimes A_{22} \end{bmatrix}^T$$

$$- [\Phi^T J^{(a,b)}(t) \quad \Psi^T J^{(a,b)}(t)] \begin{bmatrix} (P_{t_0}^\alpha)^T \otimes B_{11} & (P_{t_0}^\alpha)^T \otimes B_{12} \\ (P_{t_0}^\beta)^T \otimes B_{21} & (P_{t_0}^\beta)^T \otimes B_{22} \end{bmatrix}^T$$

$$- [U^T P_{t_0}^\alpha J^{(a,b)}(t) \quad V^T P_{t_0}^\beta J^{(a,b)}(t)] - [\Phi_0^T J^{(a,b)}(t) \quad \Psi_0^T J^{(a,b)}(t)] = 0.$$

In order to form the linear combination of two parameter shifted Jacobi polynomials $J^{(a,b)}(t)$ from the above equation, we proceed the following:

$$
[X^T \quad Y^T] \begin{bmatrix} J^{(a,b)}(t) & \mathcal{O}_{m+1} \\ \mathcal{O}_{m+1} & J^{(a,b)}(t) \end{bmatrix} - [X^T \quad Y^T] \begin{bmatrix} (P_{t_0}^{\alpha})^T \otimes A_{11} & (P_{t_0}^{\beta})^T \otimes A_{21} \\ (P_{t_0}^{\alpha})^T \otimes A_{12} & (P_{t_0}^{\beta})^T \otimes A_{22} \end{bmatrix}
$$

$$
\times \begin{bmatrix} J^{(a,b)}(t) & \mathcal{O}_{m+1} \\ \mathcal{O}_{m+1} & J^{(a,b)}(t) \end{bmatrix}
$$

$$
- [\Phi^T \quad \Psi^T] \begin{bmatrix} (P_{t_0}^{\alpha})^T \otimes B_{11} & (P_{t_0}^{\beta})^T \otimes B_{21} \\ (P_{t_0}^{\alpha})^T \otimes B_{12} & (P_{t_0}^{\beta})^T \otimes B_{22} \end{bmatrix} \begin{bmatrix} J^{(a,b)}(t) & \mathcal{O}_{m+1} \\ \mathcal{O}_{m+1} & J^{(a,b)}(t) \end{bmatrix}
$$

$$
- [U^T P_{t_0}^{\alpha} \quad V^T P_{t_0}^{\beta}] \begin{bmatrix} J^{(a,b)}(t) & \mathcal{O}_{m+1} \\ \mathcal{O}_{m+1} & J^{(a,b)}(t) \end{bmatrix}
$$

$$
- [\Phi_0^T \quad \Psi_0^T] \begin{bmatrix} J^{(a,b)}(t) & \mathcal{O}_{m+1} \\ \mathcal{O}_{m+1} & J^{(a,b)}(t) \end{bmatrix} = 0, \tag{4.12}
$$

where $\mathcal{O}_{m+1}$ is a zero vector of order $m + 1$ with all entries are equal to zero. Well known that the shifted Jacobi polynomials are orthogonal, the equation (4.12) becomes

$$
[X^T \quad Y^T] \left\{ I - \begin{bmatrix} (P_{t_0}^{\alpha})^T \otimes A_{11} & (P_{t_0}^{\beta})^T \otimes A_{21} \\ (P_{t_0}^{\alpha})^T \otimes A_{12} & (P_{t_0}^{\beta})^T \otimes A_{22} \end{bmatrix} \right\}
$$

$$
= [\Phi^T \quad \Psi^T] \begin{bmatrix} (P_{t_0}^{\alpha})^T \otimes B_{11} & (P_{t_0}^{\beta})^T \otimes B_{21} \\ (P_{t_0}^{\alpha})^T \otimes B_{12} & (P_{t_0}^{\beta})^T \otimes B_{22} \end{bmatrix} + [U^T P_{t_0}^{\alpha} \quad V^T P_{t_0}^{\beta}] + [\Phi_0^T \quad \Psi_0^T],
$$

$$
[X^T \quad Y^T] = \left\{ [\Phi^T \quad \Psi^T] \begin{bmatrix} (P_{t_0}^{\alpha})^T \otimes B_{11} & (P_{t_0}^{\beta})^T \otimes B_{21} \\ (P_{t_0}^{\alpha})^T \otimes B_{12} & (P_{t_0}^{\beta})^T \otimes B_{22} \end{bmatrix} + [U^T P_{t_0}^{\alpha} \quad V^T P_{t_0}^{\beta}] + [\Phi_0^T \quad \Psi_0^T] \right\}
$$

$$
\times \left\{ I - \begin{bmatrix} (P_{t_0}^{\alpha})^T \otimes A_{11} & (P_{t_0}^{\beta})^T \otimes A_{21} \\ (P_{t_0}^{\alpha})^T \otimes A_{12} & (P_{t_0}^{\beta})^T \otimes A_{22} \end{bmatrix} \right\}^{-1}. \tag{4.13}
$$

By using (4.3) with obtained values of the coefficient matrix $\begin{bmatrix} X \\ Y \end{bmatrix}$ from (4.13) and shifted Jacobi polynomials (4.2), one can find the approximate solution $\begin{bmatrix} x(t) \\ y(t) \end{bmatrix}$, $t \in [t_0, t_0 + \tau]$.

Case 2: For the interval $t_0 + \tau \leq t \leq t_0 + 2\tau$, applying the similar argument of Case 1, the integral solution of the system (4.10) is

$$x(t) = x(t_0 + \tau) + A_{11} I_{\tau+}^{\alpha} x(t) + A_{12} I_{\tau+}^{\alpha} y(t) + B_{11} I_{\tau+}^{\alpha} x(t - \tau) + B_{12} I_{\tau+}^{\alpha} y(t - \tau) + I_{\tau+}^{\alpha} u(t),$$

$$X^T J^{(a,b)}(t) = X_\tau^T J^{(a,b)}(t) + A_{11} X^T P_{\tau+}^{\alpha} J^{(a,b)}(t) + A_{12} Y^T P_{\tau+}^{\alpha} J^{(a,b)}(t)$$

$$+ B_{11} X^T H_\tau P_{\tau+}^{\alpha} J^{(a,b)}(t) + B_{12} Y^T H_\tau P_{\tau+}^{\alpha} J^{(a,b)}(t) + U^T P_{\tau+}^{\alpha} J^{(a,b)}(t).$$

$$y(t) = y(t_0 + \tau) + A_{21} I_{\tau+}^{\beta} x(t) + A_{22} I_{\tau+}^{\beta} y(t) + B_{21} I_{\tau+}^{\beta} x(t - \tau) + B_{22} I_{\tau+}^{\beta} y(t - \tau) + I_{\tau+}^{\beta} v(t),$$

$$Y^T J^{(a,b)}(t) = Y_\tau^T J^{(a,b)}(t) + A_{21} X^T P_{\tau+}^{\beta} J^{(a,b)}(t) + A_{22} Y^T P_{\tau+}^{\beta} J^{(a,b)}(t)$$

$$+ B_{21} X^T H_\tau P_{\tau+}^{\beta} J^{(a,b)}(t) + B_{22} Y^T H_\tau P_{\tau+}^{\beta} J^{(a,b)}(t) + V^T P_{\tau+}^{\beta} J^{(a,b)}(t).$$

Substituting the shifted Jacobi series approximations (4.3) we get the approximate solution of (4.10) for all $t \in [\tau, 2\tau]$, as follows:

$$[X^T \quad Y^T] = \left\{ \left[U^T P_\tau^{\alpha} \quad V^T P_\tau^{\beta} \right] + [X_\tau^T \quad Y_\tau^T] \right\}$$

$$\times \left\{ I - \begin{bmatrix} (P_\tau^{\alpha})^T \otimes A_{11} & (P_\tau^{\beta})^T \otimes A_{21} \\ (P_\tau^{\alpha})^T \otimes A_{12} & (P_\tau^{\beta})^T \otimes A_{22} \end{bmatrix} \right.$$

$$\left. - \begin{bmatrix} (H_\tau P_\tau^{\alpha})^T \otimes B_{11} & (H_\tau P_\tau^{\beta})^T \otimes B_{21} \\ (H_\tau P_\tau^{\alpha})^T \otimes B_{12} & (H_\tau P_\tau^{\beta})^T \otimes B_{22} \end{bmatrix} \right\}^{-1}.$$

Similar procedure can be applied to $2\tau \leq t \leq 3\tau, \ldots, (n-1)\tau \leq t \leq n\tau$ such that $n\tau \leq t_f, n = \left[\frac{t_f - t_0}{\tau} \right]$ and final step solution is obtained for the interval $t_0 + n\tau \leq t \leq t_f$. Hence we get the numerical approximate solution of the coupled system (4.10) with initial condition (4.11).

4.4.2. System Identification of Coupled FDDEs

Let us consider the vectors $x(t), y(t), u(t), v(t)$ and initial functions $\phi(t), \psi(t)$ are known values defined in the interval $[t_0, t_0 + \tau]$ and $[t_0 + \tau, t_f]$ of the coupled system (4.10) and its coefficient matrices $A_{ij}, B_{ij}, i, j = 1,2$ are unknown. The identification problem in this section is to estimate $2(n_1 + n_2)^2$ elements of $A_{ij}, B_{ij}, i, j = 1,2$ by using the shifted Jacobi polynomials.

From the integral equation of the system (4.10) with $t \in [t_0, t_0 + \tau]$, apply the approximation of shifted Jacobi polynomials are defined in above Section 4.1, we have

$$X^T = \Phi_0^T + A_{11}X^T P_{t_0}^\alpha + A_{12}Y^T P_{t_0}^\alpha + B_{11}\Phi^T P_{t_0}^\alpha + B_{12}\Psi^T P_{t_0}^\alpha + U^T P_{t_0}^\alpha, \tag{4.14}$$

$$Y^T = \Psi_0^T + A_{21}X^T P_{t_0}^\beta + A_{22}Y^T P_{t_0}^\beta + B_{21}\Phi^T P_{t_0}^\beta + B_{22}\Psi^T P_{t_0}^\beta + V^T P_{t_0}^\beta. \tag{4.15}$$

Applying the Kronecker product for the matrices in (4.14), we have

$$\begin{bmatrix} \begin{cases} x_{10} - \hat\phi_{10} \\ x_{11} \\ \vdots \\ x_{1m} \end{cases} \\ \vdots \\ \begin{cases} x_{n_10} - \hat\phi_{n_10} \\ x_{n_11} \\ \vdots \\ x_{n_1m} \end{cases} \end{bmatrix} - U^T P_{t_0}^\alpha = \left[(X^T P_{t_0}^\alpha)^T \otimes I_{n_1}\right]\begin{bmatrix} a_1^{11} \\ a_2^{11} \\ \vdots \\ \vdots \\ a_{n_1n_1}^{11} \end{bmatrix} + \left[(Y^T P_{t_0}^\alpha)^T \otimes I_{n_1}\right]\begin{bmatrix} a_1^{12} \\ a_2^{12} \\ \vdots \\ \vdots \\ a_{n_1n_2}^{12} \end{bmatrix}$$

$$+\left[(\Phi^T P_{t_0}^\alpha)^T \otimes I_{n_1}\right]\begin{bmatrix} b_1^{11} \\ b_2^{11} \\ \vdots \\ \vdots \\ b_{n_1n_1}^{11} \end{bmatrix} + \left[(\Psi^T P_{t_0}^\alpha)^T \otimes I_{n_1}\right]\begin{bmatrix} b_1^{12} \\ b_2^{12} \\ \vdots \\ \vdots \\ b_{n_1n_2}^{12} \end{bmatrix}. \tag{4.16}$$

Similarly, the equation (4.15) is expanded in the following form

$$\begin{bmatrix} \begin{cases} y_{10} - \hat{\psi}_{10} \\ y_{11} \\ \vdots \\ y_{1m} \end{cases} \\ \vdots \\ \begin{cases} y_{n_20} - \hat{\psi}_{n_20} \\ y_{n_21} \\ \vdots \\ y_{n_2m} \end{cases} \end{bmatrix} - V^T P_{t_0}^{\beta} = \left[(X^T P_{t_0}^{\beta})^T \otimes I_{n_2} \right] \begin{bmatrix} a_1^{21} \\ a_2^{21} \\ \vdots \\ \vdots \\ a_{n_2n_1}^{21} \end{bmatrix} + \left[(Y^T P_{t_0}^{\beta})^T \otimes I_{n_2} \right] \begin{bmatrix} a_1^{22} \\ a_2^{22} \\ \vdots \\ \vdots \\ a_{n_2n_2}^{22} \end{bmatrix}$$

$$+ \left[(\Phi^T P_{t_0}^{\beta})^T \otimes I_{n_2} \right] \begin{bmatrix} b_1^{21} \\ b_2^{21} \\ \vdots \\ \vdots \\ b_{n_2n_1}^{21} \end{bmatrix} + \left[(\Psi^T P_{t_0}^{\beta})^T \otimes I_{n_2} \right] \begin{bmatrix} b_1^{22} \\ b_2^{22} \\ \vdots \\ \vdots \\ b_{n_2n_2}^{22} \end{bmatrix}, \qquad (4.17)$$

For the case $t_0 + \tau \le t \le t_f$, the integral solution of the system (4.10) is expressed as in the following approximation via shifted Jacobi polynomials,

$$X^T = X_{\tau}^T + A_{11}X^T P_{\tau}^{\alpha} + A_{12}Y^T P_{\tau}^{\alpha} + B_{11}X^T H_{\tau}P_{\tau}^{\alpha} + B_{12}Y^T H_{\tau}P_{\tau}^{\alpha} + U^T P_{\tau}^{\alpha}, \quad (4.18)$$

$$Y^T = Y_{\tau}^T + A_{21}X^T P_{\tau}^{\beta} + A_{22}Y^T P_{\tau}^{\beta} + B_{21}X^T H_{\tau}P_{\tau}^{\beta} + B_{22}Y^T H_{\tau}P_{\tau}^{\beta} + V^T P_{\tau}^{\beta}. \quad (4.19)$$

Taking Kronecker product on both sides of the equation (4.18), gives

$$\begin{bmatrix} \begin{cases} x_{10} - \hat{x}_{1\tau} \\ x_{11} \\ \vdots \\ x_{1m} \end{cases} \\ \vdots \\ \begin{cases} x_{n_10} - \hat{x}_{n_1\tau} \\ x_{n_11} \\ \vdots \\ x_{n_1m} \end{cases} \end{bmatrix} - U_1^T P_{\tau}^{\alpha} = \left[(X^T P_{\tau}^{\alpha})^T \otimes I_{n_1} \right] \begin{bmatrix} a_1^{11} \\ a_2^{11} \\ \vdots \\ \vdots \\ a_{n_1n_1}^{11} \end{bmatrix} + \left[(Y^T P_{\tau}^{\alpha})^T \otimes I_{n_1} \right] \begin{bmatrix} a_1^{12} \\ a_2^{12} \\ \vdots \\ \vdots \\ a_{n_1n_2}^{12} \end{bmatrix}$$

$$+\left[(X^T H_\tau P_\tau^\alpha)^T \otimes I_{n_1}\right]\begin{bmatrix} b_1^{11} \\ b_2^{11} \\ \vdots \\ \vdots \\ \vdots \\ b_{n_1 n_1}^{11} \end{bmatrix} + \left[(Y^T H_\tau P_\tau^\alpha)^T \otimes I_{n_1}\right]\begin{bmatrix} b_1^{12} \\ b_2^{12} \\ \vdots \\ \vdots \\ \vdots \\ b_{n_1 n_2}^{12} \end{bmatrix}. \qquad (4.20)$$

Similarly, the equation (4.19) can be written as

$$\begin{bmatrix} \begin{Bmatrix} y_{10} - \hat{y}_\tau \\ y_{11} \\ \vdots \\ y_{1m} \end{Bmatrix} \\ \vdots \\ \begin{Bmatrix} y_{n_2 0} - \hat{y}_{n_2 \tau} \\ y_{n_2 1} \\ \vdots \\ y_{n_2 m} \end{Bmatrix} \end{bmatrix} - U_2^T P_\tau^\beta = \left[(X^T P_\tau^\beta)^T \otimes I_{n_2}\right]\begin{bmatrix} a_1^{21} \\ a_2^{21} \\ \vdots \\ \vdots \\ \vdots \\ a_{n_2 n_1}^{21} \end{bmatrix} + \left[(Y^T P_\tau^\beta)^T \otimes I_{n_2}\right]\begin{bmatrix} a_1^{22} \\ a_2^{22} \\ \vdots \\ \vdots \\ \vdots \\ a_{n_2 n_2}^{22} \end{bmatrix}$$

$$+\left[(X^T H_\tau P_\tau^\beta)^T \otimes I_{n_2}\right]\begin{bmatrix} b_1^{21} \\ b_2^{21} \\ \vdots \\ \vdots \\ \vdots \\ b_{n_2 n_1}^{21} \end{bmatrix} + \left[(Y^T H_\tau P_\tau^\beta)^T \otimes I_{n_2}\right]\begin{bmatrix} b_1^{22} \\ b_2^{22} \\ \vdots \\ \vdots \\ \vdots \\ b_{n_2 n_2}^{22} \end{bmatrix}. \qquad (4.21)$$

Let L and $\hat{L}$ be the $2n_1(n_1 + n_2)$ and $2n_2(n_1 + n_2)$ vectors formed from the elements of matrices $A_{11}, A_{12}, B_{11}, B_{12}$ and $A_{21}, A_{22}, B_{21}, B_{22}$ respectively are defined as

$$L = \begin{bmatrix} a_1^{11} \dots a_{n_1 n_1}^{11} & a_1^{12} \dots a_{n_1 n_2}^{12} & b_1^{11} \dots b_{n_1 n_1}^{11} & b_1^{12} \dots b_{n_1 n_2}^{12} \end{bmatrix}^T,$$

$$\hat{L} = \begin{bmatrix} a_1^{21} \dots a_{n_2 n_1}^{21} & a_1^{22} \dots a_{n_2 n_2}^{22} & b_1^{21} \dots b_{n_2 n_1}^{21} & b_1^{22} \dots b_{n_2 n_2}^{22} \end{bmatrix}^T.$$

Therefore, the equations (4.16) (4.17), (4.20) and (4.21) are rewritten in the following matrix form,

$$K_{2(m+1)n_1 \times 2n_1(n_1+n_2)} L_{2n_1(n_1+n_2) \times 1} = M_{2(m+1)n_1 \times 1},$$

$$\widehat{K}_{2(m+1)n_2 \times 2n_2(n_1+n_2)} \widehat{L}_{2n_2(n_1+n_2)\times 1} = \widehat{M}_{2(m+1)n_2\times 1}.$$

where,

$$K = \begin{bmatrix} (X^T P_{t_0}^\alpha)^T \otimes I_{n_1} & (Y^T P_{t_0}^\alpha)^T \otimes I_{n_1} & (\Phi^T P_{t_0}^\alpha)^T \otimes I_{n_1} & (\Psi^T P_{t_0}^\alpha)^T \otimes I_{n_1} \\ (X^T P_{\tau}^\alpha)^T \otimes I_{n_1} & (Y^T P_{\tau}^\alpha)^T \otimes I_{n_1} & (X^T H_\tau P_{\tau}^\alpha)^T \otimes I_{n_1} & (Y^T H_\tau P_{\tau}^\alpha)^T \otimes I_{n_1} \end{bmatrix},$$

$$\widehat{K} = \begin{bmatrix} (X^T P_{t_0}^\beta)^T \otimes I_{n_2} & (Y^T P_{t_0}^\beta)^T \otimes I_{n_2} & (\Phi^T P_{t_0}^\beta)^T \otimes I_{n_2} & (\Psi^T P_{t_0}^\beta)^T \otimes I_{n_2} \\ (X^T P_{\tau}^\beta)^T \otimes I_{n_2} & (Y^T P_{\tau}^\beta)^T \otimes I_{n_2} & (X^T H_\tau P_{\tau}^\alpha)^T \otimes I_{n_2} & (Y^T H_\tau P_{\tau}^\alpha)^T \otimes I_{n_2} \end{bmatrix},$$

$$M = \begin{bmatrix} (x_{10} - \hat{\phi}_{10}\, x_{11} \ldots x_{1m})^T & \cdots\cdots & (x_{n_10} - \hat{\phi}_{n_10}\, x_{n_11} \ldots x_{n_1m})^T \\ (x_{10} - \hat{x}_{\tau}\, x_{11} \ldots x_{1m})^T & \cdots\cdots & (x_{n_10} - \hat{x}_{n_1\tau}\, x_{n_11} \ldots x_{n_1m})^T \end{bmatrix} - \begin{bmatrix} U^T P_{t_0}^\alpha \\ U^T P_{\tau}^\alpha \end{bmatrix},$$

$$\widehat{M} = \begin{bmatrix} (y_{10} - \hat{\psi}_0\, y_{11} \ldots y_{1m})^T & \cdots\cdots & (y_{n_20} - \hat{\psi}_{n_20}\, y_{n_21} \ldots y_{n_2m})^T \\ (y_{10} - \hat{\psi}_\tau\, y_{11} \ldots y_{1m})^T & \cdots\cdots & (y_{n_20} - \hat{y}_{n_2\tau}\, y_{n_21} \ldots y_{n_2m})^T \end{bmatrix} - \begin{bmatrix} V^T P_{t_0}^\beta \\ V^T P_{\tau}^\beta \end{bmatrix}.$$

Here, the elements of matrices $K, \widehat{K}, M$ and $\widehat{M}$ are known values. The entries of matrix L and $\widehat{L}$ are obtained by using the least square method (see [17, 15]). Therefore, the least square estimation of L and $\widehat{L}$ are written as

$$L_{2n_1(n_1+n_2)\times 1} = (K^T K)^{-1} K^T M. \tag{4.22}$$

$$\widehat{L}_{2n_2(n_1+n_2)\times 1} = \left(\widehat{K}^T \widehat{K}\right)^{-1} \widehat{K}^T \widehat{M}. \tag{4.23}$$

By using the above equations, we estimate the required coefficient matrices $A_{ij}, B_{ij}, i, j = 1,2$ for the system (4.10).

4.5. NUMERICAL EXAMPLES

To illustrate the obtained theoretical rate of convergence, the numerical results are obtained with different α, β and Jacobi polynomial parameters a, b, m values. The value of absolute error for the considered problems have been calculated by applying the following formula (4.24). Since there is no exact solution, a numerical approximation for higher degrees of shifted Jacobi polynomials are used to error estimate $E(t)$.

$$E_x(t) = |x^m(t) - x^{2m}(t)|, t \in [t_0, t_f], \tag{4.24}$$

where m and $2m$ are the shifted Jacobi polynomial degrees with $x^m \simeq \sum_{i=0}^{m} x_i J_i^{(a,b)}(t)$.

Problem 1 *Consider the coupled FDDEs* (4.10) *of two different orders α and β with the following values: $n_1 = 1, n_2 = 1, \tau = 1$, the initial condition*

$$\begin{bmatrix} x(t) \\ y(t) \end{bmatrix} = \begin{bmatrix} 1 + t \\ t \end{bmatrix}, -1 \leq t \leq 0.$$

The coefficient matrices A_{ij} and B_{ij} are

$$A_{11} = -1, A_{12} = -3, A_{21} = 1, A_{22} = 2,$$

$$B_{11} = 1, B_{12} = -1, B_{21} = -1, B_{22} = 2.$$

Also, $u(t) = t^2 + 3t - 2, v(t) = 3t - 2, t \in [0,1]$.

Applying the procedure given in Subsection 4.1 with the shifted Jacobi parameters $a = 0, b = 0, m = 5$ and $\alpha = 0.6, \beta = 0.85$, we find the approximate solution $x(t)$ and $y(t)$ as

$$x(t) = 5.5857t^4 - 18.106t^3 + 19.229t^2 - 1.5289t + 0.83559,$$

$$y(t) = -0.021067t^4 + 2.0862t^3 + 0.15517t^2 - 5.0936t - 0.0238. \tag{4.25}$$

Fig. (**4.1**) shows that the approximate solution $x(t), y(t)$ from (4.25) and the error curve E_x and E_y from (4.24) with respect to the states $x(t), y(t), t \in [0,1]$ are calculated by applying the shifted Jacobi polynomial degree $m = 5$. Figs. (**4.2** and **4.3**) show that the numerical approximate solution of different values of α, β and the shifted Jacobi parameters a, b, m. The corresponding absolute error is obtained and listed in Table **4.1** that also shown in Fig. (**4.4**).

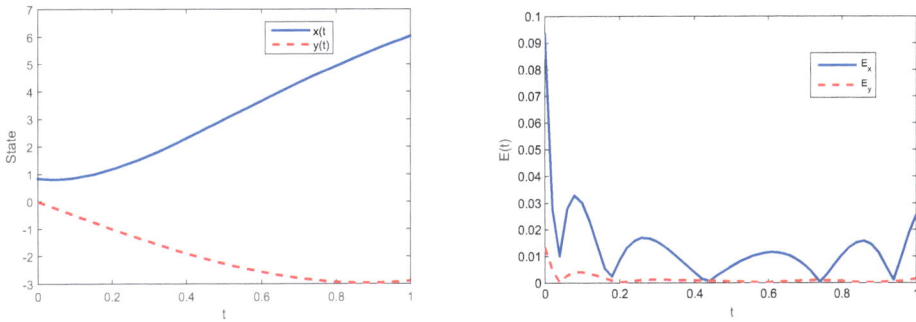

Fig. (4.1). (left) Approximate solution (right) Absolute error of Problem 1 with $\alpha = 0.6, \beta = 0.85, a = 0, b = 0, m = 5$.

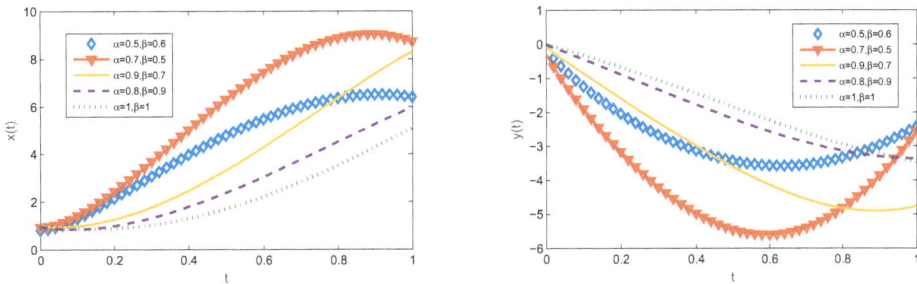

Fig. (4.2). Approximate solution of Problem 1 with $a = 1, b = 1, m = 5$.

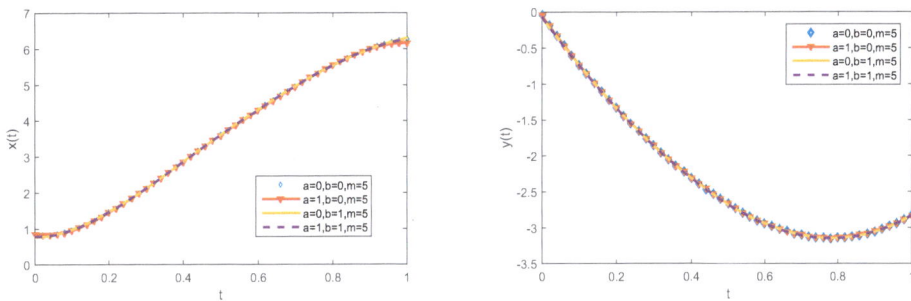

Fig. (4.3). Approximate solution of the Problem 1 with $\alpha = 0.6, \beta = 0.5$ and $m - 4$.

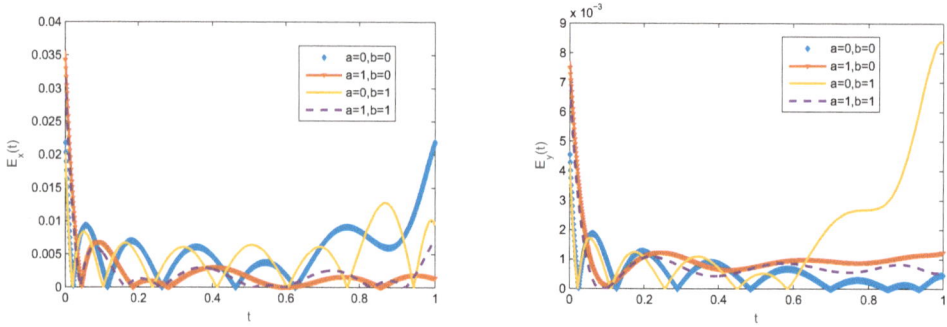

Fig. (4.4). Absolute error of states for the Problem 1 with $\alpha = 0.6, \beta = 0.5$ and $m = 4$.

Table 4.1. Absolute error approximation of Problem 1 with $\alpha = 0.6, \beta = 0.5, m = 4$ and the shifted Jacobi polynomial parameters (a, b).

Values	$a = 0, b = 0$		$a = 1, b = 0$		$a = 0, b = 1$		$a = 1, b = 1$	
t	$E_x(t)$	$E_y(t)$	$E_x(t)$	$E_y(t)$	$E_x(t)$	$E_y(t)$	$E_x(t)$	$E_y(t)$
0.0	0.0232	0.0048	0.0356	0.0077	0.0204	0.00425	0.0316	0.0069
0.2	0.0065	0.0012	0.0006	0.0010	0.0047	0.0010	0.0013	0.0010
0.4	0.0047	0.0008	0.0031	0.0006	0.0012	0.0007	0.0028	0.0004
0.6	0.0019	0.0006	0.0001	0.0009	0.0004	0.0002	0.0001	0.0008
0.8	0.0085	0.0002	0.0013	0.0009	0.005	0.0026	0.0011	0.0006
1.0	0.0219	0.0004	0.0014	0.0012	0.0094	0.0083	0.0072	0.0005

Problem 2. *Consider the coupled FDDEs (4.10) of $\alpha = 0.75$ and $\beta = 0.5$ with the following values: $n_1 = 1, n_2 = 1, \tau = 1$, the initial condition*

$$\begin{bmatrix} x(t) \\ y(t) \end{bmatrix} = \begin{bmatrix} 1 + t \\ t \end{bmatrix}, -1 \leq t \leq 0.$$

Also, the known vectors are

$$x(t) = -0.057819t^7 + 0.21716t^6 - 0.33157t^5 + 0.26505t^4$$

$$-0.11983t^3 - 1.9673t^2 - 0.000036526t + 1.0001,$$

$$y(t) = 0.046656t^7 - 0.17044t^6 + 0.24976t^5 - 0.18635t^4$$

$$+0.072856t^3 - 0.011139t^2 + 2.0056t - 0.00012332,$$

$$u(t) = 261.39t - 301.157t^2 - 1.76522t^{(5/4)} + 40.5733,$$

$$v(t) = 2.25676t^{(1/2)} - 280.7t^2 - 261.068t + 260.353.$$

To identify the system with given values of $x(t), y(t), u(t), v(t), \phi(t)$ and $\psi(t)$ from (4.22) and (4.23), one can easily estimate the corresponding unknown coefficients $A_{ij}, B_{ij}, i, j = 1,2$. The obtained approximate values of these coefficients are given in Table 2 with different values of a, b and m.

Table 4.2. Approximated values with $\alpha = 0.75, \beta = 0.5, a = 0$ and $b = 1$.

m	A_{11}	A_{12}	A_{21}	A_{22}	B_{11}	B_{12}	B_{21}	B_{22}
6	-150.58	-131.55	-140.36	120.73	110.74	-110.01	-99.30	119.99
7	-150.58	-130.63	-140.35	120.54	110.00	-110.01	-100.01	120.00
8	-150.58	-132.88	-140.35	121.44	110.39	-110.09	-100.81	120.02
9	-150.58	-132.55	-140.35	120.77	110.72	-110.01	-100.50	120.01

Remark *In order to validate the estimates of unknown coefficients in Problem 2, the identified system will be used to apply the same procedure of Problem 1 to get the approximate values of $x(t), y(t), u(t)$ and $v(t)$ which almost equal to the corresponding given values of Problem 2.*

4.6. CONCLUSION

In this chapter, shifted Jacobi polynomial approximation technique has been successfully applied to the numerical solution of coupled FDDEs. Also, the system identification with known state and input variables has been done by approximation and least square method. The effectiveness of the proposed theory has been illustrated with numerical examples.

CONSENT FOR PUBLICATION

Not applicable.

CONFLICT OF INTEREST

The authors declare no conflict of interest, financial or otherwise.

ACKNOWLEDGEMENTS

This work was supported by Science Engineering Research Board (SERB), DST, Govt. of India under YSS Project F.No:YSS/2014/000447 dated 20.11.2015. The first author was supported by the Council of Scientific and Industrial Research (CSIR) under Senior Research Fellowship (SRF), New Delhi, India (F. No. 05/715 (0017)/2016- EMR-I dated 11-04-2017).

REFERENCES

[1]   C. John, and J. J. Loiseau, *Applications of time delay systems*. Springer-Verlag: Berlin Heidelberg, 2007.

[2]   Y. Ren, Y. Qin, and R. Sakthivel, " Existence results for fractional order semilinear integro-differential evolution equations with infinite delay," *Integral Equ. Oper. Theory,* vol. 67, pp. 19–33, May 2010.

[3]   K. Diethelm, *The analysis of fractional differential equations*. Springer: London, 2010.

[4]   A. Kilbas, M. H. Srivastava, and J. J. Trujillo, *Theory and application of fractional differential equations*. Elsevier: Amsterdam, 2006.

[5]   I. Podlubny, *Fractional Differential Equations, vol. 198 of Mathematics in Science and Engineering*. Academic Press: San Diego, USA, 1999.

[6]   S. Kazem, "An integral operational matrix based on Jacobi polynomials for solving fractional-order differential equations," *Appl. Math. Model.,* vol. 37, pp. 1126–1136, Feb. 2013.

[7]   V. Daftardar-Gejji, Y. Sukale, and S. Bhalekar, "Solving fractional delay differential equations: A new approach," *Fract. Calc. Appl. Anal.,* vol. 18, pp. 400–418, Apr. 2015.

[8]   S. Bhalekar, and V. Daftardar-Gejji, "A predictor-corrector scheme for solving non- linear delay differential equations of fractional order," *J. Fractional Calc. & Appl.,* vol. 1, pp. 1–8, Jul. 2011.

[9]   M. N. Sherif, "Numerical solution of system of fractional delay differential equations using polynomial spline functions," *Appl. Math.,* vol. 7, pp. 518–526, Mar. 2016.

[10] Z. Wang, "A numerical method for delayed fractional-order differential equations," *J. Appl. Math.,* vol. 2013, pp. 1–7, Article ID 256071, May, 2013.

[11] B. P. Moghaddam, and Z. S. Mostaghim. "A numerical method based on finite difference for solving fractional delay differential equations," *J. Taibah Univ. Sci.,* vol. 7, pp. 120–127, Jul. 2013.

[12] H. Khalil, and R. A. Khan, "The use of Jacobi polynomials in the numerical solution of coupled system of fractional differential equations," *Int. J. Comput. Math.,* vol. 92, pp. 1452–1472, Jul. 2015.

[13] A. H. Bhrawy, M. M. Tharwat, and M. A. Alghamdi, "A new operational matrix of fractional integration for shifted Jacobi polynomials," *Bull. Malays. Math. Sci. Soc.,* vol. 37, pp. 983–995, Oct. 2014.

[14] P. Muthukumar, and B. Ganesh Priya, "Numerical solution of fractional delay differential equation by shifted Jacobi polynomials," *Int. J. Comput. Math.,* vol. 94, pp. 471–492, Dec. 2015.

[15]   I. R. Hornd, and J. H. Chou, "Analysis and parameter identification of time delay systems via shifted Jacobi polynomials," *Int. J. Control,* vol. 44, pp. 935–942, Oct. 1986.

[16]   S. H. Nasehi, M. Samavat, and M. A. Vali. "Analysis and parameter identification of time-delay systems using the Chebyshev wavelets," *J. Inform. Math. Sci.,* vol. 4, pp. 51–64, 2012.

[17]   B. A. Ardekani, M. Samavat, and H. Rahmani, "Parameter identification of time-delay systems via exponential: Fourier series," *Int. J. Syst. Sci.,* vol. 22, pp. 1301–1306, Jul. 1991.

[18]   F. C. Kung, and H. Lee, "Solution and parameter estimation in linear time-invariant delayed systems using Laguerre polynomial expansion," *J. Dynam. Syst. Meas. Control,* vol. 105, pp. 297-301, Dec. 1983.

[19]   T. S. Chihara, *An introduction to orthogonal polynomials.* Dover publication: New York, 2011.

Monotone Iteration Principle in the Theory of Hadamard Fractional Delay Differential Equations

Kishor D. Kucche[*]

Department of Mathematics, Shivaji University, Kolhapur, 416004 India

Abstract: This chapter deals with the existence, uniqueness and approximations of solutions of Hadamard fractional delay differential equations and neutral Hadamard fractional delay differential equations. The results are acquired *via* monotone iteration principle and hybrid fixed point theorems established by Dhage (2014) in partially ordered normed linear spaces (PONLS) by imposing weak conditions on the functions involved in the equations like mixed partial continuity and partial compactness or partial Lipschitz conditions.

Keywords: Hadamard fractional differential equations, Existence and uniqueness, Hybrid fixed point theorems, Monotone iteration principle.

AMS Subject Classification: 34A08, 34A12, 34C12,47H10, 47H07.

5.1. INTRODUCTION

Fractional differential equation (FDE) is originated as a vital and fascinating field of investigation because of its extensive applications in various branches of applied sciences and engineering. One of the important reason behind over popularity of this field is that the nonlocal characteristic of fractional order derivatives that considers memory and hereditary properties of various materials and processes. Because of this advantage mathematical formulation of many physical phenomena involving fractional order derivatives is viewed as more realistic and practical than the corresponding mathematical formulation with classical integer-order derivatives. FDEs arises in mathematical modelling of systems and processes in many fields of engineering and sciences [1 - 8], for instance, aerodynamics, fractal theory, nuclear reactors, viscoelasticity and diffusion processes, polymer rheology, electrical circuit, nonlinear oscillation of earthquake, electrodynamics of complex medium and so on. A more detailed

---

[*]**Corresponding author Kishor D. Kucche:** Department of Mathematics, Shivaji University, Kolhapur 416004 India; Tel/Fax: +91-0231-2609218; E-mail: kdkucche@gmail.com

Sachin Bhalekar (Ed.)

exposition on theory of FDEs and its applications can be found in the well known research monographs [9-13].

FDEs that involve Caputo or Riemann-Liouville fractional derivative have extensively been studied by many researchers. Many research papers devoted to investigate existence, uniqueness and various qualitative properties of solution of nonlinear FDEs, a few of them are [14-23].

Hadamard ([24], 1982) defined a new fractional derivative which involves the function $\left(\log \frac{t}{s}\right)$ in the integral kernel as a replacement of $(t - s)$ which is included in the integral kernel of both Riemann-Liouville and Caputo derivatives. Further, the Hadamard derivative involves the operator $\left(t\frac{d}{dt}\right)^n$ instead of $\left(\frac{d}{dt}\right)^n$ which is included in the definition of the Riemann-Liouville and Caputo derivative (for more details please see [25]). Though the idea of Hadamard derivative is very old, the literature survey demonstrates that the theory for Hadamard-type nonlinear FDEs is not studied to the extent to which the nonlinear FDEs involving Riemann-Liouville and Caputo derivatives are studied. For the details on calculus of Hadamard fractional derivative and the theory of FDEs involving Hadamard fractional derivative, we refer to [13, 26-36].

Ahamd *et al.* [33], [36, Chapter~2] have proved results pertaining to existence and uniqueness of solution for Hadamard-type fractional functional differential equations and neutral functional differential equations. To obtain these results, Banach contraction principle and nonlinear alternative of Leray-Schauder are used that require the nonlinear functions involved in the equations must satisfy a strong Lipschitz and compactness conditions.

Recently, Dhage [37-40] proved that the existence and uniqueness of the solutions for nonlinear differential and integrodifferential equations can also be obtained under weaker mixed partial continuity and partial compactness or partial Lipschitz conditions. In [41-44], with the same tools and techniques, Dhage investigated existence, uniqueness and approximation of the solutions for nonlinear Volterra fractional integral equations and nonlinear fractional differential equations that involves Riemann-Liouville and Caputo derivative.

Motivated by the work of Dhage [41-44] and Ahamd *et al.* [33], [36, Chapter~2], in this chapter, we establish the existence, uniqueness and approximations of solutions of Hadamard fractional delay differential equations and neutral Hadamard fractional delay differential equations assuming the weak mixed partial continuity and partial compactness or partial Lipschitz conditions on the functions involved in the equations and utilising the monotone iteration principle and hybrid fixed point theorems established by Dhage in PONLS.

The important aspect of the method employed in this chapter gives existence of solution and also gives additionally the monotone convergence of the sequence of successive approximations to the solution of Hadamard fractional delay differential equations that started with lower or upper solution of the equation as the initial approximation.

The chapter is organized as follows. In Section 2, we recall some definitions and useful results about Hadamard fractional derivatives, partially normed spaces and state hybrid fixed point theorems that are utilized in this chapter. Section 3 deals with the existence and uniqueness results of Hadamard fractional delay differential equations. Section 4 establishes the existence of solutions for hybrid Hadamard fractional delay differential equations with linear perturbation of first type. In Section 5, existence results proved for neutral Hadamard fractional delay differential equations.

5.2. PRELIMINARIES

5.2.1. Hadamard Fractional Derivative

Definition 5.1. ([9]). The Hadamard fractional integral of order q for a continuous function g is defined as

$$_H I^q g(t) = \frac{1}{\Gamma(q)} \int_1^t (\log \frac{t}{s})^{q-1} \frac{g(s)}{s} ds, q > 0.$$

Definition 5.2. ([9]). The Hadamard derivative of fractional order q for a continuous function $g: [1, \infty) \to \mathbb{R}$ is defined as

$$_H D^q g(t) = \frac{1}{\Gamma(n-q)} (t \frac{d}{dt})^n \int_1^t (\log \frac{t}{s})^{n-q-1} \frac{g(s)}{s} ds, n - 1 < q < n, n = [q] + 1,$$

where $[q]$ denotes the integer part of the real number q and $\log(\cdot) = \log_e(\cdot)$.

Lemma 5.1 ([p. 213]Kilbas). *Let $q > 0$, $n = -[-q]$ and $0 \leq \gamma < 1$. Let G be an open set in $\mathbb{R}$ and let $f: (a, b] \times G \to \mathbb{R}$ be a function such that: $f(x, y) \in C_{\gamma, \log}[a, b]$ for any $y \in G$, then the problem*

$$_H D^q y(t) = f(t, y(t)), q > 0, \tag{5.1}$$

$$_H I^{q-k} y(a+) = b_k, b_k \in \mathbb{R}, (k = 1, \dots, n, n = -[-q]), \tag{5.2}$$

satisfies the Volterra integral equation

$$y(t) = \sum_{j=1}^{n} \frac{b_j}{\Gamma(q-j+1)} (\log \tfrac{t}{a})^{q-j} + \frac{1}{\Gamma(q)} \int_a^t (\log \tfrac{t}{s})^{q-1} f(s, y(s)) \tfrac{ds}{s}, \tag{5.3}$$

for $t > a > 0$; *i.e.,* $y(t) \in C_{n-q,\log}[a, b]$ *satisfies the relations* (5.1)-(5.2) *if and only if it satisfies the Volterra integral equation* (5.3).

In particular, if $0 < q \le 1$, *problem* (5.1)-(5.2) *is equivalent to the equation*

$$y(t) = \frac{b}{\Gamma(q)} (\log \tfrac{t}{a})^{q-1} + \frac{1}{\Gamma(q)} \int_a^t (\log \tfrac{t}{s})^{q-1} f(s, y(s)) \tfrac{ds}{s}, s > a > 0. \tag{5.4}$$

More details about the Hadamard fractional calculus and theory related to Hadamard type Cauchy problem can be found in [9].

5.2.2. Partially Ordered Normed Linear Spaces (PONLS)

We review some fundamental ideas of PONLS and state the hybrid fixed point theorems which are used further in this chapter. For more detail please see Dhage [41-46].

Definition 5.3. Let $\mathcal{X}$ be real vector space. A relation $\preceq$ in $\mathcal{X}$ is said to be partially order if it fulfills the following conditions:

(a)   Reflexivity: $x \preceq x$ for all $x \in \mathcal{X}$.

(b)   Antisymmetry: for all $x, y \in \mathcal{X}$, if $x \preceq y$ and $y \preceq x$, then $x = y$.

(c)   Transitivity: for all $x, y, z \in \mathcal{X}$, if $x \preceq y$ and $y \preceq z$, then $x \preceq z$.

(d)   Order linearity:

     (i)   for all $x_1, x_2, y_1, y_2 \in \mathcal{X}$, if $x_1 \preceq y_1$ and $x_2 \preceq y_2$, then $x_1 + x_2 \preceq y_1 + y_2$;

     (ii)   for all $x, y \in \mathcal{X}$ and scalar $t \ge 0$, if $x \preceq y$, then $tx \preceq ty$.

A vector space $\mathcal{X}$ equipped with partial order $\preceq$ on it is called partially ordered vector space (POVS).

Definition 5.4. Two elements x and y in POVS $(\mathcal{X}, \preceq)$ are said to *comparable* if either $x \preceq y$ or $y \preceq x$. A non-empty subset C of $\mathcal{X}$ is called *chain* or *totally ordered* if all elements of C are comparable.

Definition 5.5. A POVS $(\mathcal{X}, \preceq)$ is said to be *regular* if $\{x_n\}$ is a nondecreasing (resp. nonincreasing) sequence in $\mathcal{X}$ such that $x_n \to x^*$ as $n \to \infty$, then $x_n \preceq x^*$ (resp. $x_n \succeq x^*$) for all $n \in \mathbb{N}$.

Definition 5.6. A POVS $(\mathcal{X}, \preceq)$ with the norm $\|\cdot\|$ defined on it is called *partially ordered normed linear space (PONLS)*.

Throughout this paper we denote by $\mathcal{X} = (\mathcal{X}, \preceq, \|\cdot\|)$ a partially ordered real normed linear space with an order relation $\preceq$ and the norm $\|\cdot\|$.

Definition 5.7. (Dhage [46]). A mapping $\mathcal{T} \colon \mathcal{X} \to \mathcal{X}$ is called *isotone* or *monotone nondecreasing* if $x \preceq y$ implies $\mathcal{T}x \preceq \mathcal{T}y$ for all $x, y \in \mathcal{X}$. Similarly, $\mathcal{T}$ is called *monotone nonincreasing* if $x \preceq y$ implies $\mathcal{T}x \succeq \mathcal{T}y$ for all $x, y \in \mathcal{X}$.

A mapping $\mathcal{T} \colon \mathcal{X} \to \mathcal{X}$ is called *monotonic* if it is either monotone nondecreasing or monotone nonincreasing.

Definition 5.8. (Dhage [46]). An operator $\mathcal{T}$ on a normed linear space $\mathcal{X}$ into itself is called *compact* if $\mathcal{T}(\mathcal{X})$ is a relatively compact subset of $\mathcal{X}$. $\mathcal{T}$ is called *totally bounded* if for any bounded subset S of $\mathcal{X}$, $\mathcal{T}(S)$ is a relatively compact subset of $\mathcal{X}$. If $\mathcal{T}$ is continuous and totally bounded, then it is called *completely continuous* on $\mathcal{X}$.

Definition 5.9. (Dhage, [42, 46]). A mapping $\mathcal{T} \colon \mathcal{X} \to \mathcal{X}$ is called *partially continuous* at a point $a \in \mathcal{X}$ if for $\varepsilon > 0$ there exists a $\delta > 0$ such that $\| \mathcal{T}x - \mathcal{T}a \| < \varepsilon$ whenever x is comparable to a and $\| x - a \| < \delta$. $\mathcal{T}$ called partially continuous on $\mathcal{X}$ if it is partially continuous at every point of it. It is clear that if $\mathcal{T}$ is partially continuous on $\mathcal{X}$, then it is continuous on every chain C contained in $\mathcal{X}$.

Definition 5.10. *A non-empty subset S of the PONLS $\mathcal{X}$ is called* partially bounded *if every chain C in S is bounded.*

Definition 5.11. (Dhage, [41, 42]). An operator $\mathcal{T}$ on a partially normed linear space $\mathcal{X}$ into itself is called *partially bounded* if $\mathcal{T}(C)$ is bounded for every chain

C in $\mathcal{X}$. $\mathcal{T}$ is called *uniformly partially bounded* if for all chains C in $\mathcal{X}$, $\mathcal{T}(C)$ in $\mathcal{X}$ are bounded by a unique constant.

Definition 5.12. *A non-empty subset S of the PONLS $\mathcal{X}$ is called* partially compact *if every chain C in S is compact.*

Definition 5.13. (Dhage, [41, 42, 49]). An operator $\mathcal{T}$ is called *partially compact* if $\mathcal{T}(C)$ is a relatively compact subset of $\mathcal{X}$ for all totally ordered sets or chains C in $\mathcal{X}$. $\mathcal{T}$ is called *uniformly partially compact* if $\mathcal{T}$ is uniformly partially bounded and partially compact operator on $\mathcal{X}$. $\mathcal{T}$ is called *partially totally bounded* if for any totally ordered and bounded subset C of $\mathcal{X}$, $\mathcal{T}(C)$ is a relatively compact subset of $\mathcal{X}$. If $\mathcal{T}$ is partially continuous and partially totally bounded, then it is called *partially completely continuous* on $\mathcal{X}$.

Remark 1. Every compact mapping on a partially normed linear space is partially compact and every partially compact mapping is partially totally bounded, however the reverse implications do not hold. Again, every completely continuous mapping is partially completely continuous and every partially completely continuous mapping is partially continuous and partially totally bounded, but the converse may not be true.

Definition 5.14. (Dhage, [41, 48]). The order relation $\preceq$ and the metric d on a non-empty set $\mathcal{X}$ are said to be *compatible* if $\{x_n\}$ is a monotone, that is, monotone nondecreasing or monotone nondecreasing sequence in $\mathcal{X}$ and if a subsequence $\{x_{n_k}\}$ of $\{x_n\}$ converges to x^* implies that the whole sequence $\{x_n\}$ converges to x^*. Similarly, given a PONLS $(\mathcal{X}, \preceq, \|\cdot\|)$, the order relation $\preceq$ and the norm $\|\cdot\|$ are said to be compatible if $\preceq$ and the metric d defined through the norm $\|\cdot\|$ are compatible.

Example 5.1. The set $\mathbb{R}$ of real numbers with usual order relation $\leq$ and the norm defined by the absolute value function possesses the compatibility property.

Example 5.2. The finite dimensional Euclidean space $\mathbb{R}^n$ with usual component-wise order relation and the standard norm possesses the compatibility property.

Definition 5.15. (Dhage [41]). A mapping $\psi: \mathbb{R}_+ \to \mathbb{R}_+$ is called a dominating function or, in short, $\mathcal{D}$-function if it is an upper semi-continuous and monotonic nondecreasing function satisfying $\psi(0) = 0$.

Note that, if $\psi_1, \psi_2 \colon \mathbb{R}_+ \to \mathbb{R}_+$ are two $\mathcal{D}$-functions then $\psi_1 + \psi_2$ and $k\psi_1, k > 0$ are $\mathcal{D}$-functions. Few examples of $\mathcal{D}$-functions provided in [42, 43, 44, 41] are: $\psi_1(r) = kr, \psi_2(r) = \dfrac{kr}{1+lr}, \psi_3(r) = \tan^{-1}r, \psi_4(r) = \log(1 + r), \psi_5(r) = e^r - 1$, where $k, l > 0$.

Definition 5.16. (Dhage [41]). Let $(\mathcal{X}, \preceq, \|\cdot\|)$ be a PONLS. A mapping $\mathcal{T} \colon \mathcal{X} \to \mathcal{X}$ is called *partially nonlinear $\mathcal{D}$-Lipschitz* if there $\mathcal{D}$-function $\psi \colon \mathbb{R}_+ \to \mathbb{R}_+$ such that

$$\| \mathcal{T}x - \mathcal{T}y \| \leq \psi(\| x - y \|) \tag{5.5}$$

for all comparable elements $x, y \in \mathcal{X}$. If $\psi(r) = kr$, $k > 0$, then $\mathcal{T}$ is called a partially Lipschitz with a Lipschitz constant k. If $k < 1$, $\mathcal{T}$ is called a partially contraction with contraction constant k. Finally, $\mathcal{T}$ is called nonlinear $\mathcal{D}$-contraction if it is a nonlinear $\mathcal{D}$-Lipschitz with $0 < \psi(r) < r$ for $r > 0$.

Following hybrid fixed theorems established by Dhage [41, 45-47] plays important role in our main results.

Theorem 5.1. (Dhage, [46, 47]). *Let $(\mathcal{X}, \preceq, \|\cdot\|)$ be a regular partially ordered complete normed linear space such that the order relation $\preceq$ and the norm $\|\cdot\|$ in $\mathcal{X}$ are compatible in every compact chain C of $\mathcal{X}$. Let $\mathcal{T} \colon \mathcal{X} \to \mathcal{X}$ be a partially continuous, nondecreasing and partially compact operator. If there exists an element $x_0 \in \mathcal{X}$ such that $x_0 \preceq \mathcal{T}x_0$ or $x_0 \succeq \mathcal{T}x_0$, then the operator equation $\mathcal{T}x = x$ has a solution x^* in $\mathcal{X}$ and the sequence $\{\mathcal{T}^n x_0\}$ of successive iterations converges monotonically to x^*.*

Remark 2. The regularity of $\mathcal{X}$ in the Theorem 5.1 may be replaced with a stronger continuity condition of the operator $\mathcal{T}$ on $\mathcal{X}$ which is a result proved in [45]

Theorem 5.2. (Dhage [41, 46]). *Let $(\mathcal{X}, \preceq, \|\cdot\|)$ be a partially ordered Banach space and let $\mathcal{T} \colon \mathcal{X} \to \mathcal{X}$ be a nondecreasing and partially nonlinear $\mathcal{D}$-contraction. Suppose that there exists an element $x_0 \in \mathcal{X}$ such that $x_0 \preceq \mathcal{T}x_0$ or $x_0 \succeq \mathcal{T}x_0$. If $\mathcal{T}$ is continuous or $\mathcal{X}$ is regular, then $\mathcal{T}$ has a fixed point x^* in $\mathcal{X}$ and the sequence $\{\mathcal{T}^n x_0\}$ of successive iterations converges monotonically to x^*. Moreover, the fixed point x^* is unique if every pair of elements in $\mathcal{X}$ has a lower and an upper bound.*

Theorem 5.3. (Dhage, [45, 47]). *Let $(\mathcal{X}, \preceq, \|\cdot\|)$ be a regular partially ordered complete normed linear space such that the order relation $\preceq$ and the norm $\|\cdot\|$ are compatible in every compact chain C of $\mathcal{X}$. Let $\mathcal{A}, \mathcal{B}: \mathcal{X} \to \mathcal{X}$ be two nondecreasing operators such that*

(a)   *$\mathcal{A}$ is partially bounded and partially nonlinear $\mathcal{D}$-contraction,*

(b)   *$\mathcal{B}$ is partially continuous and partially compact, and*

(c)   *there exists an element $x_0 \in \mathcal{X}$ such that $x_0 \preceq \mathcal{A}x_0 + \mathcal{B}x_0$ or $x_0 \succeq \mathcal{A}x_0 + \mathcal{B}x_0$.*

Then the operator equation $\mathcal{A}x + \mathcal{B}x = x$ has a solution x^* in $\mathcal{X}$ and the sequence $\{x_n\}$ of successive iterations defined by $x_{n+1} = \mathcal{A}x_n + \mathcal{B}x_n$, $n = 0, 1, \ldots$, converges monotonically to x^*.

Remark 3. The compatibility of the order relation $\preceq$ and the norm $\|\cdot\|$ in every compact chain of $\mathcal{X}$ holds if every partially compact subset of $\mathcal{X}$ possesses the compatibility property with respect to $\preceq$ and $\|\cdot\|$.

5.3. HADAMARD FRACTIONAL DELAY DIFFERENTIAL EQUATIONS

In this section, we consider Hadamard fractional delay differential equations (HFDDE) of the form:

$$^H D^q y(t) = f(t, y_t), t \in J = [1, T], 0 < q < 1, \tag{5.6}$$

$$y(t) = \phi(t), t \in [1 - r, 1], \tag{5.7}$$

$$_H I^{1-q} y(t)|_{t=1} = 0, \tag{5.8}$$

where $^H D^q$ is the Hadamard fractional derivative, $f: J \times C([-r, 0], \mathbb{R}) \to \mathbb{R}$ is a given function and $\phi \in C([1 - r, 1], \mathbb{R})$ with $\phi(1) = 0$. For any function $y: [1 - r, T] \to \mathbb{R}$ and any $t \in J$, we denote by y_t the element of $C([-r, 0], \mathbb{R})$ and is defined by

$$y_t(\theta) = y(t + \theta), \theta \in [-r, 0].$$

Consider the space $C = C([-r, 0], \mathbb{R}) = \{\varphi : \varphi : [-r, 0] \to \mathbb{R} \text{ is continuous}\}$ with the norm

$$\| \varphi \|_C = \sup_{t \in [-r,0]} |\varphi(t)|.$$

Let $B = C([1 - r, T], \mathbb{R}) = \{y : y : [1 - r, T] \to \mathbb{R} \ is \ continuous \}$ be the Banch space with the norm

$$\| y \|_B = \sup_{t \in [1-r,T]} |y(t)|. \tag{5.9}$$

Define the order relation $\preceq$ in B by

$$x \preceq y \Leftarrow x(t) \leq y(t), \forall t \in [1 - r, T]. \tag{5.10}$$

Note that, $B = (B, \preceq, \|\cdot\|_B)$ is a partially ordered Banach space with the norm defined in (5. 9) and partially order relation $\preceq$ defined in (5.10). Further, for any $t \in [1, T]$ and $\theta \in [-r, 0]$ we have $t + \theta \in [1 - r, T]$. Therefore, for any $t \in [1, T]$,

$$x, y \in B, x \preceq y \Rightarrow x(t + \theta) \leq y(t + \theta), \forall \theta \in [-r, 0]$$

$$\Rightarrow x_t(\theta) \leq y_t(\theta), \forall \theta \in [-r, 0]$$

$$\Rightarrow x_t \preceq y_t \ in \ C.$$

Thus we have

$$x, y \in B, x \preceq y \Rightarrow x_t \preceq y_t \ in \ C, for \ any \ t \in [1, T]. \tag{5.11}$$

Lemma 5.2. *Let* $(B, \preceq, \|\cdot\|_B)$ *be a partially ordered Banach space with the norm* $\|\cdot\|_B$ *and the order relation* $\preceq$ *defined by (5.9) and (5.10) respectively. Then* $\|\cdot\|_B$ *and* $\preceq$ *are compatible in every partially compact subset of B.*

The proof of the Lemma 5.2 can be completed on similar line with very little modification in the proof of the Lemma given in [43, 44] and hence we omit the details.

Definition 5.17. A function $y \in B = C([1 - r, T], \mathbb{R})$ that satisfies (5.6)-(5.8) is called the solution of the initial value problem (5.6)-(5.8).

Definition 5.18. *A function* $u \in B$ *is said to be lower solution of the HFDDE* (5.6)-(5.8) *if it satisfies the inequations*

$$^{H}D^{q}y(t) \leq f(t, y_t), t \in J = [1, T], 0 < q < 1,$$

$$y(t) \leq \phi(t), t \in [1 - r, 1],$$

$$_{H}I^{q}y(t)|_{t=1} = 0.$$

Definition 5.19. *A function $v \in B$ is called an upper solution of the HFDDE (5.6)-(5.8) if it satisfies the inequations*

$$^{H}D^{q}y(t) \geq f(t, y_t), t \in J = [1, T], 0 < q < 1,$$

$$y(t) \geq \phi(t), t \in [1 - r, 1],$$

$$_{H}I^{q}y(t)|_{t=1} = 0.$$

5.3.1. An Existence Result for HFDDE

In this section, results about existence of solution to the HFDDE (5.6)-(5.8) is proved. We list the following hypotheses.

(H1) There exist a constant $L_f > 0$ such that $|f(t, \psi)| \leq L_f, for\ all\ t \in J$ and $\psi \in C$.

(H2) The mapping $f(t, \psi)$ is monotone nondecreasing in $\psi \in C$ for each $t \in J$.

(H3) The HFDDE (5.6)-(5.8) has lower solution $u \in B$.

Theorem 5.4. If the hypotheses (H1)–(H3) are satisfied, then the HFDDE (5.6)-(5.8) has a solution $\tilde{y} : [1 - r, T] \to \mathbb{R}$ in B and the sequence $\{y_n\}_{n=0}^{n=\infty}$ of successive approximation defined by

$$y_0 = u, y_{n+1}(t) = \begin{cases} \phi(t), \text{if } t \in [1 - r, 1], \\ \frac{1}{\Gamma(q)} \int_1^t \left(\log \frac{t}{s} \right)^{q-1} f\left(s, y_{n_s} \right) \frac{ds}{s}, \text{if } t \in [1, T]. \end{cases} \quad (5.12)$$

converges monotonically to $\tilde{y}$.

Proof. In the view of the Lemma 5.1, the HFDDE (5.6)-(5.7) is equivalent to the fractional Volterra integral equations

$$y(t) = \begin{cases} \phi(t), \text{if } t \in [1-r,1], \\ \frac{1}{\Gamma(q)} \int_1^t \left(\log \frac{t}{s}\right)^{q-1} f(s,y_s) \frac{ds}{s}, \text{if } t \in [1,T]. \end{cases} \tag{5.13}$$

By Lemma 5.2, every compact chain in B has the compatibility property with respect to the $\|\cdot\|_B$ and the order relation $\preceq$ in B.

Define operator $\mathcal{F}: B \to B$ by

$$\mathcal{F}y(t) = \begin{cases} \phi(t), \text{if } t \in [1-r,1], \\ \frac{1}{\Gamma(q)} \int_1^t \left(\log \frac{t}{s}\right)^{q-1} f(s,y_s) \frac{ds}{s}, \text{if } t \in [1,T]. \end{cases} \tag{5.14}$$

Then the equivalent fractional Volterra integral equation to the HFDDE (5.6)-(5.8) is written as

$$x(t) = \mathcal{F}x(t), t \in [1-r,T].$$

We prove that the operator $\mathcal{F}$ verifies the conditions of Theorem 5.1. We give the proof step wise.

Step 1. $\mathcal{F}: B \to B$ nondecreasing.

Let any $x, y \in B$ with $x \preceq y$. Then for any $t \in [1,T]$ we have $x_t \le y_t$ in C. By hypothesis (H2), for any $t \in J = [1,T]$, we have

$$\mathcal{F}x(t) = \frac{1}{\Gamma(q)} \int_1^t \left(\log \frac{t}{s}\right)^{q-1} f(s,x_s) \frac{ds}{s}$$

$$\le \frac{1}{\Gamma(q)} \int_1^t \left(\log \frac{t}{s}\right)^{q-1} f(s,y_s) \frac{ds}{s}$$

$$= \mathcal{F}y(t)$$

and

$$\mathcal{F}x(t) = \phi(t) = \mathcal{F}y(t), t \in [1-r,1].$$

Therefore

$$\mathcal{F}x(t) \le \mathcal{F}y(t), for \ all \ t \in [1-r,T] \Rightarrow \mathcal{F}x \preceq \mathcal{F}y.$$

This proves that $\mathcal{F}: B \to B$ nondecreasing operator.

Step 2. $\mathcal{F}: B \to B$ partially continuous.

Let C be a chain in B. Let $\{y_n\} \subseteq C$ such that $y_n \to y$. Then for each $t \in J$ we have $y_{n_t} \to y_t$. By dominated convergence theorem, for each $t \in J$

$$\lim_{n \to \infty} \mathcal{F} y_n(t) = \lim_{n \to \infty} \frac{1}{\Gamma(q)} \int_1^t \left(\log \frac{t}{s}\right)^{q-1} f(s, y_{n_s}) \frac{ds}{s}$$

$$= \frac{1}{\Gamma(q)} \int_1^t \left(\log \frac{t}{s}\right)^{q-1} \left\{\lim_{n \to \infty} f(s, y_{n_s})\right\} \frac{ds}{s}$$

$$= \frac{1}{\Gamma(q)} \int_1^t \left(\log \frac{t}{s}\right)^{q-1} f(s, y_s) \frac{ds}{s}$$

$$= \mathcal{F} y(t).$$

Also for each $t \in [1 - r, 1]$,

$$\lim_{n \to \infty} \mathcal{F} y_n(t) = \lim_{n \to \infty} \phi(t) = \phi(t) = \mathcal{F} y(t).$$

Therefore

$$\lim_{n \to \infty} \mathcal{F} y_n(t) = \mathcal{F} y(t) \ for \ all \ t \in [1 - r, T].$$

This proves $\{\mathcal{F} y_n\} \to \mathcal{F} y$ pointwise on $[1 - r, T]$.

Next, to prove the sequence $\{\mathcal{F} y_n\}$ is equicontinuous in B we proceed as follows.

Case-I : Let any $t_1, t_2 \in J$ with $t_1 < t_2$.

Then,

$$|\mathcal{F} y_n(t_2) - \mathcal{F} y_n(t_1)|$$

$$= \left| \frac{1}{\Gamma(q)} \int_1^{t_2} \left(\log \frac{t_2}{s}\right)^{q-1} f(s, y_{n_s}) \frac{ds}{s} - \frac{1}{\Gamma(q)} \int_1^{t_1} \left(\log \frac{t_1}{s}\right)^{q-1} f(s, y_{n_s}) \frac{ds}{s} \right|$$

$$= \frac{1}{\Gamma(q)} \left| \int_1^{t_1} \left\{ \left(\log\frac{t_2}{s} \right)^{q-1} - \left(\log\frac{t_1}{s} \right)^{q-1} \right\} f(s, y_{n_s}) \frac{ds}{s} \right|$$

$$+ \frac{1}{\Gamma(q)} \left| \int_{t_1}^{t_2} \left(\log\frac{t_2}{s} \right)^{q-1} f(s, y_{n_s}) \frac{ds}{s} \right|$$

$$\leq \frac{L_f}{\Gamma(q)} \left| \int_1^{t_1} \left\{ \left(\log\frac{t_2}{s} \right)^{q-1} - \left(\log\frac{t_1}{s} \right)^{q-1} \right\} \frac{ds}{s} \right| + \frac{L_f}{\Gamma(q)} \left| \int_{t_1}^{t_2} \left(\log\frac{t_2}{s} \right)^{q-1} \frac{ds}{s} \right|$$

$$= \frac{L_f}{\Gamma(q)} \left| \frac{(\log t_2)^q}{q} - \frac{(\log t_1)^q}{q} \right| + \frac{L_f}{\Gamma(q)} \frac{\left(\log\frac{t_2}{t_1} \right)^q}{q}$$

$$= \frac{L_f}{\Gamma(q+1)} \left\{ |(\log t_2)^q - \log t_1)^q| + \left(\log\frac{t_2}{t_1} \right)^q \right\}$$

Case-II : If $t_1 < t_2 \leq 1$ then $|\mathcal{F}y_n(t_2) - \mathcal{F}y_n(t_1)| = |\phi(t_2) - \phi(t_1)|$.

Case-III : Let $t_1 \leq 1 \leq t_2$. Then

$$|\mathcal{F}y_n(t_2) - \mathcal{F}y_n(t_1)| = \left| \frac{1}{\Gamma(q)} \int_1^{t_2} \left(\log\frac{t_2}{s} \right)^{q-1} f(s, y_{n_s}) \frac{ds}{s} - \phi(t_1) \right|$$

$$\leq \left| \frac{1}{\Gamma(q)} \int_1^{t_2} \left(\log\frac{t_2}{s} \right)^{q-1} f(s, y_{n_s}) \frac{ds}{s} - \phi(1) \right| + |\phi(1) - \phi(t_1)|$$

From the above three cases it follows that for any $t_1, t_2 \in [1 - r, T]$,

$$|\mathcal{F}y_n(t_2) - \mathcal{F}y_n(t_1)| \to 0 \ as \ |t_1 - t_2| \to 0 \ uniformly \ for \ all \ n \in \mathbb{N}.$$

Hence the convergence $\mathcal{F}y_n \to \mathcal{F}y$ is uniform and therefore $\mathcal{F}: B \to B$ is a partially continuous.

Step 3. $\mathcal{F}: B \to B$ is partially compact.

Let C be any chain in B. Let any $y \in C$. Then for any $t \in J = [1, T]$,

$$|\mathcal{F}y(t)| \leq \frac{1}{\Gamma(q)} \left| \int_1^t \left(\log\frac{t}{s} \right)^{q-1} f(s, y_s) \frac{ds}{s} \right|$$

$$\leq \frac{1}{\Gamma(q)} \int_1^t \left(\log\frac{t}{s}\right)^{q-1} |f(s,y_s)| \frac{ds}{s}$$

$$\leq \frac{L_f}{\Gamma(q)} \int_1^t \left(\log\frac{t}{s}\right)^{q-1} \frac{ds}{s} = \frac{L_f}{\Gamma(q+1)} (\log t)^q \leq \frac{L_f}{\Gamma(q+1)} (\log T)^q.$$

Also, for any $t \in [1-r, 1]$, $|\mathcal{F}y(t)| = |\phi(t)| \leq \|\phi\|_{[1-r,1]}$, where $\|\phi\|_{[1-r,1]} = \sup_{s\in[1-r,1]} |\phi(s)|$. Therefore

$$\|\mathcal{F}y\|_B = \sup_{t\in[1-r,T]} |\mathcal{F}y(T)| \leq \max\left\{\frac{L_f}{\Gamma(q+1)} (\log T)^q, \| \phi \|_{[1-r,1]}\right\}$$

Hence $\mathcal{F}(C)$ is uniformly bounded subset of B.

Next, to prove $\mathcal{F}(C)$ is an equicontinuous subset in B, let any $y \in C$ and $t_1, t_2 \in J$ with $t_1 < t_2$. Then,

$$|\mathcal{F}y(t_2) - \mathcal{F}y(t_1)|$$

$$= \left|\frac{1}{\Gamma(q)} \int_1^{t_2} \left(\log\frac{t_2}{s}\right)^{q-1} f(s,y_{n_s}) \frac{ds}{s} - \frac{1}{\Gamma(q)} \int_1^{t_1} \left(\log\frac{t_1}{s}\right)^{q-1} f(s,y_s) \frac{ds}{s}\right|$$

$$= \frac{1}{\Gamma(q)} \left|\int_1^{t_1} \left\{\left(\log\frac{t_2}{s}\right)^{q-1} - \left(\log\frac{t_1}{s}\right)^{q-1}\right\} f(s,y_s) \frac{ds}{s}\right|$$

$$+ \frac{1}{\Gamma(q)} \left|\int_{t_1}^{t_2} \left(\log\frac{t_2}{s}\right)^{q-1} f(s,y_s) \frac{ds}{s}\right|$$

$$\leq \frac{L_f}{\Gamma(q)} \left|\int_1^{t_1} \left\{\left(\log\frac{t_2}{s}\right)^{q-1} - \left(\log\frac{t_1}{s}\right)^{q-1}\right\} \frac{ds}{s}\right| + \frac{L_f}{\Gamma(q)} \left|\int_{t_1}^{t_2} \left(\log\frac{t_2}{s}\right)^{q-1} \frac{ds}{s}\right|$$

$$= \frac{L_f}{\Gamma(q+1)} \left\{|(\log t_2)^q - \log t_1)^q| + \left(\log\frac{t_2}{t_1}\right)^q\right\}.$$

Note that, in the above inequation the right hand side is not dependening on y and we have $|\mathcal{F}y(t_2) - \mathcal{F}y(t_1)| \to 0$ as $t_2 \to t_1$. Next, consider the cases $t_1 < t_2 \leq 1$ and $t_1 \leq 1 \leq t_2$. We can write

$$|\mathcal{F}y(t_2) - \mathcal{F}y(t_1)| = |\mathcal{F}y(t_2) - \phi(t_1)| \leq |\mathcal{F}y(t_2) - \phi(1)| + |\phi(1) - \phi(t_1)|.$$

Observe that the equicontinuity in these cases obtained from above inequation *via* uniform continuity of ϕ on $[-r, 0]$.

Step 4. We now prove that $\exists u \in B$ satisfying $u \leq \mathcal{F}u$.

By hypothesis (H3) there exists $u \in B$ such that

$$^H D^q u(t) \leq f(t, u_t), t \in J = [1, T], 0 < q < 1, \tag{5.15}$$

$$y(t) \leq \phi(t), t \in [1 - r, 1], \tag{5.16}$$

$$_H I^{q-1} y(t)|_{t=1} = 0, \tag{5.17}$$

In the view of Lemma 5.1 the equivalent fractional Volterra integral inequations to the fractional differential inequations (5.15)-(5.17) are

$$u(t) \leq \frac{1}{\Gamma(q)} \int_1^t \left(\log \frac{t}{s} \right)^{q-1} f(s, u_s) \frac{ds}{s}, t \in [1, T].$$

$$u(t) \leq \phi(t), t \in [1 - r, 1]$$

Using definition of operator $\mathcal{F}$, we have

$$u \leq \mathcal{F}u.$$

So from the steps 1-4, we can conclude that $\mathcal{F}$ fulfills the assumptions of the Theorem 5.1. By Theorem 5.1 and in the view of Remark 2, $\mathcal{F}$ has fixed point $\tilde{y} \colon [-r, T] \to \mathbb{R}$ in B which is solution of HFDDE (5.6)-(5.8). Furthermore, $y_n \to \tilde{y}$ in B monotonically. The proof of the theorem is completed.

Remark 4. We may acquire the existence result by replacing the assumption (H3) in the Theorem 5.4 by

(*H*3)' The HFDDE (5.6)-(5.8) has upper solution $u \in B$.

5.3.2. An Uniqueness Result for HFDDE

We now derive the uniqueness result for HFDDE (5.6)-(5.8) *via* weak Lipschitz condition:

(H4) There exist a $\mathcal{D}-$ function Ω that satisfy

$$0 \leq f(t, x) - f(t, y) \leq \Omega(\|x_t - y_t\|_C), \forall x, y \in B \text{ with } x \geq y \text{ and } t \in [1, T].$$

Moreover, $0 < \frac{(\log T)^q}{\Gamma(q+1)} \Omega(r) < r$ for each $r > 0$.

Theorem 5.5. If the hypotheses (H2), (H3) and (H4) are hold. Then the HFDDE (5.6)-(5.8) has a unique solution $\tilde{y}: [1 - r, T] \to \mathbb{R}$ in B and the sequence $\{y_n\}_{n=0}^{n=\infty}$ of successive approximation defined by (5.12) converges monotonically to $\tilde{y}$.

Proof. Note that, $B = C([1 - r, T], \mathbb{R})$ is a lattice with the relation $\preceq$ given in (5.10) and hence every pair of elements in B has lower and upper bounds. Consider $\mathcal{F}: B \to B$ defined in (5.14). Then we can write equivalent integral equation of initial value problem (5.6)-(5.8) as

$$x(t) = \mathcal{F}x(t), x \in B, t \in [1 - r, T].$$

We have already proved that $\mathcal{F}$ is continuous and nondecreasing. Thus to prove $\mathcal{F}$ verifies the conditions of Theorem 5.2, it only remains to prove $\mathcal{F}: B \to B$ is a nonlinear $\mathcal{D}$ − contraction.

Let any $x, y \in B$ with $x \geq y$. Then, by using hypothesis (H4), for any $t \in [1, T]$,

$$|\mathcal{F}x(t) - \mathcal{F}y(t)| = \left| \frac{1}{\Gamma(q)} \int_1^t \left(\log \frac{t}{s} \right)^{q-1} |f(s, x_s) - f(s, y_s)| \frac{ds}{s} \right|$$

$$\leq \frac{1}{\Gamma(q)} \int_1^t \left(\log \frac{t}{s} \right)^{q-1} \Omega(\| x_s - y_s \|_C) \frac{ds}{s}$$

Note that for any $x \in B$ and any $t \in [1, T]$ we have

$$\|x_t\|_C = \sup_{-r \leq \theta \leq 0} |x_t(\theta)| = \sup_{-r \leq \theta \leq 0} |x(t + \theta)| \leq \sup_{1 - r \leq t + \theta \leq T} |x(t + \theta)| \leq \|x\|_B. \tag{5.18}$$

Therefore

$$|\mathcal{F}x(t) - \mathcal{F}y(t)| \leq \frac{1}{\Gamma(q)} \left\{ \int_1^t \left(\log \frac{t}{s} \right)^{q-1} \frac{ds}{s} \right\} \Omega(\| x - y \|_B)$$

$$= \frac{(\log t)^q}{\Gamma(q + 1)} \Omega(\| x - y \|_B)$$

$$\leq \frac{(\log T)^q}{\Gamma(q+1)} \Omega(\| x - y \|_B) \tag{5.19}$$

Also

$$|\mathcal{F}x(t) - \mathcal{F}y(t)| = 0, if \ t \in [1-r,1] \tag{5.20}$$

From (5.19) and (5.20),

$$|\mathcal{F}x(t) - \mathcal{F}y(t)| \leq \frac{(\log T)^q}{\Gamma(q+1)} \ \Omega(\|x-y\|_B) = \psi(\|x-y\|_B), \forall t \in [1-r,T],$$

where

$$\psi(r) = \frac{(\log T)^q}{\Gamma(q+1)} \ \Omega(r) < r, r > 0.$$

Therefore

$$\|\mathcal{F}x - \mathcal{F}y\|_B = \sup_{t \in [1-r,T]} |\mathcal{F}x(t) - \mathcal{F}y(t)| \leq \psi(\|x-y\|_B) \ for \ all \ x,y \in B, with \ x \geq y.$$

This proves $\mathcal{F}: B \rightarrow B$ is a partially nonlinear $\mathcal{D}-$ contraction. Using hypothesis (H3) and proceding as in step-4 of the proof of Theorem 5.4, $\exists u \in B$ such that $u \preceq \mathcal{F}u$. We have proved that $\mathcal{F}$ fulfills every one of the conditions of Theorem 5.2. Hence $\exists \tilde{y} \in B$ such that $\tilde{y} = \mathcal{F}\tilde{y}$, which is a unique solution of HFDDE (5.6)-(5.8). Furthermore, the sequence $\{y_n\}_{n=0}^{\infty}$ converges monotonically to $\tilde{y}$. This complete the proof.

Remark 5. We can acquire the uniqueness result by replacing the assumption (H3) in the Theorem 5.5 by

(*H3*)' The HFDDE (5.6)-(5.8) has upper solution $u \in B$.

5.4. LINEAR PERTURBATIONS OF FIRST TYPE

In some cases it may possible that the nonlinear function f involved in the HFDDE (5.6)-(5.8) neither fulfills the requirements of Theorem 5.1 nor fulfills the requirements of Theorem 5.2. However, it may possible to decompose the functions f in the form $f = f_1 + f_2$, where f_1 and f_2 fulfills the the requirements of Theorem 5.1 and Theorem 5.2 respectively. Then the existence and monotonic approximation of the solutions of such HFDDE can be obtained under suitable mixed hybrid conditions *via* hybrid fixed point Theorem 5.3. In the terminology of Dhage [41], such equations are called hybrid HFDDE with linear perturbation of the HFDDE (5.6)-(5.8) of first type.

With the same notations of the section 3, we consider the nonlinear hybrid Hadamard fractional delay differential equations (HHFDDE) of the form:

$$^H D^q y(t) = f(t, y_t) + g(t, y_t), t \in J = [1, T], 0 < q < 1, \qquad \textbf{(5.21)}$$

$$y(t) = \phi(t), t \in [1 - r, 1], \qquad \textbf{(5.22)}$$

$$_H I^{1-q} y(t)|_{t=1} = 0, \qquad \textbf{(5.23)}$$

where $f, g : J \times C([-r, 0], \mathbb{R}) \to \mathbb{R}$ are continuous function.

Definition 5.20. A function $y \in B = C([1 - r, T], \mathbb{R})$ that satisfies (5.21)-(5.23) is called the solution of the initial value problem (5.21)-(5.23).

Definition 5.21. A function $u \in B$ is said to be a lower solution of HHFDDE (5.21)-(5.23) if it satisfies

$$^H D^q y(t) \leq f(t, y_t) + g(t, y_t), t \in J = [1, T], 0 < q < 1,$$

$$y(t) \leq \phi(t), t \in [1 - r, 1],$$

$$_H I^{1-q} y(t)|_{t=1} = 0.$$

Definition 5.22. A function $u \in B$ is called an upper solution of HHFDDE (5.21)-(5.23) if it satisfies

$$^H D^q y(t) \geq f(t, y_t) + f(t, x_t), t \in J = [1, T], 0 < q < 1,$$

$$y(t) \geq \phi(t), t \in [1 - r, 1],$$

$$_H I^{1-q} y(t)|_{t=1} = 0.$$

Theorem 5.6. Assume that the hypotheses (H1) and (H4) are satisfied. In addition, we assume

(H5) $\exists L_g > 0$ such that $|g(t, \psi)| \leq L_g, \forall t \in J$ and $\psi \in C$;

(H6) the mapping $g(t, \psi)$ is monotone nondecreasing in $\psi \in C$ for each $t \in J$;

(H7) the HHFDDE (5.21)-(5.23) has lower solution $u \in B$.

Then HHFDDE (5.21)-(5.23) has a solution $\tilde{y} \colon [1 - r, T] \to \mathbb{R}$ in B and the sequence $\{y_n\}_{n=0}^{n=\infty}$ of successive approximation defined by

$$y_0 = u, y_{n+1}(t) = \begin{cases} \phi(t), \text{if } t \in [1 - r, 1], \\ \frac{1}{\Gamma(q)} \int_1^t \left(\log\frac{t}{s}\right)^{q-1} \{f(s, y_{n_s}) + g(s, y_{n_s})\}\frac{ds}{s}, \text{if } t \in [1, T]. \end{cases} \quad (5.24)$$

converges monotonically to $\tilde{y}$.

Proof. Already we have disccused in the Theorem 5.4, each compact chain in B has compatibility property with respect to $\|\cdot\|_B$ and $\preceq$ in B.

Further, by Lemma 5.1, the HHFDDE (5.21)-(5.23) is equivalent to the fractional Volterra integral equations

$$y(t) = \begin{cases} \phi(t), \text{if } t \in [1 - r, 1], \\ \frac{1}{\Gamma(q)} \int_1^t \left(\log\frac{t}{s}\right)^{q-1} f(s, y_s)\frac{ds}{s} + \frac{1}{\Gamma(q)} \int_1^t \left(\log\frac{t}{s}\right)^{q-1} g(s, y_s)\frac{ds}{s}, \text{if } t \in [1, T]. \end{cases} \quad (5.25)$$

Define operator $\tilde{\mathcal{F}}, \tilde{\mathcal{G}} \colon B \to B$ by

$$\tilde{\mathcal{F}}y(t) = \begin{cases} 0, \text{if } t \in [1 - r, 1], \\ \frac{1}{\Gamma(q)} \int_1^t \left(\log\frac{t}{s}\right)^{q-1} f(s, y_s)\frac{ds}{s}, \text{if } t \in [1, T]. \end{cases} \quad (5.26)$$

and

$$\tilde{\mathcal{G}}y(t) = \begin{cases} \phi(t), \text{if } t \in [1 - r, 1], \\ \frac{1}{\Gamma(q)} \int_1^t \left(\log\frac{t}{s}\right)^{q-1} g(s, y_s)\frac{ds}{s}, \text{if } t \in [1, T]. \end{cases} \quad (5.27)$$

Then the equivalent fractional Volterra integral equations (5.25) can be written as

$$\tilde{\mathcal{F}}y(t) + \tilde{\mathcal{G}}y(t) = y(t), y \in B, t \in [1 - r, T]. \quad (5.28)$$

Following the similar arguments as in the proofs of Theorem 5.4 and 5.5, with suitable modification, one can easily prove that $\tilde{\mathcal{F}}$ is partially bounded and partially nonlinear $\mathcal{D}$-contraction and $\tilde{\mathcal{G}}$ is partially continuous and partially compact.

By using assumption (H7) there exists $u \in B$ such that

$$^H D^q u(t) \leq f(t, u_t) + g(t, u_t), t \in J = [1, T], 0 < q < 1,$$

$$u(t) \leq \phi(t), t \in [1 - r, 1],$$

$$_H I^{1-q} u(t)|_{t=1} = 0.$$

By Lemma 5.1, above fractional differential inequations is equivalent to the inequations

$$u(t) \leq \phi(t), \text{if } t \in [1 - r, 1]$$

$$u(t) \leq \frac{1}{\Gamma(q)} \int_1^t \left(\log\frac{t}{s}\right)^{q-1} f(s, u_s) \frac{ds}{s} + \frac{1}{\Gamma(q)} \int_1^t \left(\log\frac{t}{s}\right)^{q-1} g(s, u_s) \frac{ds}{s}, \text{if } t \in [1, T].$$

Using the definition of the operators $\tilde{\mathcal{F}}$ and $\tilde{\mathcal{G}}$ and the above fractional integral inequations, we have $u \in B$ such that

$$u \leq \tilde{\mathcal{F}}u + \tilde{\mathcal{G}}u.$$

Applying Theorem 5.3, the operator equation

$$\tilde{\mathcal{F}}y + \tilde{\mathcal{G}}y = y$$

has fixed point $\tilde{y} : [1 - r, T] \to \mathbb{R}$ in B which is solution of HHFDDE (5.21)-(5.23). Further, the sequence $\{y_n\}_{n=0}^{n=\infty}$ defined by (5.24) converges to $\tilde{y}$.

Remark 6. We may acquire the existence result by replacing the assumption (H7) in the Theorem 5.6 by

(H7)' HHFDDE (5.21)-(5.23) has upper solution $u \in B$.

5.5.   NEUTRAL   HADAMARD   FRACTIONAL   DIFFERENTIAL EQUATIONS (NHFDE)

We estblish the existence results *via* fixed point Theorem 5.3 for NHFDE of the form:

$$^H D^q (y(t) - e(t, y_t)) = f(t, y_t), t \in J = [1, T], 0 < q < 1, \qquad \textbf{(5.29)}$$

$$y(t) = \phi(t), t \in [1 - r, 1], \qquad \textbf{(5.30)}$$

$$_HI^{1-q}y(t)|_{t=1} = 0, \tag{5.31}$$

where f and ϕ are as given in the problem (5.6)-(5.8) and $e: J \times C([1 - r, 0], \mathbb{R}) \to \mathbb{R}$ staisfy $e(1, \phi) = 0$.

Theorem 5.7. Asssume that (H1) and (H2) holds. In addition, assume that

(N1) (i) There exist $L_e > 0$ such that

$$|e(t, \psi)| \le L_e, t \in J, \psi \in C;$$

and $e(t, \psi)$ is monotone nondecreasing in $\psi \in C$ for each $t \in J$;

(N2) There exists $\mathcal{D} -$ function ψ_e such that

$$0 \le e(t, x_t) - e(t, x_t) \le \psi_e(\|x_t - y_t\|_C),$$

for all $x, y \in B$ with $x \succeq y$ and $t \in [1, T]$, where $\psi_e(r) < r$, for $r > 0$;

(N3) The NHFDE (5.29)-(5.31) has lower solution $u \in B$.

Then NHFDE (5.29)-(5.31) has a solution $\tilde{y}: [1 - r, T] \to \mathbb{R}$ in B and the sequence $\{y_n\}_{n=0}^{n=\infty}$ of successive approximation defined by

$$y_0 = u, y_{n+1}(t) = \begin{cases} \phi(t), \text{if } t \in [1 - r, 1], \\ e(t, y_{n_t}) + \frac{1}{\Gamma(q)} \int_1^t \left(\log\frac{t}{s}\right)^{q-1} f(s, y_{n_s})\frac{ds}{s}, \text{if } t \in [1, T]. \end{cases} \tag{5.32}$$

converges monotonically to $\tilde{y}$.

Proof. By Lemma 5.1, the equivalent integral inequations to NHFDE (5.29)-(5.31) is given by

$$y(t) = \begin{cases} \phi(t), \text{if } t \in [1 - r, 1], \\ e(t, y_t) + \frac{1}{\Gamma(q)} \int_1^t \left(\log\frac{t}{s}\right)^{q-1} f(s, y_s)\frac{ds}{s}, \text{if } t \in [1, T]. \end{cases} \tag{5.33}$$

Consider $\mathcal{F}: B \to B$ defined by

$$\mathcal{F}y(t) = \begin{pmatrix} \phi(t), \text{if } t \in [1-r, 1], \\ \dfrac{1}{\Gamma(q)} \displaystyle\int_1^t \left(\log\frac{t}{s}\right)^{q-1} f(s, y_s) \dfrac{ds}{s}, \text{if } t \in [1, T], \end{pmatrix}$$

which is same as defined in the proof of Theorem 5.4. Further, define $\mathcal{E} : B \to B$ by

$$\mathcal{E}y(t) = \begin{pmatrix} 0, \text{if } t \in [1-r, 1], \\ e(t, x_t), \text{if } t \in J. \end{pmatrix} \tag{5.34}$$

Then the equivalent integral (5.33) of NHFDE (5.29)-(5.31) framed as the operator equation

$$\mathcal{F}y + \mathcal{E}y = y.$$

We have already proved that $\mathcal{F} : B \to B$ is nondecreasing, partially continuous and partially compact the operator. The proof of $\mathcal{E} : B \to B$ is nondecreasing, partially bounded and nonlinear $\mathcal{D} -$ contraction is given stepwise.

Step 1. $\mathcal{E} : B \to B$ nondecreasing.

Let any $x, y \in B$ with $x \preceq y$. Then for any $t \in [1, T]$ we have $x_t \preceq y_t$ in C. By hypothesis (N1) (ii), we have

$$\mathcal{E}x(t) = e(t, x_t) \leq e(t, y_t) = \mathcal{E}y(t), t \in J = [1, T].$$

Also

$$\mathcal{E}x(t) = 0 = \mathcal{E}y(t), t \in [1-r, 1].$$

Therefore

$$\mathcal{E}x(t) \leq \mathcal{E}y(t), t \in [1-r, T] \Rightarrow \mathcal{E}x \preceq \mathcal{E}y.$$

This prove $\mathcal{E} : B \to B$ is nondecreasing.

Step 2. $\mathcal{E} : B \to B$ is partially bounded.

For any $x \in B$,

$$|\mathcal{E}x(t)| = |e(t, x_t)| \leq L_e, t \in [1, T]$$

and

$$|\mathcal{E}x(t)| = 0, t \in [1, T].$$

Therefore

$$|\mathcal{E}x(t)| \le L_e, \forall t \in [1-r, T] \Rightarrow \|\mathcal{E}x\|_B = \sup_{t \in [1-r, T]} |\mathcal{E}x(t)| \le L_e.$$

Hence $\mathcal{E}: B \to B$ is partially bounded.

Step 3. $\mathcal{E}: B \to B$ is non linear $\mathcal{D}$ − contraction.

Let any $x, y \in B$ with $x \succeq y$. Then $x_t \succeq y_t$ in C. Using hypothesis (N2) and using inequation (5.18), we obtain

$$|\mathcal{E}x(t) - \mathcal{E}y(t)| = e(t, x_t) - e(t, y_t) \le \psi_e(\|x_t - y_t\|_C) \le \psi_e(\|x - y\|_B), t \in [1, T]. \quad \textbf{(5.35)}$$

Further,

$$|\mathcal{E}x(t) - \mathcal{E}y(t)| = 0, t \in [1-r, 1] \tag{5.36}$$

From (5.35) and (5.36) we have

$$|\mathcal{E}x(t) - \mathcal{E}y(t)| \le \psi_e(\|x - y\|_B), \forall t \in [1-r, T].$$

Therefore

$$\|\mathcal{E}x - \mathcal{E}y\|_B = \sup_{t \in [1-r, T]} |\mathcal{E}x(t) - \mathcal{E}y(t)| \le \psi_e(\|x - y\|_B) \le \|x - y\|_B,$$

where $0 < \psi_e(r) < r$ for $r > 0$. Hence $\mathcal{E}: B \to B$ is non linear $\mathcal{D}$ − contraction.

Finally, using the hypothesis (N3), there exists $u \in B$ that satisfies the inequations

$$^HD^q(y(t) - e(t, y_t)) \le f(t, y_t), t \in J = [1, T], 0 < q < 1,$$

$$y(t) \le \phi(t), t \in [1-r, 1],$$

$$_HI^{1-q}y(t)|_{t=1} = 0.$$

and hence satisfies the equivalent fractional integral inequations

$$y(t) \le \phi(t), \text{if } t \in [1-r, 1], \tag{5.37}$$

$$y(t) \leq e(t, y_t) + \frac{1}{\Gamma(q)} \int_1^t \left(\log \frac{t}{s}\right)^{q-1} f(s, y_s) \frac{ds}{s}, \text{if } t \in [1, T]. \quad (5.38)$$

Using the definitions of the operators $\mathcal{F}$ and $\mathcal{E}$, and the fractional integral inequations (5.37)-(5.38), there exists $u \in B$ satisfying

$$u \leq \mathcal{F}u + \mathcal{E}u.$$

Fullfilled every conditions of Theorem 3, the operator equation $\mathcal{F}y + \mathcal{E}y = y$ has a fixed point $\tilde{y} \in B$. This $\tilde{y} \in B$ is the solution of (5.29)-(5.31). Moreover, the sequence $\{y_n\}_{n=0}^{n=\infty}$ of successive approximation defined by (5.32) converges monotonically to $\tilde{y}$.

5.6. CONCLUSION

Similar arguments of [39, 41, 46, 47] hold for HFDDEs and NHFDDEs. The existence results obtained *via* fixed theorems of Schauder and nonlinear alternative of Laray- Schauder involves construction of closed, convex subset of B which is not needed in the methods employed here. Existence and uniqueness results for HFDDEs *via* hybrid fixed point theorems are obtained under weaker conditions on the functions involved in the equations. As discussed in [39, 43], here aslo Theorems 5.4, 5.5 and 5.6 does not make use of any Lipschitz type condition on the nonlinear functions involved in (5.6)-(5.8) and (5.29)-(5.31), but even then the monotone convergence of successive approximations to the solutions is proved for HFDDE (5.6)-(5.8) and NHFDDE (5.29)-(5.31).

The main constraint in the application of the methods utilized here is that there is no way to acquire the rate of convergence of sequence of successive approximations. Further, from Theorem 5.4, Theorem 5.6 and Theorem 5.7 it can be seen that to obtain the results the boundedness condtion is required on the functions involved in the equations. This condition restrict the study to a particular class of HFDDE and NHFDDE. Thus it will be very interesting if one can obtain the same types of results by droping boundedness conditions.

We remark that the method employed in this chapter can be extended with suitable modifications to investigate the existence and approximations of solutions for Hadamard fractional delay differential equations with nonlocal conditions or, then again with different sorts of boundary conditions.

CONSENT FOR PUBLICATION

Not applicable.

CONFLICT OF INTEREST

The authors declare no conflict of interest, financial or otherwise.

ACKNOWLEDGEMENTS

Declared none.

REFERENCES

[1]   N. Laskin, "Fractional market dynamics," *Physica A*, vol. 28, pp. 7482–492, 2000.

[2]   I. Petras, and R. L. Magin, "Simulation of drug uptake in a two compartmental fractional model for a biological system," *Commun. Nonlinear Sci. Numer. Simul.*, vol. 16, no. 12, pp. 4588–4595, Dec. 2011.

[3]   J. Cao, C. Ma, Z. Jiang, and S. Liu, "Nonlinear dynamic analysis of fractional order rub-impact rotor system," *Commun. Nonlinear Sci. Numer. Simul.*, vol. 16, no. 3, pp. 1443–1463, Mar. 2011.

[4]   Y. Luo, Y. Q. Chen, and Y. Pi, "Experimental study of fractional order proportional derivative controller synthesis for fractional order systems," *Mechatronics*, vol. 21, no. 1, pp. 204–214, Feb. 2011.

[5]   X.-J. Yang, and H. M. Srivastava, "An asymptotic perturbation solution for a linear oscillator of free damped vibrations in fractal medium described by local fractional derivatives," *Commun. Nonlinear Sci. Numer. Simul.*, vol. 29, no. 1, pp. 499–504, Dec. 2015.

[6]   X.-J. Yang, H. M. Srivastava, J. He, and D. Baleanu, "Cantor-type cylindrical-coordinate method for differential equations with local fractional derivatives," *Phys. Lett. A,* vol. 377, no. 28, pp. 1696–1700, Oct. 2013.

[7]   X. Zhao, H. T. Yang, and Y. Q. He, "Identification of constitutive parameters for fractional viscoelasticity," *Commun. Nonlinear Sci. Numer. Simul.*, vol. 19, no. 1, pp. 311–322, Jan. 2014.

[8]   N. Ozdemir, and D. Avci, "Optimal control of a linear time-invariant space time fractional diffusion process," *J. Vib. Control*, vol. 30, no. 3, pp. 370–380, Feb. 2014.

[9]   A. A. Kilbas, H. M. Srivastava, and J. J. Trujillo, *Theory and Applications of Fractional Differential Equations*. Elsevier: Amsterdam, 2006.

[10]   V. Lakshmikantham, S. Leela, and J. Vaaundhara Devi, *Theory of Fractional Dynamic Systems*. Cambridge Scientific Publishers: UK, 2009.

[11]   K. S. Miller, and B. Ross, *An Introduction the the Fractional Calclus and Fractional Equations*. Wiley: New York, 1993.

[12]   I. Podlubny, *Fractional Differential Equations*. Academic Press: New York, 1993.

[13]   D. Baleanu , J. A. T. Machado, and A. C. J. Luo, *Fractional Dynamics and Control*. Springer: New York, 2012.

[14]   R. P. Agarwal, M. Belmekki, and M. Benchohra, "A survey on semilinear differential equations and inclusions involving Riemann-Liouville fractional derivative," *Adv. Diff. Equ.*, vol 2009, no. 1, Article ID 981728, Mar. 2009.

[15]   R. P. Agarwal, M. Benchohra, and S. Hamani, "A survey on existence results for boundary value problems of nonlinear fractional differential equations and inclusions," *Acta Appl. Math.*, vol. 109, no. 3, pp. 973–1033, Mar. 2010.

[16]   D. Delbosco, and L. Rodino, "Existence and uniqueness for a nonlinear fractional differential equation," *J. Math. Anal.Appl.*, vol. 204, no. 2, pp. 609–625, Dec. 1996.

[17] K. Diethelm, and N. J. Ford, "Analysis of fractional differential equations," *J. Math. Anal. Appl.*, vol. 265, no. 2, pp. 229–248, Jan. 2002.

[18] V. Lakshmikantham, and A. S. Vatsala, "Basic theory of fractional differential equations," *Nonlinear Anal.*, vol. 69, no. 8, pp. 2677–2682, Oct. 2008.

[19] Y. Zhou, "Existence and uniqueness of solutions for a system of fractional differential equations," *Fract. Calc. Appl. Anal.*, vol. 12, no. 2, pp. 195–204, 2003.

[20] V. Daftardar-Gejji, and A. Babakhani, "Analysis of a system of fractional differential equations," *J. Math. Anal. Appl.*, vol. 293, no. 2, pp. 511–522, May 2004.

[21] V. Daftardar-Gejji, and H. Jafari, "Analysis of a system of nonautonomous fractional differential equations involving Caputo derivatives," *J. Math. Anal. Appl.*, vol. 328, no. 2, pp. 1026–1033, Apr. 2007.

[22] K. D. Kucche, J. J. Nieto, and V. Venktesh, "Theory of nonlinear implicit fractional differential equations," *Differ. Equ. Dyn. Syst.* DOI 10.1007/s12591-016-0297-7.

[23] K. D. Kucche, and J. J. Trujillo, "Theory of System of Nonlinear Fractional Differential Equations", *Progress in Fractional Differentiation and Applications*, vol. 3, no. 1, pp. 7–18, Jan. 2017.

[24] J. Hadamard, Essai sur l'etude des fonctions donnees par leur developpment de Taylor, *J. Mat. Pure Appl. Ser.*, vol. 8, pp. 101–186, 1892.

[25] S. G. Samko, A. A. Kilbas, and O. I. Marichev, *Fractional Integrals and Derivatives: Theory and Applications*. Gordon and Breach Science Publishers: Switzerland, 1993.

[26] P. L. Butzer, A. A. Kilbas, and J. J. Trujillo, "Compositions of Hadamard-type fractional integration operators and the semigroup property," *J. Math. Anal. Appl.*, vol. 269, no. 2, pp. 387–400, May 2002.

[27] P. L. Butzer, A. A. Kilbas, and J. J. Trujillo, "Fractional calculus in the Mellin setting and Hadamard-type fractional integrals," *J. Math. Anal. Appl.*, vol. 269, no. 1, pp. 1–27, May 2002.

[28] P. L. Butzer, A. A. Kilbas, and J. J. Trujillo, "Mellin transform analysis and integration by parts for Hadamard-type fractional integrals," *J. Math. Anal. Appl.*, vol. 270, no. 1, pp. 1–15, Jun. 2002.

[29] A. A. Kilbas, "Hadamard-type fractional calculus," *J. Korean Math. Soc.*, vol. 38, no. 6, pp. 1191–1204, 2001.

[30] A. A. Kilbas, and J. J. Trujillo, "Hadamard-type integrals as G-transforms," *Integr. Transf. Spec. F.*, vol. 14, no. 5, pp. 413–427, Oct. 2003.

[31] B. Ahmad, and S. K. Ntouyas, "On Hadamard fractional integrodifferential boundary value problems," *J. Appl. Math. Comput.*, vol. 47, no. 1–2, pp. 119–131, Feb. 2015.

[32] B. Ahmad, and S. K. Ntouyas, "A fully Hadamard type integral boundary value problem of a coupled system of fractional differential equations," *Fract. Calc. Appl. Anal.*, vol. 17, no. 2, pp. 348–360, Jun. 2014.

[33] B. Ahmad, and S. K. Ntouyas, "Initial value problems of fractional order Hadamard-type functional differential equations," *Electron. J. Differ. Eq.*, vol. 2015, no. 77, pp. 1-9, Mar. 2015.

[34] M. Li, and J. Wang, "Analysis of nonlinear Hadamard fractional differential equations *via* properties of Mittagâ€"Leffler functions," *J. Appl. Math. Comput.*, vol. 51, no. 1–2, pp. 487–508, Jun. 2016.

[35] J. Wang, and Z. Lin, "Ulam's type stability of Hadamard type fractional integral equations," *Filomat*, vol. 28, no. 7, pp. 1323–1331, 2014.

[36] B. Ahmad, A. Alsaedi, S. K. Ntouyas, and J. Tariboon, *Hadamard-Type Fractional Differential Equations, Inclusions and Inequalities*. Springer International Publishing: New York, 2017.

[37] B. C. Dhage, and S. B. Dhage, "Approximating positive solutions of PBVPs of nonlinear first order ordinary quadratic differential equations," *Appl. Math. Lett.*, vol. 46, pp. 133–142, Aug. 2015.

[38] B. C. Dhage, and S. B. Dhage, "Approximating solutions of nonlinear pbvps of hybrid differential equations *via* hybrid fixed point theory," *Indian J. Math.*, vol. 57, no. 1, pp. 103–119, 2015.

[39] B. C. Dhage, and S. B. Dhage, "Approximating solutions of nonlinear PBVP of second order differential equations *via* hybrid fixed point theory," *Electron. J. Differ. Eq.*, vol. 2015, no. 20, pp. 1–10, Jan. 2015.

[40] B. C. Dhage, S. B. Dhage, and S. K. Ntouyas, "Approximating solutions of nonlinear hybrid differential equations," *Appl. Math. Lett.*, vol. 34, pp. 76–80, Aug. 2014.

[41] B. C. Dhage, "Hybrid fixed point theory in partially ordered normed linear spaces and applications to fractional integral equations," *Differ. Equ. Appl.*, vol.5, no.2, pp. 155-184, 2013.

[42]   B. C. Dhage, "Global attractivity results for comparable solutions of nonlinear hybrid fractional integral equations," *Differ. Equ. Appl.*, vol. 6, no. 2, pp. 165–186, 2014.

[43]   B. C. Dhage, "A new monotone iteration principle in the theory of nonlinear fractional differential equations," *Int. J. Anal. Appl.*, vol. 8, no. 2, pp. 130–143, Aug. 2015.

[44]   B. C. Dhage, S. B. Dhage, and S. K. Ntouyas, "Existence and approximate solutions for fractional differential equations with nonlocal conditions," *J. Fract. Calc. Appl.*, vol.7, no.1, pp. 24–35, Jan. 2016.

[45]   B. C. Dhage, "Partially condensing mappings in partially ordered normed linear spaces and applications to functional integral equations," *Tamkang J. Math.*, vol.45, no.4, pp. 397–426, Dec. 2014.

[46]   B. C. Dhage, "Operator theoretic techniques in the theory of nonlinear ordinary hybrid differential equations," *Nonlinear Analysis Forum*, vol.20, no.1, pp. 15–31, 2015.

[47]   B. C. Dhage, S. B. Dhage, and J. R. Graef, "Dhage iteration method for initial value problems for nonlinear first order hybrid integrodifferential equations," *J. Fixed Point Theory Appl.*, vol. 18, pp. 309–326, Jun. 2016.

[48]   B. C. Dhage, and S. B. Dhage "Approximating positive solutions of nonlinear first order ordinary quadratic differential equations," *Cogent Mathematics*, vol. 2, no. 1, Article ID 1023671, pp. 1–10, Dec. 2015.

[49]   B. C. Dhage, S. B. Dhage, and S. K. Ntouyas, "Approximating solutions of nonlinear second order ordinary differential equations *via* Dhage iteration principle," *Malaya J. Mat.*, vol. 4, no.1, pp. 8–18, 2016.

[50]   B. Ahmad, and S. K. Ntouyas, "Initial-value problems for hybrid Hadamard fractional differential equations," *Electron. J. Differ. Eq.*, vol. 2014 , no. 161, pp. 1–8, Jul. 2014.

CHAPTER 6

Dynamics of Fractional Order Modified Bhalekar-Gejji System

Sachin Bhalekar*

Department of Mathematics, Shivaji University, Kolhapur 416004, India

Abstract: In the present chapter, we study the dynamics of modified Bhalekar-Gejji system. We discuss the stability, symmetry, dissipativity and chaos in the proposed system. We control the chaos using linear feedback control and synchronize the system with new chaotic system using active control.

Keywords: Chaos, fractional order, Bhalekar-Gejji system, synchronization.

AMS Subject Classification: 26A33, 65P20, 34L30.

6.1. INTRODUCTION

Fractional differential equations are the models involving non-integer order derivatives [1, 2, 3]. The order of the derivative in such equations may be a rational number, irrational number, complex number or even some function. These derivatives are usually defined in terms of an integral (with suitable kernel) and hence non-local unlike integer order derivative. One has to consider all the history of the function starting from initial point while evaluating fractional derivative. This results some complications in numerical computation of fractional derivative. The process of numerical solution of fractional differential equations (FDE) takes longer than that of integer order differential equations (IODE).

On the other hand, this nonlocal operator is proved very useful while modeling memory and hereditary properties of the natural system. Further, there are some systems with intermediate behavior (*e.g.* viscoelastic systems) which can be modeled only using fractional derivatives. We list few applications of fractional calculus: Fractional diffusion equation is used to model various phenomena [4, 5, 6, 7, 8, 9]. Applications to viscoelasticity are discussed by Mainardi in his

*Corresponding author Sachin Bhalekar: Department of Mathematics, Shivaji University, Kolhapur 416004, India; Tel: +91 231 2609218; E-mail: sbb_maths@unishivaji.ac.in, sachin.math@yahoo.co.in

monograph [10]. Magin devoted [11] to the applications in bioengineering. Few more applications include signal processing [12], image processing [13], image encryption [14], cryptography [15], control theory [16], thermodynamics [17] and nonlinear dynamics [18]. Chaotic fractional order systems can be used to generate reliable cryptographic schemes [19].

Various inequivalent definitions of fractional derivative are given in the literature [1, 2, 3]. Each such definition has its own importance. The classical one is Riemann-Lioville (RL) definition. It has importance in Mathematical Analysis. However, it is not suitable in modeling some physical systems. Further, the RL derivative of a constant is not zero. On the other hand, the definition proposed by Caputo involves initial values at integer order derivatives of the function and hence more suitable to model real life problems. The Caputo derivative of constant is zero as required.

Natural systems are usually modeled using (nonlinear) differential equations. Solutions of such equations may be constant (equilibrium points), converging (in the neighborhood of asymptotically stable equilibrium), unbounded (moving away from unstable equilibrium) or oscillating periodically. Further, some nonlinear systems of order greater than or equal to three may exhibit aperiodic oscillations for all the time. Such oscillations are bounded and very sensitive to initial conditions. Lyapunov exponents of these systems are positive and the solutions are said to be chaotic [20]. There are various systems exhibiting chaos viz. Lorenz system, Rossler system, Chua system and so on. Fractional order counterparts of these systems can also produce chaotic oscillations even for system order $\bar{\alpha} = \alpha_1 + \alpha_2 + \alpha_3$ less than three. However, there is always a threshold value of $\bar{\alpha}$ for which chaos disappears. This value is called Minimum Effective Dimension (MED) of that system. Examples of these fractional order systems include Lorenz system [21], Chen system [22], Lü system [23], Rossler system [24]and Liu [25] system.

In the present article, we modify fractional order Bhalekar-Gejji system and explore its rich dynamics. The chapter is organized as below:

In Section 6.2, we take a brief review about Bhalekar-Gejji (BG) system. Some basic definitions related to fractional calculus are described in Section 6.3. Numerical methods for solving fractional differential equations are discussed in Section 6.4. Section 6.5 is about stability results of fractional order differential equations. The modified BG system is presented in Section 6.6. The proposed system is analyzed in Section 6.7. Chaos in modified BG system is studied in

Section 6.8. Section 6.9 deals with chaos control in this system. Hybrid synchronization between modified BG system and new system is presented in Section 6.10. Conclusions are summarized in Section 6.11.

6.2. RELATED WORK

Bhalekar-Gejji (BG) system is a new chaotic dynamical system proposed by Bhalekar and Daftardar-Gejji [26] in 2011. It comprises of following system of nonlinear ordinary differential equations

$$\dot{x} = \omega x - y^2$$

$$\dot{y} = \mu(z - y)$$

$$\dot{z} = ay - bz + xy. \tag{6.1}$$

It is shown that the system is chaotic for a range of parameter values. Numerical bifurcation analysis is done by these authors for the parameters b, ω and μ. Various limit cycles and chaotic attractors are obtained for different values of parameters. Further, synchronization of this system with itself is also presented in this paper. Fig. (**6.1**) shows the chaotic attractor generated by the BG system with $\omega = -4$, $\mu = 9$, $a = 30$ and $b = 1$.

By adding a constant gain in third equation of BG system, Bhalekar [27] shown that the two-scroll attractor in this system is formed from two one-scroll attractors. Figs. (**6.2** and **6.3**) show the left and right one-scroll attractors for gain $+47$ and -47 respectively.

Singh, Singh and Roy discussed synchronization and anti-synchronization of BG system and Lu system using sliding mode control [28] and nonlinear active control [29, 30]. Lyapunov stability theory is used by these researchers to design the control in this procedure. Further, the authors have discussed the application of synchronization of BG system in secure communication [31] also.

Synchronization between unified chaotic systems (Chen system, Lorenz system *etc.*) and BG system in presence of uncertainties and disturbances is achieved by Vargas *et al.* [32] using adaptive scheme.

Daftardar-Gejji, Sukale and Bhalekar [33] proposed fractional order analog of BG system and solved it by using new predictor-corrector method.

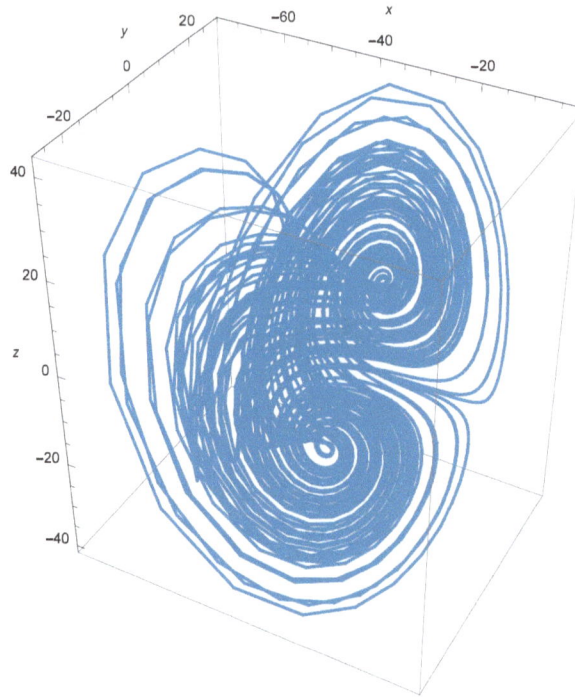

Fig. (6.1). Chaotic attractor generated by BG system.

Borah, Singh and Roy also considered fractional order analog of BG system in [34]. They have shown that the bandwidth of this system is broader than various chaotic systems (Lorenz system, Lu system, Chen system, Volta system and so on). Thus the fractional order BG system is more suitable for secure communication applications. Further, the electronic circuit for this system is also designed in this paper.

Aqeel and Ahmad [35] characterized Hopf bifurcation in BG system using numerical continuation. Analytical results are used to prove the existence of Hopf bifurcation and to study various properties of the system. Deshpande, Daftardar-Gejji and Sukale [36] proved existence of Hopf bifurcation in fractional order BG system. Hopf critical curves for various values of fractional order are presented in this work.

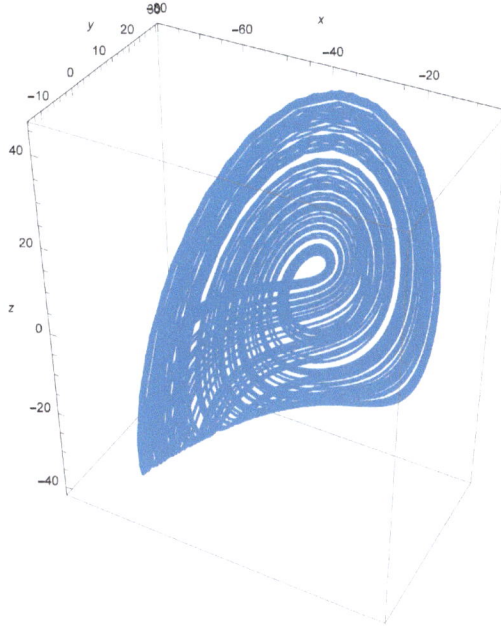

Fig. (6.2). Chaotic attractor generated by BG system.

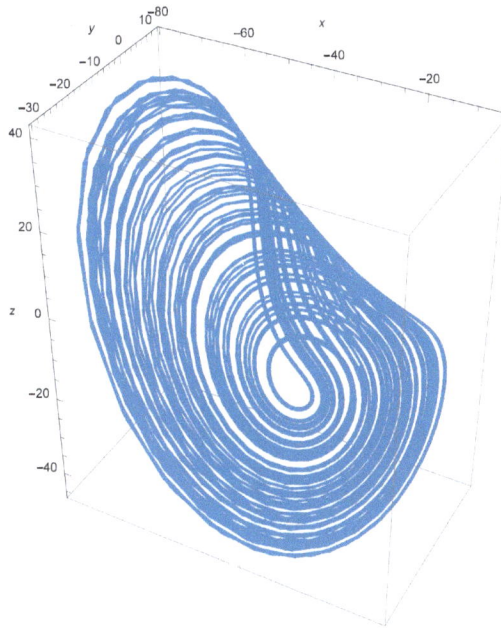

Fig. (6.3). Chaotic attractor generated by BG system.

6.3. PRELIMINARIES

In this section, we discuss some basic definitions and analytical results regarding fractional calculus.

Definition 6.1. *[1, 3] Riemann-Liouville fractional integration of order α is defined as*

$$I^\alpha f(t) = \frac{1}{\Gamma(\alpha)} \int_0^t (t-y)^{\alpha-1} f(y)\, dy, t > 0. \tag{6.2}$$

Definition 6.2. *[1, 3] Caputo fractional derivative of order α is defined as*

$$D^\alpha f(t) = I^{m-\alpha}\left(\frac{d^m f(t)}{dt^m}\right), 0 \le m-1 < \alpha \le m. \tag{6.3}$$

Note that for $0 \le m-1 < \alpha \le m, a \ge 0$ and $\gamma > -1$

$$I^\alpha (t-a)^\gamma = \frac{\Gamma(\gamma+1)}{\Gamma(\gamma+\alpha+1)}(t-a)^{\gamma+\alpha}, \tag{6.4}$$

$$(I^\alpha D^\alpha f)(t) = f(t) - \sum_{k=0}^{m-1} f^{(k)}(0)\frac{t^k}{k!}. \tag{6.5}$$

6.4. NUMERICAL METHODS FOR FDDEs

6.4.1. Fractional Adams Method (FAM)

Fractional Adams method (FAM) proposed by Diethelm *et al.* [37, 38, 39] is widely used numerical scheme to solve FDEs. In this Section, we describe this algorithm.

Consider for $\alpha \in (m-1, m]$ the initial value problem (IVP)

$$D^\alpha y(t) = f(t, y(t)), 0 \le t \le T, \tag{6.6}$$

$$y^{(k)}(0) = y_0^{(k)}, k = 0, 1, \cdots, m-1. \tag{6.7}$$

The IVP (6.6)–(6.7) is equivalent to the Volterra integral equation

$$y(t) = \sum_{k=0}^{m-1} y_0^{(k)} \frac{t^k}{k!} + \frac{1}{\Gamma(\alpha)} \int_0^t (t-\tau)^{\alpha-1} f(\tau, y(\tau)) d\tau. \tag{6.8}$$

Consider the uniform grid $\{t_n = nh/n = 0,1,\cdots,N\}$ for some integer N and $h:= T/N$. Let $y_h(t_n)$ be approximation to $y(t_n)$. Assume that we have already calculated approximations $y_h(t_j), j = 1,2,\cdots,n$ and we want to obtain $y_h(t_{n+1})$ by means of the equation

$$y_h(t_{n+1}) = \sum_{k=0}^{m-1} \frac{t_{n+1}^k}{k!} y_0^{(k)} + \frac{h^\alpha}{\Gamma(\alpha+2)} f\left(t_{n+1}, y_h^P(t_{n+1})\right) + \frac{h^\alpha}{\Gamma(\alpha+2)} \sum_{j=0}^{n} a_{j,n+1} f\left(t_j, y_n(t_j)\right) \qquad (6.9)$$

where

$$a_{j,n+1} = \begin{cases} n^{\alpha+1} - (n-\alpha)(n+1)^\alpha, & if\ j = 0, \\ (n-j+2)^{\alpha+1} + (n-j)^{\alpha+1} - 2(n-j+1)^{\alpha+1}, & if\ 1 \le j \le n, \\ 1, & if\ j = n+1. \end{cases}$$

The preliminary approximation $y_h^P(t_{n+1})$ is called predictor and is given by

$$y_h^P(t_{n+1}) = \sum_{k=0}^{m-1} \frac{t_{n+1}^k}{k!} y_0^{(k)} + \frac{1}{\Gamma(\alpha)} \sum_{j=0}^{n} b_{j,n+1} f\left(t_j, y_n(t_j)\right), \qquad (6.10)$$

where

$$b_{j,n+1} = \frac{h^\alpha}{\alpha}((n+1-j)^\alpha - (n-j)^\alpha). \qquad (6.11)$$

Error in this method is

$$max_{j=0,1,\cdots,N}|y(t_j) - y_h(t_j)| = O(h^p), \qquad (6.12)$$

where $p = min(2,1+\alpha)$.

6.4.2. New Predictor-Corrector Method (NPCM)

The FAM described in previous section is improved by Daftardar-Gejji *et al.* using an iterative method [40]. The new predictor-corrector method (NPCM) [33] is described as below:

$$x_{n+1}^p = \sum_{k=0}^{m-1} y_0^k \frac{t_{n+1}^k}{k!} + \frac{h^\alpha}{\Gamma(\alpha+2)} \sum_{j=0}^{n} a_{j,n+1} f(t_j, y_j), \qquad (6.13)$$

$$z_{n+1}^p = \frac{h^\alpha}{\Gamma(\alpha+2)} f(t_{n+1}, x_{n+1}^p), \qquad (6.14)$$

$$y^c_{n+1} = x^p_{n+1} + \frac{h^\alpha}{\Gamma(\alpha+2)} f(t_{n+1}, x^p_{n+1} + z^p_{n+1}). \qquad (6.15)$$

Here x^p_{n+1} and z^p_{n+1} are called as predictors and y^c_{n+1} is the corrector, and y_j denotes the approximate value of the solution.

6.5. STABILITY ANALYSIS

Consider the following fractional order system

$$D^{\alpha_1} x_1 = f_1(x_1, x_2, \cdots, x_n),$$

$$D^{\alpha_2} x_2 = f_2(x_1, x_2, \cdots, x_n),$$

$$\vdots$$

$$D^{\alpha_n} x_n = f_n(x_1, x_2, \cdots, x_n), \qquad (6.16)$$

where $0 < \alpha_i < 1$ are fractional orders. The system (6.16) is called as a commensurate order if $\alpha_1 = \alpha_2 = \cdots = \alpha_n$ otherwise an incommensurate order.

A point $p = (x_1^*, x_2^*, \cdots, x_n^*)$ is called an equilibrium point of system (6.16) if $f_i(p) = 0$ for each $i = 1, 2, \cdots, n$.

Theorem 6.1. *[41, 42] Consider $\alpha = \alpha_1 = \alpha_2 = \cdots = \alpha_n$ in (6.16). An equilibrium point p of the system (6.16) is locally asymptotically stable if all the eigenvalues of the Jacobian matrix*

$$J = \begin{pmatrix} \partial_1 f_1(p) & \partial_2 f_1(p) & \cdots & \partial_n f_1(p) \\ \partial_1 f_2(p) & \partial_2 f_2(p) & \cdots & \partial_n f_2(p) \\ \vdots & \vdots & \vdots & \vdots \\ \partial_1 f_n(p) & \partial_2 f_n(p) & \cdots & \partial_n f_n(p) \end{pmatrix}. \qquad (6.17)$$

evaluated at p satisfy the following condition

$$|arg(Eig(J))| > \alpha\pi/2. \qquad (6.18)$$

Theorem 6.2. *[43, 44] Consider the incommensurate fractional ordered dynamical system given by (6.16). Let $\alpha_i = v_i/u_i$, $(u_i, v_i) = 1$, u_i, v_i be positive integers. Define M to be the least common multiple of u_i's.*

Define

$$\Delta(\lambda) = diag([\lambda^{M\alpha_1}, \lambda^{M\alpha_2}, \cdots, \lambda^{M\alpha_n}]) - J \qquad (6.19)$$

where J is the Jacobian matrix as defined in (6.17) evaluated at point p. Then p is locally asymptotically stable if all the roots of the equation $det(\Delta(\lambda)) = 0$ satisfy the condition $|arg(\lambda)| > \pi/(2M)$.

This condition is equivalent to the following inequality

$$\frac{\pi}{2M} - \min_i\{|arg(\lambda_i)|\} < 0. \qquad (6.20)$$

Thus an equilibrium point p of the system (6.16) is asymptotically stable if the condition (6.20) is satisfied. The term $\frac{\pi}{2M} - \min_i\{|arg(\lambda_i)|\}$ is called as the instability measure for equilibrium points in fractional order systems (IMFOS). Hence, a necessary condition for fractional order system (6.16) to exhibit chaotic attractor is [43]

$$\text{IMFOS} \geq 0. \qquad (6.21)$$

Theorem 6.3. *[45] Consider the polynomial*

$$P(\lambda) = \lambda^3 + a_1\lambda^2 + a_2\lambda + a_3 = 0. \qquad (6.22)$$

Define the discriminant for equation (6.22) as

$$D(P) = 18a_1a_2a_3 + (a_1a_2)^2 - 4a_3(a_1)^3 - 4(a_2)^3 - 27(a_3)^2. \qquad (6.23)$$

1. If $D(P) > 0$ then all the roots of $P(\lambda)$ satisfy the condition

$$|arg(\lambda)| > \alpha\pi/2 \qquad (6.24)$$

where $0 \leq \alpha \leq 1$.

2. If $D(P) < 0$, $a_1 \geq 0$, $a_2 \geq 0$, $a_3 > 0$, $\alpha < 2/3$ then (6.24) is satisfied.

3. If $D(P) < 0$, $u_1 > 0$, $a_2 > 0$, $a_1a_2 - a_3$ then (6.24) is satisfied for all $0 \leq \alpha < 1$.

6.6. MODIFIED BHALEKAR-GEJJI SYSTEM

We consider modification in (6.1) described by following system:

$$D^{\alpha_1}x = \omega x - y^2$$

$$D^{\alpha_2}y = \mu(z - y) - \delta y e^x$$

$$D^{\alpha_3}z = ay - bz + xy, \tag{6.25}$$

where $0 < \alpha \le 1$, $\omega = -2.667$, $\mu = 10$, $a = 27.3$, $b = 1$ and $\delta = 1$. We have added the term $-\delta y e^x$ in second equation of BG system. The new model is also chaotic for some parameter values.

6.7. ANALYSIS OF BG SYSTEM

Now we discuss some dynamical properties of the system (6.25).

1. **Symmetry:** It can be observed that the transformation $(x, y, z) \to (x, -y, -z)$ does not alter the system. This shows that the system is symmetric about Z-axis.

2. **Dissipativity:** Consider

$$div(V) = \frac{\partial \dot{x}}{\partial x} + \frac{\partial \dot{y}}{\partial y} + \frac{\partial \dot{z}}{\partial z}$$

$$= -13.667 - e^x < 0. \tag{6.26}$$

Therefore, the system is dissipative.

3. **Non-generalized Lorenz system:** The generalized Lorenz system (GLS) discussed in [46] is of the form

$$\dot{X} = \begin{pmatrix} A & 0 \\ 0 & \lambda_3 \end{pmatrix} X + x \begin{pmatrix} 0 & 0 & 0 \\ 0 & 0 & -1 \\ 0 & 1 & 0 \end{pmatrix} X, \tag{6.27}$$

where $X = [x, y, z]^T$, $\lambda_3 \in \Re$ and A is a 2×2 matrix with eigenvalues $\lambda_1, \lambda_2 \in \Re$ such that $-\lambda_2 > \lambda_1 > -\lambda_3 > 0$. Since our system

$$\dot{X} = \begin{pmatrix} \omega & 0 & 0 \\ 0 & -\mu & \mu \\ 0 & a & -b \end{pmatrix} X + y \begin{pmatrix} 0 & -1 & 0 \\ 0 & 0 & 0 \\ 1 & 0 & 0 \end{pmatrix} X + \begin{pmatrix} 0 \\ -\delta y e^{x} \\ 0 \end{pmatrix} \quad \textbf{(6.28)}$$

is having different structure than the system (6.3), it is not a GLS.

4. **Equilibrium points and their stability:** The constant solutions of the dynamical system are called equilibrium points. The modified system (6.25) has three equilibrium points viz. $O = (0,0,0)$ and $E_{\pm} = (-26.3, \pm 8.37509, \pm 8.37509)$. Stability of equilibrium depends upon the eigenvalues of the Jacobian matrix evaluated at that point. In this case, the Jacobian is given by the matrix

$$J(X) = \begin{pmatrix} -2.667 & -2y & 0 \\ -ye^{x} & -e^{x} - 10 & 10 \\ y & 27.3 + x & -1 \end{pmatrix}. \quad \textbf{(6.29)}$$

The eigenvalues of $J(0)$ are -23.2627, 11.2627 and -2.667. Therefore, O is a saddle point of index 1. It is useful in connecting scrolls of chaotic attractor. The eigenvalues of $J(E_{\pm})$ are $1.59719 \pm 8.9804\iota$ and -16.8614. Thus, $E_{\pm}$ are saddle points of index 2. The scrolls of chaotic attractor are usually about these equilibrium points.

6.8. CHAOS IN MODIFIED BG SYSTEM OF FRACTIONAL ORDER

It is shown in the literature that the system (6.1) is chaotic for some values of parameters. In this section, we discuss chaos in (6.25) for commensurate as well as incommensurate order cases.

6.8.1. Commensurate Order System

Consider the system (6.25) with $\alpha_1 = \alpha_2 = \alpha_3 = \alpha$. In view of condition (6.18), the equilibrium point of the system is asymptotically stable if $\alpha < \frac{2}{\pi}|arg(Eig(J))|$. The system will not exhibit chaotic oscillations if at least one equilibrium is asymptotically stable. It can be checked that $\frac{2}{\pi}|arg(Eig(J))| = 0.887947$ for both the eigenvalues $E_{\pm}$. The Figs. (**6.4**, **6.5** and **6.6**) show the stable solutions x, y and z respectively for $\alpha = 0.88$. The chaotic trajectories and attractors for $\alpha = 0.89$ are shown in Figs. (**6.7**, **6.8**, **6.9** and **6.10**).

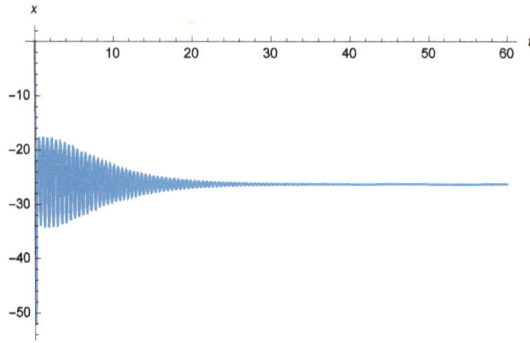

Fig. (6.4). Stable time series $x(t)$ for $\alpha = 0.88$.

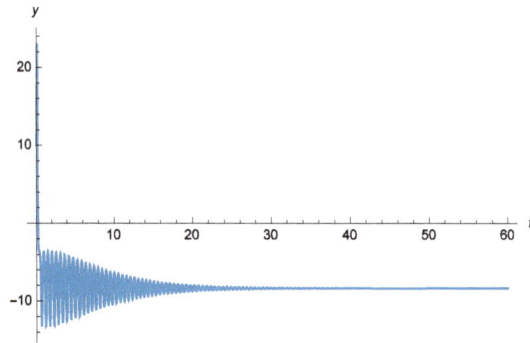

Fig. (6.5). Stable time series $y(t)$ for $\alpha = 0.88$.

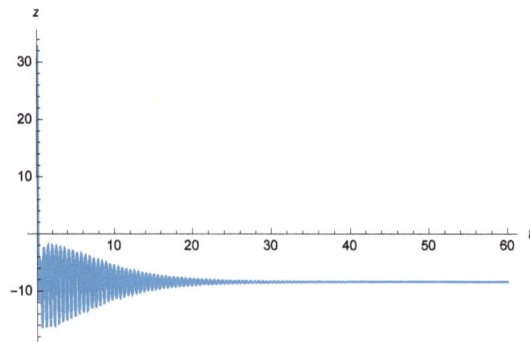

Fig. (6.6). Stable time series $z(t)$ for $\alpha = 0.88$.

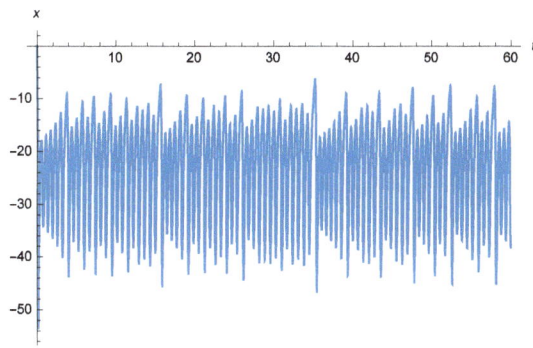

Fig. (6.7). Chaotic time series $x(t)$ for $\alpha = 0.89$.

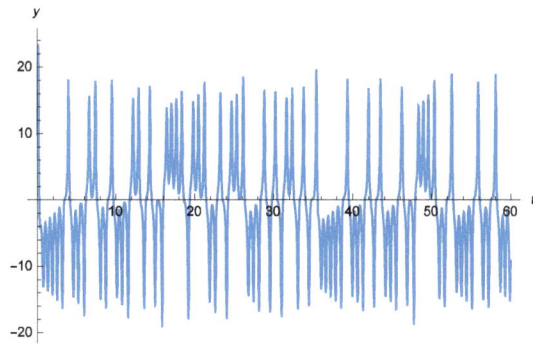

Fig. (6.8). Chaotic time series $y(t)$ for $\alpha = 0.89$.

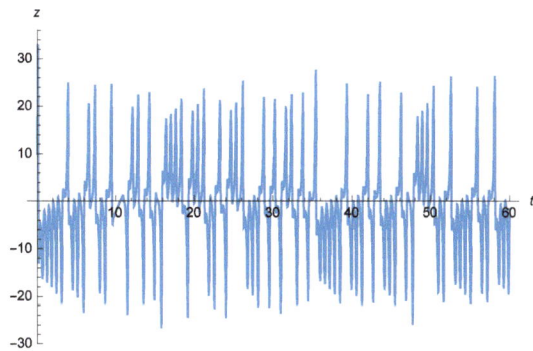

Fig. (6.9). Chaotic time series $z(t)$ for $\alpha = 0.89$.

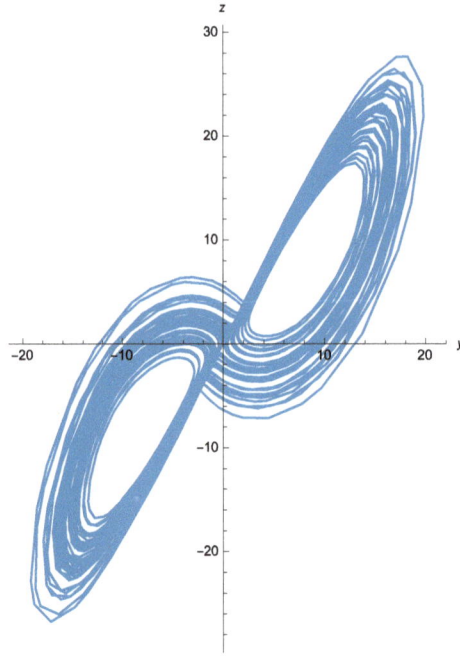

Fig. (6.10). Chaotic attractor in $y - z$ plane for $\alpha = 0.89$.

6.8.2. Incommensurate Order System

Consider $\alpha_1 = 0.65 = 13/20$, $\alpha_2 = 1$ and $\alpha_3 = 1$. In this case, $M = 20$ and IMFOS= -0.000984947. The system is stable in this case (cf. Fig. **6.11**). Now, consider $\alpha_1 = 0.69 = 69/100$, $\alpha_2 = 1$ and $\alpha_3 = 1$. It can be checked that the IMFOS= 0.0000332647. The system exhibit chaotic oscillations as shown in Fig. **(6.12)**.

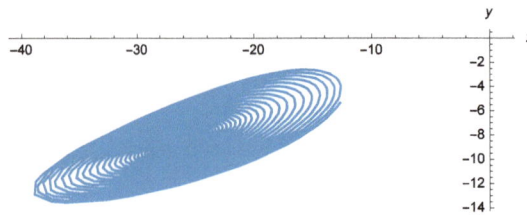

Fig. (6.11). Stable solution for $\alpha_1 = 0.65$, $\alpha_2 = 1$ and $\alpha_3 = 1$.

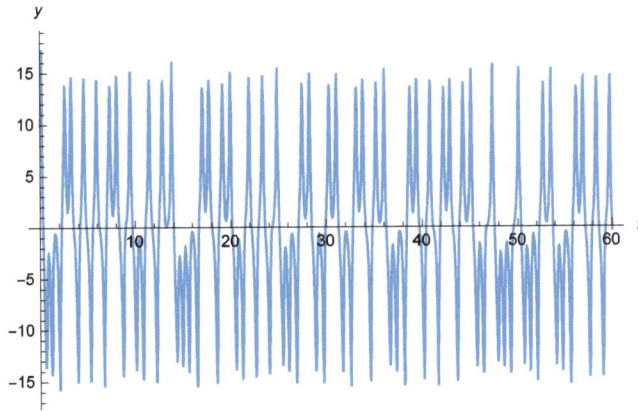

Fig. (6.12). Chaotic solution for $\alpha_1 = 0.69$, $\alpha_2 = 1$ and $\alpha_3 = 1$.

6.9. CHAOS CONTROL

Chaos in the system can be suppressed by applying a proper control. In this section, we use linear feedback control to suppress the chaotic oscillations in the modified BG system. Consider the controlled system

$$D^\alpha x = \omega x - y^2 + u$$

$$D^\alpha y = \mu(z - y) - \delta y e^x$$

$$D^\alpha z = ay - bz + xy, \tag{6.30}$$

where u is linear feedback control. We take $u = kz$, where k is real number chosen properly so as the system (6.30) satisfies the stability condition (6.18). Equilibrium points of this system are $O = (0,0,0)$ and $E_1^\pm = \big(-26.3, 0.5k \pm 0.5\sqrt{280.568 + k^2}, 0.5k \pm 0.5\sqrt{280.568 + k^2}\big)$. If $k^2 \geq 0.056288$ then there are two more equilibrium points viz.
$E_2^\pm = \left(5.77052, 16.5353k \pm 0.5\sqrt{1093.66k^2 - 61.5599}, \dfrac{546.83k \pm}{16.5353\sqrt{1093.66k^2 - 61.5599}}\right)$. To suppress chaos, it is sufficient to stabilize a single equilibrium point. The eigenvalues of Jacobian matrix of system (6.30) evaluated at E_1^- are described as

$$\lambda_1 = -(0.0661417 + 0.114561\iota)\big(-1572k^2 + 1572\sqrt{k^2 + 280.6}k$$

$$+ \Big(864(-k^2 + \sqrt{k^2 + 280.6}k - 65.86\Big)^3$$

$$+\left(-1572k^2 + 1572\sqrt{k^2 + 280.568}k - 314990\right)^2\right)^{1/2} - 314990\right)^{1/3}$$

$$+\left((0.629961 - 1.09112\iota)\left(-k^2 + \sqrt{k^2 + 280.568}k - 65.8506\right)\right)$$

$$/\left(-1572k^2 + 1572\sqrt{k^2 + 280.6}k + \left(864\left(k^2 + \sqrt{k^2 + 280.6}k - 65.86\right)^3\right.\right.$$

$$+\left(-1572k^2 + 1572\sqrt{k^2 + 280.6}k - 314990\right)^2\right)^{1/2} - 314990\right)^{1/3} - 4.556$$

$$\lambda_2 = -(0.0661417 - 0.114561\iota)\left(-1572k^2 + 1572\sqrt{k^2 + 280.6}k\right.$$

$$+\left(864\left(-k^2 + \sqrt{k^2 + 280.6}k - 65.86\right)^3\right.$$

$$+\left(-1572k^2 + 1572\sqrt{k^2 + 280.568}k - 314990\right)^2\right)^{1/2} - 314990\right)^{1/3}$$

$$+\left((0.629961 + 1.09112\iota)\left(-k^2 + \sqrt{k^2 + 280.568}k - 65.8506\right)\right)$$

$$/\left(-1572k^2 + 1572\sqrt{k^2 + 280.6}k + \left(864\left(k^2 + \sqrt{k^2 + 280.6}k - 65.86\right)^3\right.\right.$$

$$+\left(-1572k^2 + 1572\sqrt{k^2 + 280.6}k - 314990\right)^2\right)^{1/2} - 314990\right)^{1/3} - 4.556$$

$$\lambda_3 = 0.132283\left(-1572k^2 + 1572\sqrt{k^2 + 280.6}k\right.$$

$$+\left(864\left(-k^2 + \sqrt{k^2 + 280.6}k - 65.86\right)^3\right.$$

$$+\left(-1572k^2 + 1572\sqrt{k^2 + 280.568}k - 314990\right)^2\right)^{1/2} - 314990\right)^{1/3}$$

$$-\left(1.25992\left(-k^2 + \sqrt{k^2 + 280.568}k - 65.8506\right)\right)$$

$$/\left(-1572k^2 + 1572\sqrt{k^2 + 280.6}k + \left(864\left(k^2 + \sqrt{k^2 + 280.6}k - 65.86\right)^3\right.\right.$$

$$+\left(-1572k^2 + 1572\sqrt{k^2 + 280.6}k - 314990\right)^2\Big)^{1/2} - 314990\right)^{1/3} - 4.556.$$

We have plotted absolute value of arguments of these eigenvalues and the curves $\alpha\pi/2$ for different values of fractional order α in Figs. (**6.13** and **6.14**). Note that the figures match for λ_2 and λ_3. These figures give estimate for the value of control parameter k used to stabilize the system. For example, in case of $\alpha = 0.92$, we should take $k > 2$ to control the chaos in system (cf. Fig. **6.15** with $k = 2.2$).

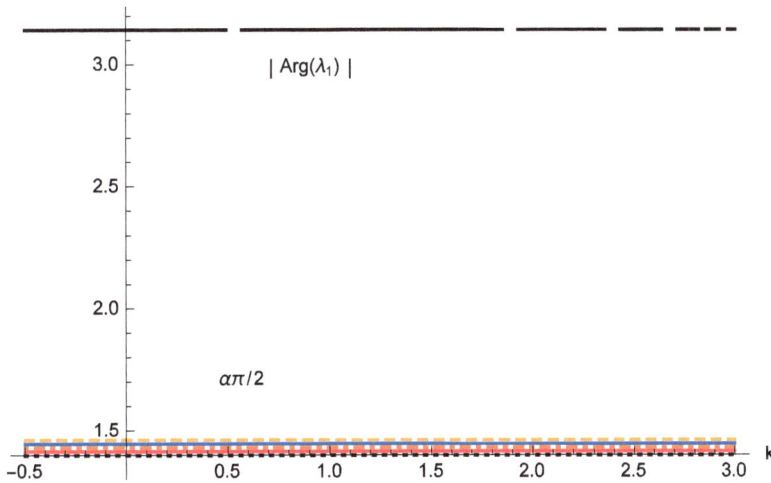

Fig. (6.13). Estimate for control parameter k with eigenvalue λ_1.

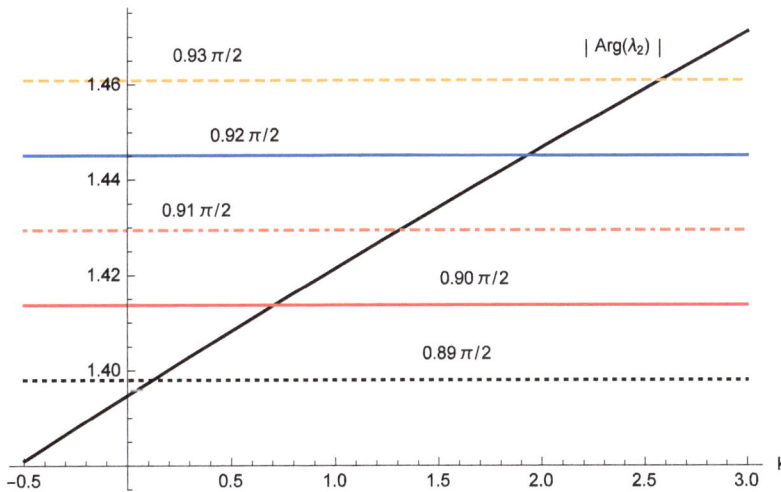

Fig. (6.14). Estimate for control parameter k with eigenvalue λ_2 and λ_3.

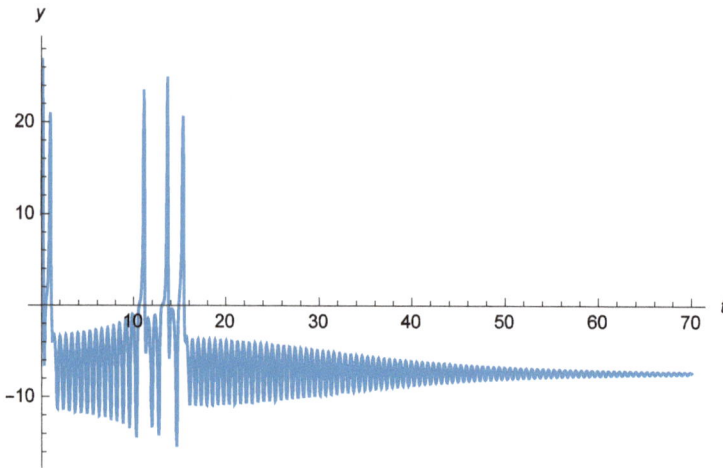

Fig. (6.15). Controlled orbit for $\alpha = 0.92$ with $k = 2.2$.

6.10. HYBRID SYNCHRONIZATION

In the presence of proper control terms, the two different chaotic signals can oscillate together [47, 48]. This phenomenon is termed as chaos synchronization and has applications in secure communications [49] and cryptography [50]. There are various technique to synchronize chaotic signals. The phenomenon is called antisynchronization if the value of one signal is negation of the other (after some time period). In this case, the trajectories are mirror images of each other. In contrast, the complete synchronization means that both the trajectories have same value. In hybrid synchronization, there is one pair of signals which gives rise to complete synchronization and the other one exhibit antisynchronization.

In this section, we utilize active control method to present hybrid synchronization in modified BG system and fractional order version of new chaotic system [51].

Modified BG system

$$D^\alpha x_1 = -2.667x_1 - y_1^2$$

$$D^\alpha y_1 = 10(z_1 - y_1) - y_1 e^{x_1}$$

$$D^\alpha z_1 = 27.3y_1 - z_1 + x_1 y_1 \qquad (6.31)$$

is taken as the drive system and the controlled new system

$$D^\alpha x_2 = -1.5x_2 + y_2 z_2 + u_1,$$

$$D^\alpha y_2 = -\sinh(y_2) + x_2 z_2 + u_2,$$

$$D^\alpha z_2 = z_2 - x_2 y_2 + u_3 \tag{6.32}$$

as response system.

The active control terms u_i's are chosen as $u_1 = -1.167x_1 - y_2 z_2 - y_1^2 + V_1$, $u_2 = \sinh(y_2) - x_2 z_2 - 10(z_2 + y_2) + y_1 e^{x_1} + V_2$ and $u_3 = -2z_1 + 27.3y_1 + x_2 y_2 + x_1 y_1 + V_3$. We define the errors in synchronization as $e_1 = x_2 - x_1$, $e_2 = y_2 + y_1$ and $e_3 = z_2 - z_1$. Note that there will be complete synchronization in signals x and z whereas antisynchronization in the signal y. The error system can now be described as

$$D^\alpha e_1 = -1.5e_1 + V_1,$$

$$D^\alpha e_2 = -10e_2 + 10e_3 + V_2,$$

$$D^\alpha e_3 = e_3 + V_3. \tag{6.33}$$

Further, if we choose $V_1 = V_2 = 0$ and $V_3 = -2e_3$ then the system (6.33) becomes

$$\begin{pmatrix} D^\alpha e_1 \\ D^\alpha e_2 \\ D^\alpha e_3 \end{pmatrix} = \begin{pmatrix} -1.5 & 0 & 0 \\ 0 & -10 & 10 \\ 0 & 0 & -1 \end{pmatrix} \begin{pmatrix} e_1 \\ e_2 \\ e_3 \end{pmatrix}. \tag{6.34}$$

The eigenvalues of the coefficient matrix of this error system are $-1.5, -10, -1$. Thus, the error system is asymptotically stable *i.e.* the errors will tend to zero as t tends to infinity. Figs. (**6.16, 6.17** and **6.18**) show synchronized orbits (solid line = drive signals; dashed line = response signals) for fractional order $\alpha = 0.94$.

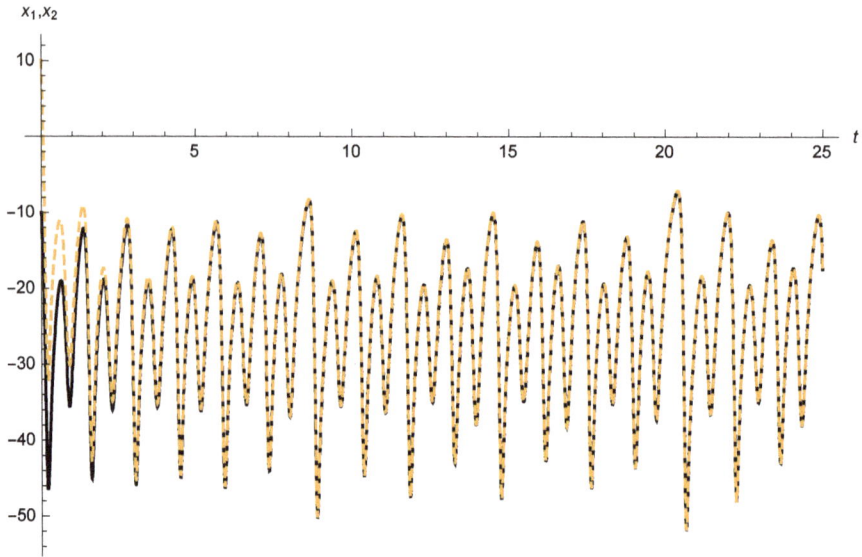

Fig. (6.16). Synchronized signals $x_1(t)$ and $x_2(t)$ for $\alpha = 0.94$.

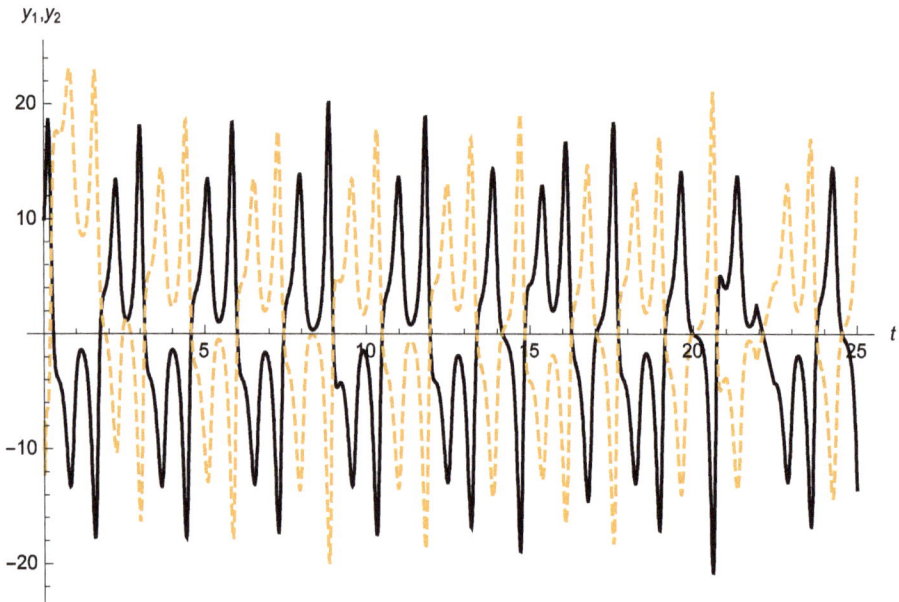

Fig. (6.17). Antisynchronized signals $y_1(t)$ and $y_2(t)$ for $\alpha = 0.94$.

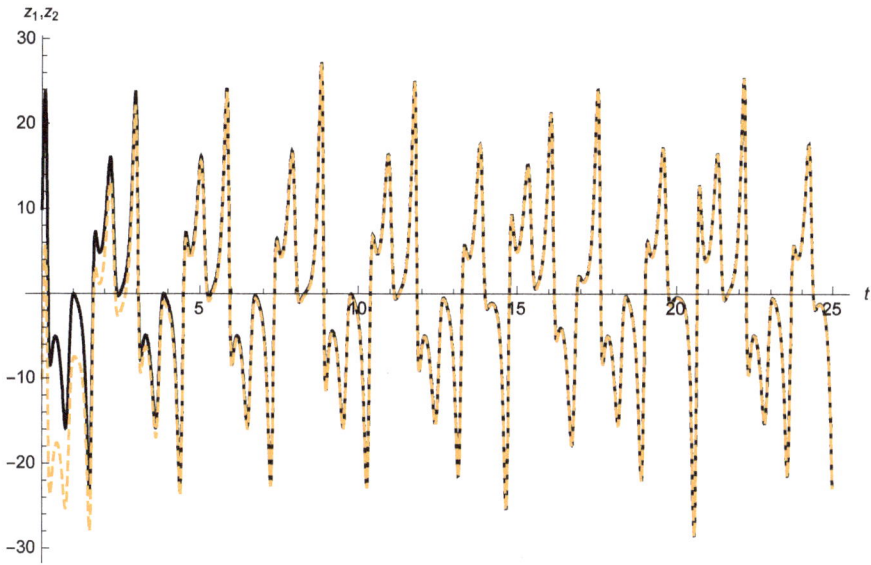

Fig. (6.18). Synchronized signals $z_1(t)$ and $z_2(t)$ for $\alpha = 0.94$.

The errors are plotted in Fig. (**6.19**).

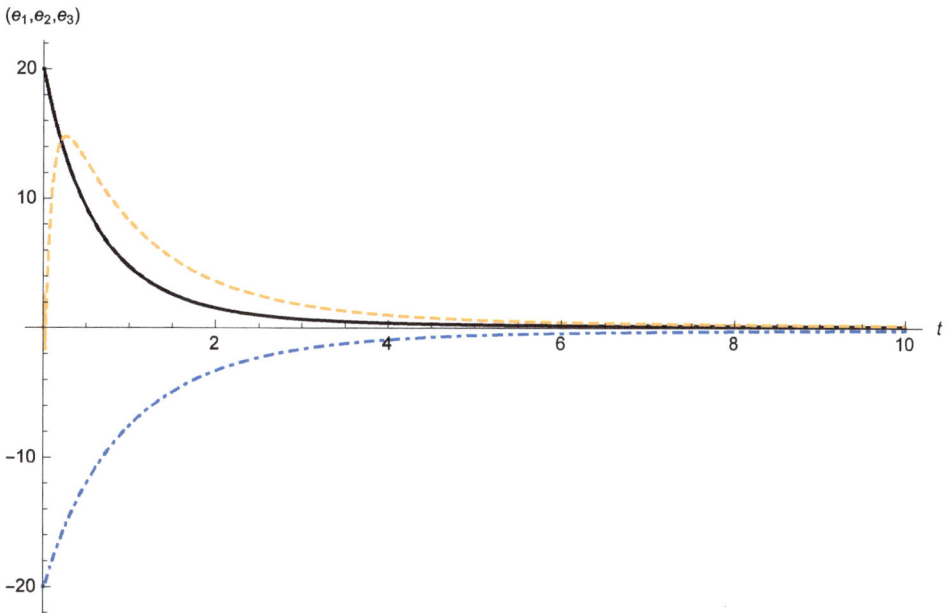

Fig. (6.19). Errors in synchronization for $\alpha = 0.94$.

6.11. CONCLUSION

In this chapter, we have proposed a modification in Bhalekar-Gejji (BG) system by adding a nonlinearity in second equation. We have taken a review of results related to BG system. The equilibrium points of modified system are obtained and their stability is discussed. The system is solved using new predictor-corrector scheme and chaos in the system is investigated. Linear feedback method is used to control chaos in modified BG system. Finally, the system is synchronized with new chaotic system by using active control.

CONSENT FOR PUBLICATION

Not applicable.

CONFLICT OF INTEREST

The authors declare no conflict of interest, financial or otherwise.

ACKNOWLEDGEMENTS

Author acknowledges CSIR, New Delhi for funding through Research Project [25(0245)/15/EMR-II].

REFERENCES

[1]   I. Podlubny, *Fractional Differential Equations*. Academic Press: San Diego, 1999.
[2]   S.G. Samko, A.A. Kilbas, and O.I. Marichev, *Fractional Integrals and Derivatives: Theory and Applications*. Gordon and Breach: Yverdon, 1993.
[3]   A.A. Kilbas, H.M. Srivastava, and J.J. Trujillo, *Theory and Applications of Fractional Differential Equations*. Elsevier: Amsterdam, 2006.
[4]   F. Mainardi, Y. Luchko and G. Pagnini, "The fundamental solution of the space-time fractional diffusion equation," *Fract. Calc. Appl. Anal.*, vol. 4, no. 2, pp. 153–192, Apr. 2001.
[5]   V. Daftardar-Gejji and S. Bhalekar, "Solving multi-term linear and non-linear diffusion-wave equations of fractional order by Adomian decomposition method," *Appl. Math. Comput.*, vol. 202, pp. 113–120, Aug. 2008.
[6]   I.S. Jesus, and J.A.T. Machado, "Fractional control of heat diffusion systems," *Nonlinear Dyn.*, vol. 54, no. 3, pp. 263–282, Nov. 2008.
[7]   W. Chen, and G. Pang, "A new definition of fractional Laplacian with application to modeling three-dimensional nonlocal heat conduction," *J. Comput. Phys*, vol. 309, pp. 350–367, Mar. 2016.
[8]   W. Chen, H. Sun, X. Zhang, and D. Korosak, "Anomalous diffusion modeling by fractal and fractional derivatives," *Comput. Math. Appl.*, vol. 59, no. 5, pp. 1754–1758, Mar. 2010.
[9]   W. Chen, Y. Liang, and X. Hei, "Structural derivative based on inverse Mittag-Leffler function for modeling ultraslow diffusion," *Fract. Calc. Appl. Anal.*, vol. 19, no. 5, pp. 1250–1261, Oct. 2016.
[10]   F. Mainardi, *Fractional calculus and waves in linear viscoelasticity: An introduction to mathematical models*. Imperial College Press: London, 2010.
[11]   R.L. Magin, *Fractional calculus in bioengineering*. Begll House Publishers: Redding, 2006.

[12]   T.J. Anastasio, "The fractional-order dynamics of Brainstem Vestibulo–Oculomotor neurons," *Biol. Cybern.*, vol. 72, pp. 69–79, Nov. 1994.

[13]   C. Tseng, and S.L. Lee, "Digital image sharpening using Riesz fractional order derivative and discrete hartley transform," In: *IEEE Asia Pacific Conference on Circuits and Systems (APCCAS)*, Ishigaki, 2014, pp. 483–486.

[14]   G.C. Wu, D. Baleanu, and Z.X. Lin, "Image encryption technique based on fractional chaotic time series," *J. Vib. Control*, vol. 22, no. 8. pp. 2092–2099, May 2016.

[15]   S. Wang, W. Sun, C.Y. Ma, D. Wang, and Z. Chen, "Secure communication based on a fractional order chaotic system," *Int. J. Security and Its Applications*, vol. 7, no. 5, pp. 205–216, Sep. 2013.

[16]   J. Sabatier, S. Poullain, P. Latteux, J. Thomas, and A. Oustaloup, "Robust speed control of a low damped electromechanical system based on CRONE control: Application to a four mass experimental test bench," *Nonlinear Dyn.*, vol. 38, pp. 383–400, Dec. 2004.

[17]   R.P. Meilanov, and R.A. Magomedov, "Thermodynamics in fractional calculus," *J. Eng. Phys. Thermophys.*, vol. 87, no. 6, pp. 1521–1531, Nov. 2014.

[18]   B. Xu, D. Chen, H. Zhang, and F. Wang, "Modeling and stability analysis of a fractional-order Francis hydro-turbine governing system," *Chaos Solitons Fractals*, vol. 75, pp. 50–61, Jun. 2015.

[19]   P. Muthukumar, and P. Balasubramaniam, "Feedback synchronization of the fractional order reverse butterfly-shaped chaotic system and its application to digital cryptography," *Nonlinear Dyn.*, vol. 74, pp. 1169–1181, Dec. 2013.

[20]   K.T. Alligood, T.D. Sauer, and J.A. Yorke, *Chaos: An Introduction to Dynamical Systems*, Springer: New York, 2008.

[21]   I. Grigorenko, and E. Grigorenko, "Chaotic dynamics of the fractional Lorenz system," *Phys. Rev. Lett.*, vol. 91, p. 034101, Jul. 2003.

[22]   C. Li, and G. Peng, "Chaos in Chen's system with a fractional order," *Chaos Solitons Fractals*, vol. 22, pp. 443–450, Oct. 2004.

[23]   J.G. Lü, "Chaotic dynamics of the fractional order Lü system and its synchronization," *Phys. Lett. A*, vol. 354, no. 4, pp. 305–311, Jun. 2006.

[24]   C. G. Li, and G. Chen, "Chaos and hyperchaos in the fractional order Rossler equations," *Phys. A: Stat. Mech. Appl.*, vol. 341, pp. 55–61, Oct. 2004.

[25]   V. Daftardar-Gejji, and S. Bhalekar, "Chaos in fractional ordered Liu system," *Comput. Math. Appl.*, vol. 59, pp. 1117–1127, Feb. 2010.

[26]   S. Bhalekar, and V. Daftardar-Gejji, "A new chaotic dynamical system and its synchronization," In: *Proceedings of the international conference on mathematical sciences in honor of Prof. AM Mathai*, 2011, pp. 3–5.

[27]   S. Bhalekar, "Forming mechanizm of Bhalekar-Gejji chaotic dynamical system," *American J. Comput. Appl. Math.*, vol. 2, no. 6, pp. 257–259, Nov. 2012.

[28]   J.P. Singh, P.P. Singh, and B.K. Roy, "Synchronization of Lu and Bhalekar-Gejji chaotic systems using sliding mode control," In: *IEEE International Conference on Information Communication and Embedded Systems (ICICES)*, 2014, pp. 1–7.

[29]   J.P. Singh, P.P. Singh, and B.K. Roy, "Hybrid synchronization of Lu and Bhalekar-Gejji chaotic systems using nonlinear active control," *IFAC Proceedings*, vol. 47, no. 1, pp. 292–296, Jan. 2014.

[30]   P.P. Singh, J.P. Singh, and B.K. Roy, "Synchronization and anti-synchronization of Lu and Bhalekarâ€"Gejji chaotic systems using nonlinear active control," *Chaos, Solitons & Fractals*, vol. 69, pp. 31–39, Dec. 2014.

[31]   P.P. Singh, J.P. Singh, and B.K. Roy, "Synchronization of chaotic systems using NAC and its application to secure communication," *Int. J. Control Th. Appl.*, vol. 8, no. 3, pp. 995–1003, Dec. 2015.

[32]   J.A. Vargas, E. Grzeidak, K.H. Gularte, and S.C. Alfaro, "An adaptive scheme for chaotic synchronization in the presence of uncertain parameter and disturbances," *Neurocomputing*, vol. 174, pp. 1038–1048, Jan. 2016.

[33]   V. Daftardar-Gejji, Y. Sukale, and S. Bhalekar, "A new predictorâ€"corrector method for fractional differential equations," *Appl. Math. Comput.*, vol. 244, pp. 158–182, Oct. 2014.

[34] M. Borah, P.P. Singh, and B.K. Roy, "Improved chaotic dynamics of a fractional-order system, its chaos-suppressed synchronisation and circuit implementation," *Circuits Systems Signal Process.*, vol. 35, no. 6, pp. 1871–1907, Jun. 2016.

[35] M. Aqeel, and S. Ahmad, "Analytical and numerical study of Hopf bifurcation scenario for a three-dimensional chaotic system," *Nonlinear Dyn.*, vol. 84, no. 2, pp. 755–765, Apr. 2016.

[36] A.S. Deshpande, V. Daftardar-Gejji, and Y. Sukale, "On Hopf bifurcation in fractional dynamical systems," *Chaos Solitons Fractals*, vol. 98, pp. 189–198, Apr. 2017.

[37] K. Diethelm, N.J. Ford, and A.D. Freed, "A predictor-corrector approach for the numerical solution of fractional differential equations," *Nonlinear Dyn.*, vol. 29, pp. 3–22, Jul. 2002.

[38] K. Diethelm, "An algorithm for the numerical solution of differential equations of fractional order," *Elec. Trans. Numer. Anal.*, vol. 5, pp. 1–6, Mar. 1997.

[39] K. Diethelm, and N.J. Ford, "Analysis of fractional differential equations," *J. Math. Anal. Appl.*, vol. 265, pp. 229–48, Jan. 2002.

[40] V. Daftardar-Gejji, and H. Jafari, "An iterative method for solving nonlinear functional equations," *J. Math. Anal. Appl.*, vol. 316, no. 2, pp. 753–763, Apr. 2006.

[41] M.S. Tavazoei, and M. Haeri, "Regular oscillations or chaos in a fractional order system with any effective dimension," *Nonlinear Dyn.*, vol. 54, no. 3, pp. 213–222, Nov. 2008.

[42] D. Matignon, "Stability results for fractional differential equations with applications to control processing," In: *Computational Engineering in Systems and Application Multiconference*, pp. 963–968, Lille: France, 1996.

[43] M.S. Tavazoei, and M. Haeri, "Chaotic attractors in incommensurate fractional order systems," *Physica D*, vol. 237, pp. 2628–2637, Oct. 2008.

[44] W. Deng, C. Li, and J. Lu, "Stability analysis of linear fractional differential system with multiple time delays," *Nonlinear Dyn.*, vol. 48, pp. 409–416, Jun. 2007.

[45] E. Ahmed, A.M.A. El-Sayed, and H.A.A. El-Saka, "On some Routh-Hurwitz conditions for fractional order differential equations and their applications in Lorenz, Rossler, Chua and Chen systems," *Phys. Lett. A*, vol. 358, pp. 1–4, Oct. 2006.

[46] S. Celikovsky, and G. Chen, "On the generalized Lorenz canonical form," *Chaos Solitons Fractals*, vol. 26, pp. 1271–1276, Dec. 2005.

[47] L.M. Pecora, and T.L. Carroll, "Synchronization in chaotic systems," *Phys. Rev. Lett.*, vol. 64, no. 8, pp. 821–825, Feb. 1990.

[48] L.M. Pecora, and T.L. Carroll, "Driving systems with chaotic signals," *Phys. Rev. A*, vol. 44, pp. 2374–2384, Aug. 1991.

[49] R. Hilfer, Ed., *Applications of Fractional Calculus in Physics*. World Scientific: Singapore, 2001.

[50] R. He, and P.G. Vaidya, "Implementation of chaotic cryptography with chaotic synchronization," *Phys. Rev. E*, vol. 57, no. 2, pp. 1532–1535, Feb. 1998.

[51] S. Bhalekar, and V. Daftardar-Gejji, "Synchronization of different fractional order chaotic systems using active control," *Commun. Nonlinear Sci. Numer. Simul.*, vol. 15, no. 11, pp. 3536–3546, Nov. 2010.

Current Developments in Mathematical Sciences, 2018, *Vol. 1*, 183-198

CHAPTER 7

Grünwald-Letnikov Derivative: Analysis in Range of First Order

Radosław Cioć*

Faculty of Transport and Electrical Engineering, Kazimierz Pulaski University of Technology and Humanities in Radom, Malczewskiego Str. 29, Radom 26-600, Poland

Abstract: This chapter analyses Grünwald-Letnikov $f^{(\eta)}(t)$ derivative in the range of first order derivative $f^{(1)}(t)$. Influence of non-integer derivative order to derivative value and relations between derivative order, increment of function value and increment of function argument are presented.

Problems arising from the use of any order of non-integer derivative and any increment of function are presented in examples. The author argues that the sign of Grünwald-Letnikov order is changing with an increment of function. He also argues that the G-L derivative does not fulfil the Mean-Value Theorem for derivatives of all orders.

Keywords: Fractional calculus, Grünwald-Letnikov differintegrals, fractional order interpretation.

AMS Subject Classification: 26A33, 28E05, 33E30.

7.1. INTRODUCTION

The beginning of the fractional calculus dates back to the letter from de l'Hospital to Leibniz in 1695. In the letter, de l'Hospital mentioned the notation $d^n y / dx^n$ invented by Leibniz and asked: "What if n be $1/2$?". Leibniz replay: "You can see by that, Sir, that one can express by an infinite series a quantity such as $d^{1/2}\overline{xy}$ or $d^{1:2}\overline{xy}$. Although infinite series and geometry are distant relations, infinite series admits only the use of exponents that are positive and negative integers, and does not, as yet, know the use of fractional exponents". In the same letter he continues: "Thus it follows that $d^{1/2}x$ will be equal $x\sqrt{dx:x}$. This is an apparent paradox from which, one day, useful consequences will be drawn."[1, 2].

*Corresponding author Radosław Cioć:** Faculty of Transport and Electrical Engineering, Kazimierz Pulaski University of Technology and Humanities in Radom, Malczewskiego Str. 29, Radom 26-600, Poland; Tel: +48 483617700, +48 483617763; Fax: +48 483617704; E-mail: r.cioc@uthrad.pl

The subject is further developed by Euler, Laplace, Lacroix, Fourier, Liouville, Grünwald, Letnikov, Nekrassov, Weyl, Nishimoto, Podlubny, Kilbas and so on. Niels Henrik Abel was the first who used Fractional Calculus in real-life problem. The problem discused by Abel is callled Tautochrone problem [3]. The detailed history of fractional calculus can be found in works of [4, 5].

There are different inequivalent definitions of fractional derivative. One of them is Grünwald-Letnikov derivative. It is defined by replacing the natural number by real number and by introducing Gamma function in the definition of derivative as a limit of function [5-10]. This operation must have an effect on the calculated value. The chapter analyses this effect for positive order of Grünwald-Letnikov derivative in range of first order derivative.

7.2. GRÜNWALD-LETNIKOV DERIVATIVE FROM THE FIRST ORDER DERIVATIVE POINT OF VIEW

Definition of real order derivative is derived as below:

$$f'(t) = f^{(1)}(t) = \frac{df(t)}{dt} = \lim_{dt \to 0} \frac{f(t+dt)-f(t)}{dt}, \tag{7.1}$$

where dt is the increment of the independent variable t and $df(t)$ is the increment of a function dependent on t.

The second derivative is:

$$f''(t) = f^{(2)}(t) = \lim_{dt \to 0} \frac{f'(t+dt)-f'(t)}{dt}$$

$$= \lim_{dt \to 0} \frac{f(t)-2f(t+dt)+f(t+2dt)}{(dt)^2}. \tag{7.2}$$

In general, the derivative of any order $n \in \mathbb{N}$ is formulated as [7, 9, 10]

$$f^{(n)}(t) = \lim_{dt \to 0} \frac{\sum_{m=0}^{n}(-1)^m \binom{n}{m}f(t-mdt)}{(dt)^n} \tag{7.3}$$

where:

$$dt = t_2 - t_1 = t_3 - t_2 = t_l - t_{l-1}$$

$$m = 0,1,\dots,l$$

$$\binom{n}{m} = \frac{n!}{m!(n-m)!} \text{ for } n \geq m.$$

Grünwald-Letnikov derivative of order $\eta \in \mathbb{R}$ is obtained by replacing factorial function with gamma function and n with η:

$$\binom{n}{m} = \frac{n!}{m!(n-m)!} = \frac{\Gamma(n+1)}{m!\Gamma(n-m+1)} = \frac{\Gamma(\eta+1)}{m!\Gamma(\eta-m+1)} \tag{7.4}$$

By inserting (7.4) to (7.3) and replacing n with η, we obtain the Grünwald-Letnikov derivative:

$$f^{(\eta)}(t) = \lim_{dt \to 0} \frac{\sum_{m=0}^{p}(-1)^m \frac{\Gamma(\eta+1)}{m!\Gamma(\eta-m+1)} f(t-mdt)}{(dt)^\eta} \tag{7.5}$$

where: $p = \left\lfloor \frac{t_l - t_0}{dt} \right\rfloor$

By substituting $p = n = 1$, we recover the definition (7.2) as

$$f^{(1)}(t) = \lim_{dt \to 0} \frac{f(t_2) - f(t_1)}{dt} = \frac{df(t)}{dt} \tag{7.6}$$

Also, if we put $p = l$ and $n = 1$ in (7.5), we get (7.5) as

$$f^{(\eta)}(t) = \lim_{dt \to 0} \frac{f(t_2) - \eta f(t_1)}{(dt)^\eta} = \frac{d^\eta f(t)}{(dt)^\eta} \tag{7.7}$$

where:

η is the independent variable (order of G-L derivative) describing increment change dt,

$(dt)^\eta$ is the increment of independent variable t described by η,

$d^\eta f(t)$ is the increment of function dependent on dt^η.

From (7.7) the result is that (7.6) is a form of (7.7) for $\eta = 1$.

Let order η be connected with increment dt by relation [11, 12]:

$$dt = \Delta T + (dt)^\eta \tag{7.8}$$

where:

$$\Delta T > 0 \text{ for } (dt)^{\eta} > dt$$

$$\Delta T = 0 \text{ for } (dt)^{\eta} = dt$$

$$\Delta T < 0 \text{ for } (dt)^{\eta} < dt.$$

Geometrical interpretations of (7.6) and (7.7) are shown in Fig. (**7.1** and **7.2**), respectively linear characteristic of $f(t_1)$ and $f(t_2)$ comes from the definition of (**7.6**). Characteristic of $f(t_{\eta})$ and $f(t_2)$ depends on G-L definition (**7.7**). Order η can be interpreted as an influence parameter of non-linearity of function $f(t)$ between $f(t_1)$ and $f(t_2)$ when function values are unknown in (t_1, t_2).

Change of η has an impact on the value of increment of dt and it estimates function value $f(t)$ in point $t = t_{\eta} = t_2 - (dt)^{\eta}$ (Fig. **7.1**):

$$f(t_{\eta}) = \eta f(t_1) \tag{7.9}$$

In (7.7), $dt \to 0$. This is a relative expression, especially in useful work. For example when $f(t)$ is a measured velocity in time t_1 and t_2 where $dt = t_2 - t_1$ is sample time and $f'(t)$ is an acceleration [8, 11]. From the sentence, we have an additional assumption:

$$0 < dt < 1 \tag{7.10}$$

From (7.10) and (7.8) we receive:

$$(dt)^{\eta} = dt \text{ for } \eta = 1 \tag{7.11}$$

$$(dt)^{\eta} < dt \text{ for } \eta > 1 \tag{7.12}$$

$$(dt)^{\eta} > dt \text{ for } \eta < 1 \tag{7.13}$$

The case (7.12) is shown in Fig. (**7.1**). For (7.11) $f(t_{\eta}) = f(t_1)$. The case (7.13) is shown in the Fig. (**7.2**).

From Figs. (**1** and **2**) it turns out that η is a factor that estimates function value $f(t)$ for $t = t_{\eta}$ based on the value $f(t_1)$ and an increase of $(dt)^{\eta}$:

$$\begin{cases} |f(t_\eta)| < |f(t_1)| \; for \; t_\eta < t_1 \\ |f(t_\eta)| = |f(t_1)| \; for \; t_\eta = t_1 \\ |f(t_\eta)| > |f(t_1)| \; for \; t_\eta > t_1 \end{cases} \tag{7.14}$$

Absolut values in (7.14) take into possible minus value of $f(t_\eta)$.

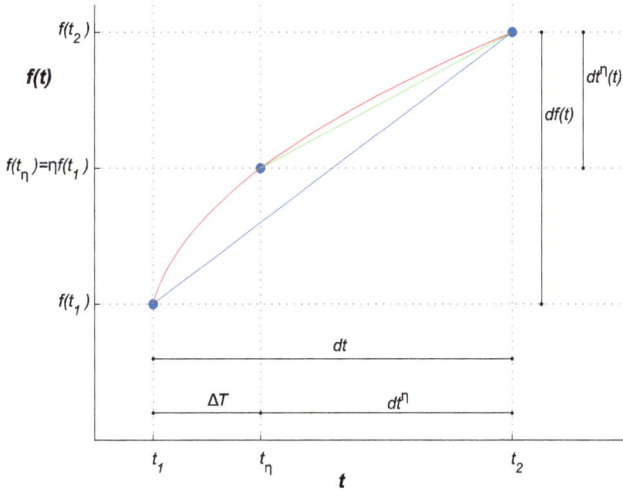

Fig. (7.1). Geometrical interpretation of G-L and first order derivative for $dt > (dt)^\eta$.

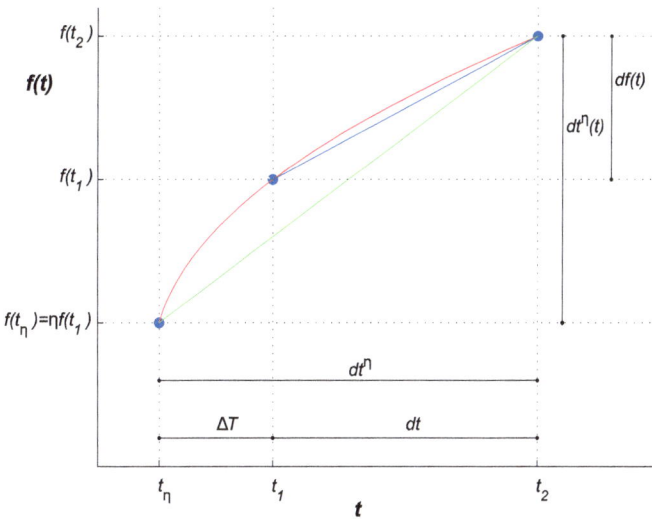

Fig. (7.2). Geometrical interpretation of G-L and first order derivative for $dt < (dt)^\eta$.

7.3. PROPERTIES of η ORDER G-L DERIVATIVE

7.3.1. Estimation of $f(t_\eta)$

From (7.8):

$$\eta = \log_{dt}(dt - \Delta T) \tag{7.15}$$

The relationship $\eta(\Delta T)$ for example values of dt within the range from $-\Delta T$ to ΔT is shown in Fig. (**7.3**).

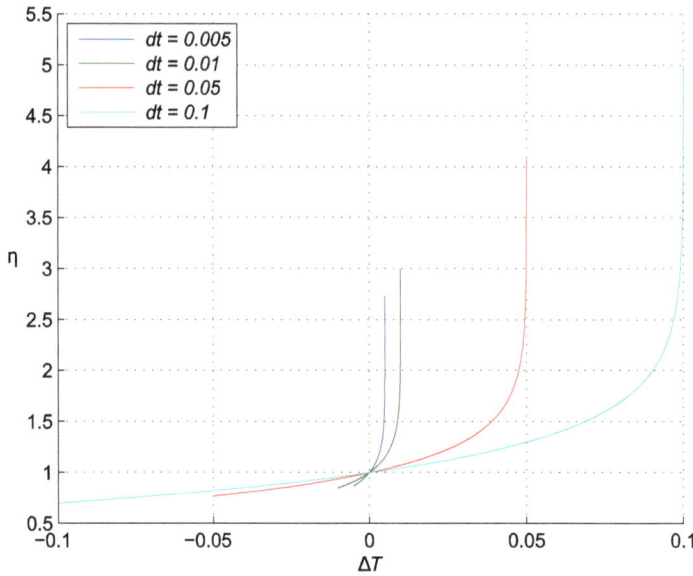

Fig. (7.3). $\eta(\Delta T)$ for example values of dt.

Fig. (**7.3**) shows that $\eta > 1$ for $\Delta T > 0$ (or in a different way for $dt > (dt)^\eta$) and $\eta < 1$ for $\Delta T < 0$ (for $dt < (dt)^\eta$). For $+\Delta T$ and $-\Delta T$ η runs have different increment for the same $|\Delta T|$. In case of large relative values of $|\Delta T/dt|$, η increases from 1 to asymptote in dt. The increase is faster if dt is smaller. That property made the values of $f(t_\eta)$ different from the values laying on a straight line between points $(t_1, f(t_1))$ and $(t_2, f(t_2))$ (Fig. **7.1**). Non-linearity of $f(t)$ is larger if $f(t_\eta)$ is bigger between the points $(t_1, f(t_1))$ and $(t_2, f(t_2))$. Increase of n is negative for $-\Delta T$ and relatively smaller than for positive η (Fig. **7.4**).

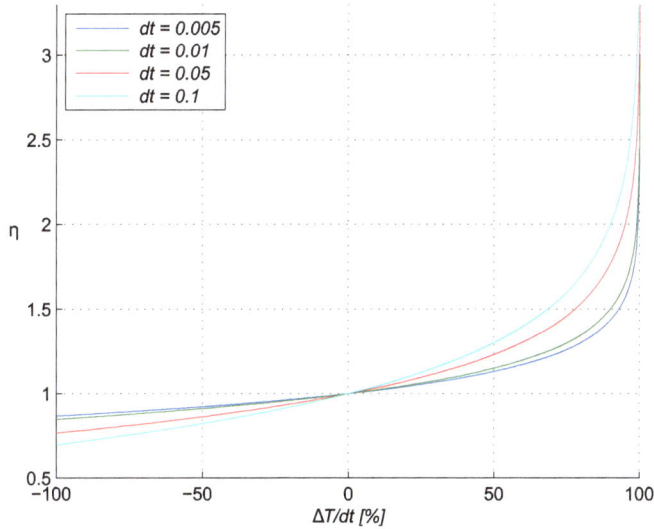

Fig. (7.4). $\eta(\Delta T/dt)$ for example values of dt.

On (7.9) and $t_\eta = t_1 + \Delta T$ (Fig. **7.1**), the result shows that $f(t_\eta)$ is estimated by $f(t_1)$ and η. But not for all t_η is true, because:

$$\forall(t_\eta \in [-t_1, t_2) \wedge \begin{cases} f(t_1) > 0 : f(t_\eta) > f(t_1) \\ f(t_1) = 0 : f(t_\eta) = f(t_1) \\ f(t_1) < 0 : f(t_\eta) < f(t_1) \end{cases} \tag{7.16}$$

and

$$\forall(t_\eta \to t_2) \wedge \begin{cases} f(t_1) > 0 : f(t_\eta) \to +\infty \\ f(t_1) = 0 : f(t_\eta) = 0 \\ f(t_1) < 0 : f(t_\eta) \to -\infty \end{cases} \tag{7.17}$$

Characteristics of $f(t_\eta) = \eta f(t_1)$ for sampled values of $f(t_1)$, $f(t_2)$ and dt are shown in Fig. (**7.5** and **7.6**).

From $f(t_\eta \to t_2) \to \pm\infty$ (7.17), it can be observed that (7.9) for order (7.15) is not a function that estimates the values of $f(t)$ for any t, known $f(t_1)$, $f(t_2)$ and dt. Estimated characteristic for $dt = 0.005$ is shown in Fig. (**7.5**). For other dt, the characteristics are not estimations.

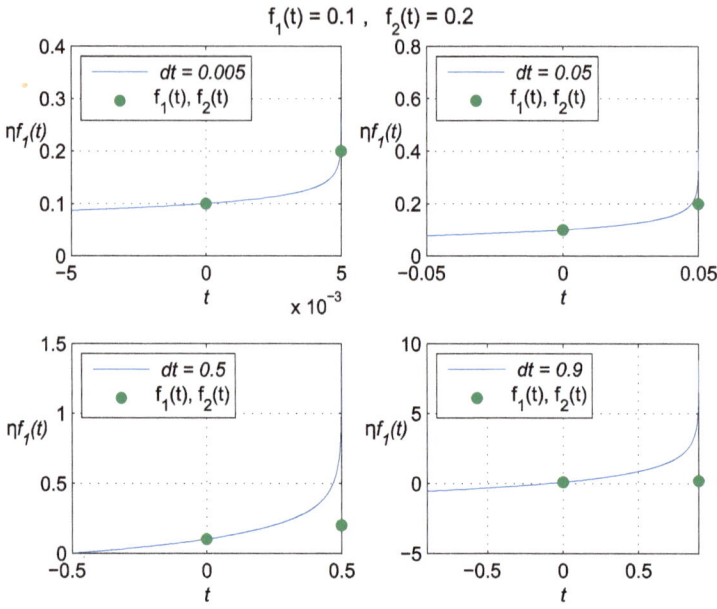

Fig. (7.5). $f(t_\eta)$ for $f(t_1) = 0.1, f(t_2) = 0.2$.

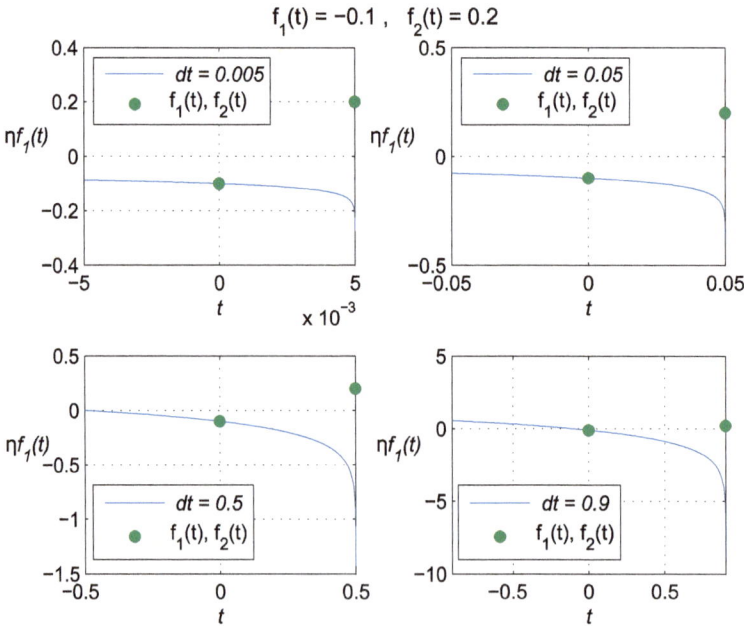

Fig. (7.6). $f(t_\eta)$ for $f(t_1) = -0.1, f(t_2) = 0.2$.

7.3.2. Derivative $f^{(\eta)}(t)$

The goal of the G-L derivative (**7.7**) is not an estimation of $f(t_\eta)$. The goal is the calculation of non-integer order derivative $f^{(\eta)}(t)$, where $t = t_\eta$ and an increment $dt = (dt)^\eta$. It is made by calculation of increment $(dt)^\eta$, where the relationship between η and dt is written as (7.15) and $f(t_\eta)$ is estimated by (7.9).

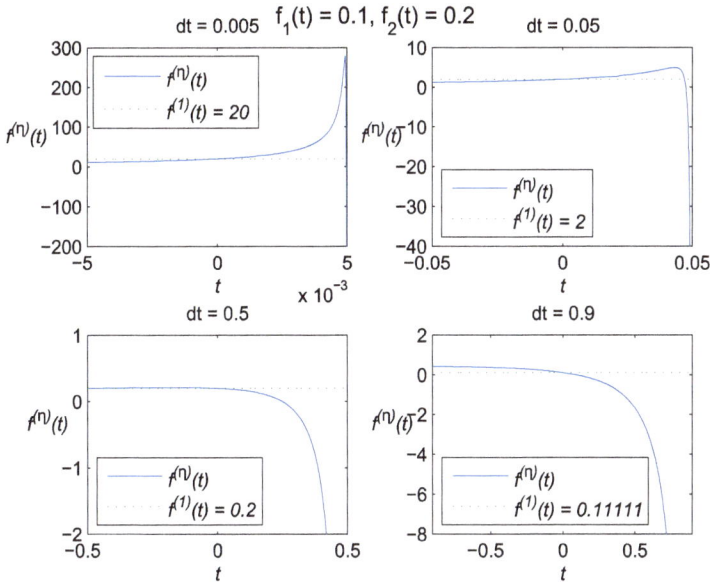

Fig. (7.7). $f^{(\eta)}(t_\eta)$ for $f(t_1 = 0) = 0.1$, $f(t_2 = t_1 + dt) = 0.2$.

Derivative (7.7) depends on $f(t_1)$, $f(t_2)$, dt and η values. Because η is connected with logarithmic relation of dt (7.15), this relationship is transmitted for shape of $f^{(\eta)}(t)$ characteristic (Fig. **7.6**). Non-integer order derivative (7.7) and first order derivative for $t_\eta \in [(t_1 - dt), t_2]$ are shown in Fig. (**7.7** and **7.8**), respectively derivative $f^{(1)}(t)$ is constant for each t_η because the values between $f(t_1)$ and $f(t_2)$ have linear change.

Characteristics $f^{(\eta)}(t)$ of η are shown in Fig. (**7.9** and **7.10**). Marginal values of η axis are values for $t_\eta = -dt$ and $t_\eta = dt$. For $dt = 0.9$ we have negative value of η. The value is the result of increment where:

$$(dt)^\eta = dt - (-\Delta T) > 1 \qquad (7.18)$$

at the assumption (7.10).

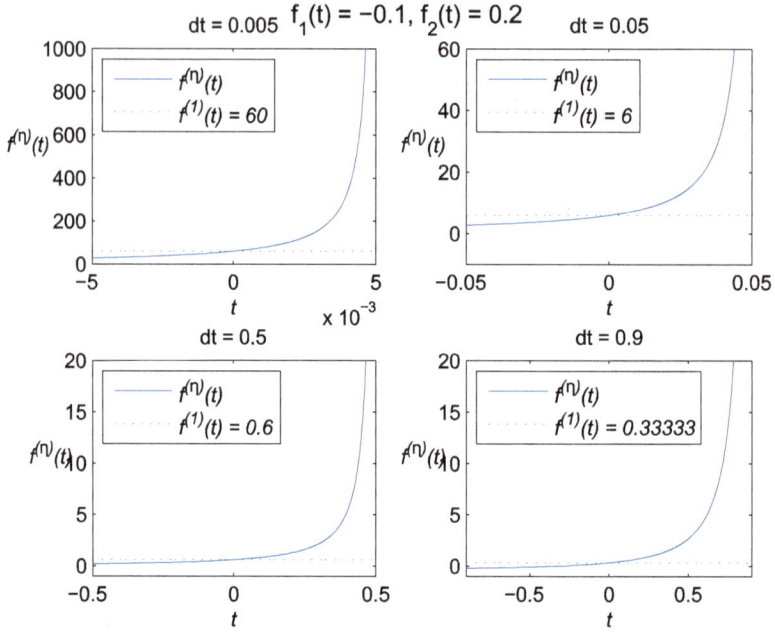

Fig. (7.8). $f^{(\eta)}(t_\eta)$ for $f(t_1 = 0) = -0.1$, $f(t_2 = t_1 + dt) = 0.2$.

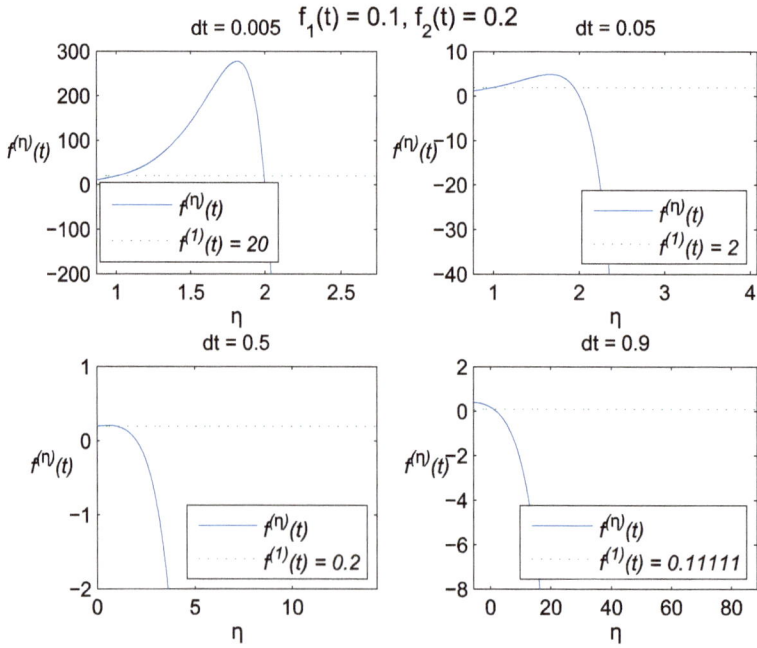

Fig. (7.9). $f^{(\eta)}(t_\eta)$ of η for $f(t_1) = 0.1$, $f(t_2) = 0.2$.

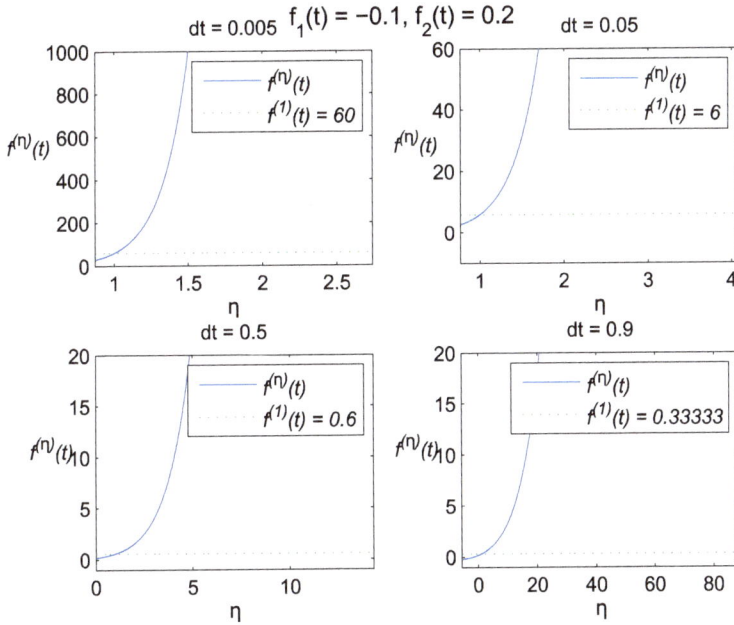

Fig. (7.10). $f^{(\eta)}(t_\eta)$ of η for $f(t_1) = -0.1$, $f(t_2) = 0.2$.

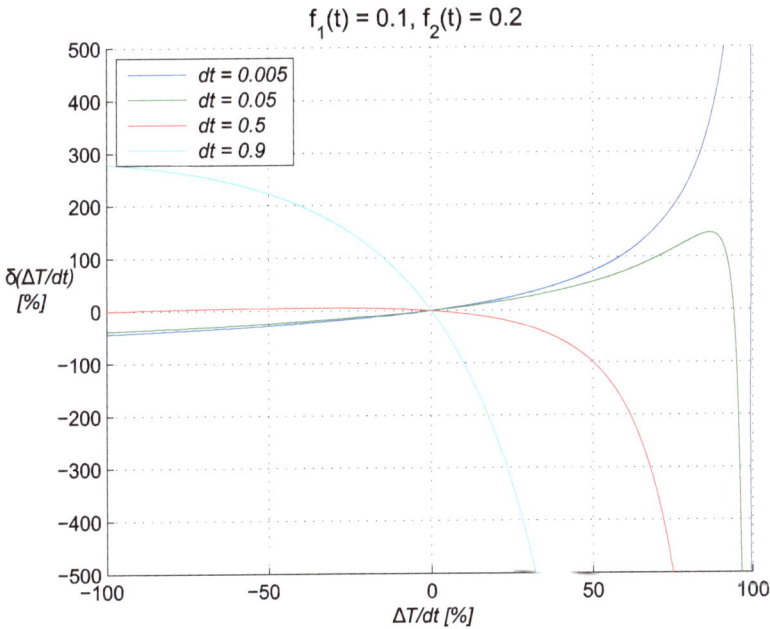

Fig. (7.11). $\delta_\eta(\Delta T/dt)$ for positive $f(t_1)$.

The relations between G-L derivative, η, dt and sign of $f(t_1)$ are shown in the Figs. (**7.7-7.10**). The relationships are nonlinear and,

$$\forall(t_\eta \to t_2) \wedge \begin{cases} f(t_1) \geq 0 : f^\eta(t_\eta) \to -\infty \\ f(t_1) < 0 : f^\eta(t_\eta) \to +\infty \end{cases} \tag{7.19}$$

(7.19) is real for minus and plus values of $f(t_2)$.

Let $\delta_\eta(\Delta T)$ be the relative difference of G-L derivative and first order derivative:

$$\delta_\eta(\Delta T) = \frac{f^{(\eta)}(t_\eta) - f^{(1)}(t_\eta)}{f^{(1)}(t_\eta)} 100\% \tag{7.20}$$

(7.20) can be named as Nonlinearity Index of η Order Derivative. $\delta_\eta(\Delta T/dt)$ is shown in Figs. (**7.11** and **7.13**) where relation of ΔT, dt and η is written by (7.15).

$\delta_\eta(\eta)$ is shown in the Figs. (**7.12** and **7.14**). Marginal values of η axis are values for $t_\eta = -dt$ and $t_\eta = dt$.

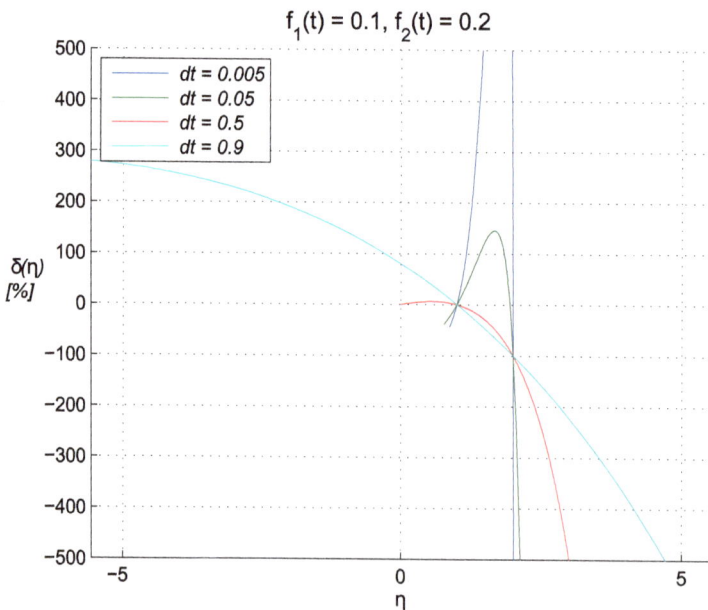

Fig. (7.12). $\delta_\eta(\eta)$ for positive $f(t_1)$.

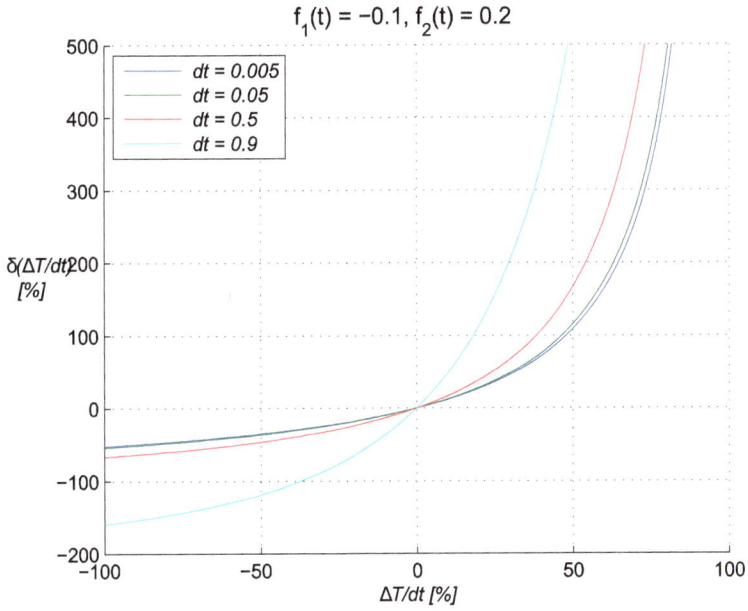

Fig. (7.13). $\delta_\eta(\Delta T/dt)$ for negative $f(t_1)$.

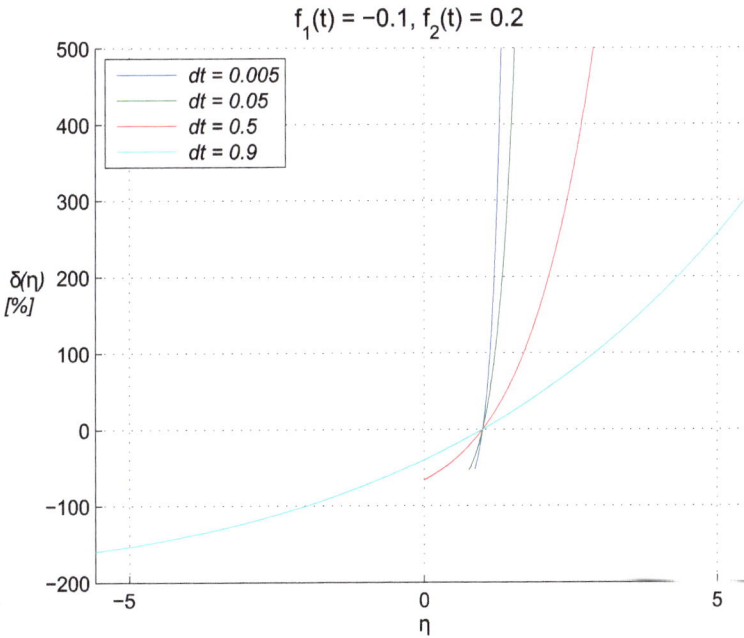

Fig. (7.14). $\delta_\eta(\eta)$ for negative $f(t_1)$.

Properties of δ_η (7.20):

$$\begin{cases} \forall(\eta < 1): \delta_\eta < 0 \\ \forall(\eta = 1): \delta_\eta = 0 \\ \forall(\eta > 1): \delta_\eta > 0 \end{cases} \tag{7.21}$$

For $\delta_\eta < 0$ and $\delta_\eta > 0$: $\eta f(t_\eta)$ is a point on nonlinear approximation of function between $f(t_1)$ and $f(t_2)$. For $\delta_\eta = 0$: $\eta f(t_\eta)$ is a point on linear approximation of function between $f(t_1)$ and $f(t_2)$.

Let $t_1 < t_\eta < t_2$ (fig. 1), G-L derivative (7.7) calculates the value of function $f(t)$ derivative in point t_η and $(dt)^\eta = t_2 - t_\eta$. Because $f(t)$ is a continuous function between $f(t_1)$ and $f(t_2)$, so it results as $f(t_\eta) = \eta f(t_1)$ (or $f(t_\eta) \cong \eta f(t_1)$) and $f(t_\eta)$ should lay on (or close to) the function run $f(t)$. η has values from 1 to $log_{dt}(dt - \Delta T)$ (positive part of ΔT in Fig. **7.3**) and for $t_1 < t_\eta < t_2$, should be high values. The result of this $f(t_\eta)$ can be much higher from $f(t_1)$ and $f(t_2)$. This difference will be larger for higher dt. In cases when the difference between $f(t_1)$ and $f(t_2)$ is small, it can make $\eta f(t_1)$ different from the real value of $f(t_\eta)$ laying on the $f(t)$ characteristic.

For $t_1 > t_\eta$ (Fig. **7.2**), t_η is not in the range $[t_1, t_2]$. In this case, η changes like in Fig. (**7.3**) for minus ΔT. If $f(t)$ is an increasing function ($f(t_1) < f(t_2)$) it can be assumed within some tolerance range that $\eta f(t_1)$ is an estimate of $f(t_\eta)$ because $f(t_\eta) < f(t_1) < f(t_2)$ and $f(t)$ in the range $[t_1, t_\eta]$ is an increasing function. From $f(t_1) > f(t_2)$ (for decreasing $f(t)$ function) and $\eta < 1$ the result is $f(t_\eta) < f(t_1) > f(t_2)$ and $f(t)$ from range $[t_\eta, t_1]$ is an increasing function in contrast to range $[t_1, t_2]$ where $f(t)$ is a decreasing function. From this the result, it can be observed that $\eta f(t_1)$ is not an estimate of $f(t)$ at t_η.

7.4. CONCLUSION

Order η depends on the change of dt increment of independent variable t in $(dt)^\eta$ and values of $f(t)$ (7.8) at a point $t = t_\eta = t_2 - (dt)^\eta$ (Fig. **7.1**) comes from the analysis of G-L derivative (**7.7**). Following this, η order describes linear (for $\eta = 1$) or nonlinear (for $\eta \neq 1$) $f(t)$ characteristic. Nonlinearity Index of η Order Derivative δ_η has been written as the relative difference between G-L derivative and first order derivative (7.20).

The way of $f(t_\eta)$ estimation (7.9) and estimation of G-L derivative (7.7) for samples of $f(t_1)$ and $f(t_2)$ has been shown. The estimation has been made by logarithmic dependence between η, dt and ΔT (7.15) without taking into consideration the values of $f(t_2)$. For that reason $f(t_\eta \to t_2) \to \pm\infty$ (7.17) and estimation of $f(t_\eta)$ are not real for all t_η.

From (7.19) comes $f^{(\eta)}(t_\eta \to t_2) \to \pm\infty$. G-L derivative should have value of $f^{(\eta)}(t_\eta \to t_2) \cong f^{(1)}(t_\eta \to t_2) \cong f^{(1)}(t)$ because $f(t_\eta \to t_2)$ lays close to the line joining $f(t_1)$ and $f(t_2)$ (Fig. **7.1**) and $f^{(1)}(t) \cong f^{(1)}(t_1)$ for $t \in (t_1, t_2)$. From this reason, G-L derivative has not obtained properly values for $t_\eta \to t_2$. That sentence is confirmed by characteristics shown in Fig. (**7.7** and **7.8**).

In the Mean-Value Theorem of Derivatives results, there exists a point $t_\eta \in (t_1, t_2)$ in which a tangent to f is parallel to a secant drawn between $(t_1, f(t_1))$ and $(t_2, f(t_2))$ [13]. The same for minimum one value of t_η there exists equality $(f(t_2) - f(t_1))/dt = f^{(\eta)}(t_\eta)$. Fig. (**7.7-7.10**) shows that it is not true for all values of $f(t_1), f(t_2)$, η and dt. In most of the cases, characteristics of G-L, derivative for $\eta \neq 1$ and the Mean-Value Theorem of Derivatives are mutually exclusive in the range of first order. The M-VT of Derivatives and G-L derivative are fully unanimous for small dt values ($dt = 0.005$ and 0.05 in the figures) and the same sign of $f(t_1)$ and $f(t_2)$ only.

CONSENT FOR PUBLICATION

Not applicable.

CONFLICT OF INTEREST

The authors declare no conflict of interest, financial or otherwise.

ACKNOWLEDGEMENTS

Declared none.

REFERENCES

[1]   G. Leibniz, "Letter from Hanover, Germany to G. F. A. l'Hospital, September 30, 1695," *Mathematische Schriften*, vol. 2, pp. 301–302, 1849.
[2]   B. Ross, "The development of fractional calculus 1695–1900," *Historia Mathematica*, vol. 4, pp. 75–89, Feb. 1977.
[3]   N. H. Abel, *Oeuvres completes*, vol. 1. Imprimerie Grøndahl & Søn, 1881.

[4]   K. Oldham and J. Spanier, *The Fractional Calculus*, vol. 111. Academic Press, New York, 1974.

[5]   K. S. Miller and B. Ross, *An Introduction To The Fractional Calculus And Fractional Differential Equations*. Wiley-Interscience, 1993.

[6]   A. Letnikov, "An explanation of fundamental notions of the theory of differentiation of fractional order," *Mat. Sb*, vol. 6, pp. 413–445, 1872.

[7]   S. Das, *Functional Fractional Calculus For System Identification And Controls*. Springer-Verlag, Berlin, 2008.

[8]   R. Herrmann, *Fractional Calculus: An Introduction For Physicists*. World Scientific, London, 2014.

[9]   I. Podlubny, *Fractional Differential Equations*, vol. 198. Academic Press, San Diego, 1998.

[10]   I. Petras, *Fractional-Order Nonlinear Systems: Modeling, Analysis And Simulation*. Springer: London, 2011.

[11]   R. Cioć, "Physical and geometrical interpretation of Grünwald-Letnikov differintegrals: measurement of path and acceleration," *Fractional Calculus and Applied Analysis*, vol. 19, pp. 161–172, Feb. 2016.

[12]   R. Cioć, "Digital fractional integrator," in *Theory and Applications of Non-integer Order Systems*, pp. 169–174, Springer, London, 2017.

[13]   T. M. Apostol, *Calculus, Vol. 1: One-Variable Calculus, with an introduction to linear algebra*. John Wiley and Sons Inc, 1967.

CHAPTER 8

GPU Computing of Special Mathematical Functions used in Fractional Calculus

Parag Patil[1], Navin Singhaniya[1], Chaitanya Jage[1], Vishwesh A. Vyawahare[1], * Mukesh D. Patil[2] and P.S.V. Nataraj[3]

[1]*Department of Electronics Engineering, Ramrao Adik Institute of Technology, D. Y. Patil Vidyanagar, Nerul, Navi Mumbai 400 706, India*
[2]*Department of Electronics and Telecommunication Engineering, Ramrao Adik Institute of Technology, D. Y. Patil Vidyanagar, Nerul, Navi Mumbai 400 706, India*
[3]*IDP in Systems and Control Engineering, Indian Institute of Technology Bombay, Mumbai 400 076, India*

Abstract: Fractional calculus is a field with growing interest for researchers due to its applications in the various fields like mathematical modeling, control systems, image processing, financial systems, *etc.* Special mathematical functions like Gamma functions, Mittag-Leffler function, Hypergeometric function, *etc.* play a crucial role in the analytical and numerical solutions of fractional differential equations. However, the numerical computation of these functions is a tedious and time-consuming task. This is due to the fact that these functions do not have straightforward definitions and are mostly represented by series expansions. This chapter attempts to exploit the parallel computing power of Graphics Processing Unit (GPU) for computing some of the well-known special functions used in fractional calculus. Using the numerical computational platform of MATLAB and its parallel computing toolbox, the chapter reports the use of GPU hardware to compute these functions using their series definitions. It is shown with the help of various case studies and for different parameter combinations that the implementation of parallel computation of special functions reduces the execution time. A comparative study showing the effect of function parameters on their parallel computation is also presented.

Keywords: Fractional Calculus, Special Functions, Parallel computation, GPU Computing.

AMS Subject Classification: 26A33, 65P20, 34L30.

8.1. INTRODUCTION

This chapter deals with the implementation of special mathematical functions used in fractional calculus on Graphics Processing Unit (GPU). The numerical

*Corresponding author **Vishwesh A. Vyawahare:** Department of Electronics Engineering, Ramrao Adik Institute of Technology, D. Y. Patil Vidyanagar, Nerul, Navi Mumbai 400 706, India; Tel: +91 22 2770 9574; Fax: +91 22 2770 9573; E-mail: vishwesh.vyawahare@rait.ac.in

methods for computation of mathematical quantities which do not have any general definition, can be described by using special functions of fractional calculus. Fractional calculus is a more than 300 years old field of mathematical analysis which describes the dynamics of a system more precisely than conventional integer order methods [1]. It studies the possibility of taking real number powers or complex number powers of differentiation operator and integration operator. The use of arbitrary or fractional order instead of integer order results into more compact and realistic models of various physical, engineering, atomization, biological, biomedical, chemical, economic phenomena [3]. As mentioned in the book "Numerical methods for special functions" by authors Gil, A. and Segura, J. and Temme, N.M., the ability of a mathematical function to be useful in applications and to satisfy certain special properties is what makes it a "special function" [2]. Some commonly used special functions include Gamma function, Mittag-Leffler function, Gauss Hypergeometric function, Dawson function, *etc.*

Most of the fractional differential and integral equations can be solved using special functions of fractional calculus. Special functions appear in the solution of fractional differential equations and the integral equations. Therefore, special functions like Gamma function, Mittag-Leffler function, Confluent hypergeometric function, Gauss hypergeometric function, Bessel Wright function, Dawsons function are necessary tools related to differentiation and integration of fractional order and also fractional calculus appliactions. The importance of these special functions can be understood by following examples. Mittag-Leffler function is used to calculate the step response of linear Fractional-order systems [5]. Sergei Rogosin in his paper mentioned that Abel integral equation used in the study of heat and mass transfer, the solar or a planetary atmosphere can be solved using Mittag-Leffler function [4]. Herbert Buchholz in his book titled "The Confluent Hypergeometric Function with Special Emphasis on its Applications" discussed solution of the wave equation for problems involving parabolic mirrors or parabolic antennas using confluent hypergeometric function [6] and according to James B. Seaborn [7] most of the functions encountered in applied mathematics can be expressed in terms of hypergeometric function.

The time required to obtain the solution of these fractional integro-differential equations is high [5], when long time vectors are considered while implementing. In this study an attempt is made to overcome this problem by the use of parallel computing approach. In parallel computing, many arithmetic and logical operations are carried out simultaneously using multicore processors.

Traditionally, parallel computing has been carried out using Central Processing Unit (CPU) with handful of cores. But, we are using the capabilities of graphics processor for parallel computing. Even though original purpose of GPU is graphics rendering, in recent years the GPU is increasingly being used for scientific computations. This practice of using GPU for general purpose computations is called General Purpose Computing on Graphics Processing Unit (GPGPU) Technology. In paper [50] a parallel GPU solution of the Caputo fractional reaction-diffusion equation in one spatial dimension with explicit finite difference approximation has been provided. The optimized GPU solution obtained is faster than the optimized parallel CPU solution. So the power of parallel computing on GPU for solving fractional applications can be recognized. In [51], the development of parallel algorithms to solve fractional differential equations using a numerical approach has been considered. The methodology adopted is to convert the existing numerical schemes and to develop prototype parallel programs using the MATLAB Parallel Computing Toolbox. In this parallel implementation of the Diethelm-Chern Algorithm, Fractional Adams Method, Lubich's Fractional Multistep Method has been done on MPCT. See also [52-55] which describe the various applications of fractional calculus. Unlike CPU, Graphics Processing Unit has arrays of hundreds of smaller processors which function in parallel. We have utilised this parallel architecture of GPU to accelerate the execution of above mentioned special functions. To harness the power of GPU, we have used Parallel Computing Toolbox (PCT) of MATLAB.

The organization of the chapter is as follows: In section 2 various special functions and their numerical methods are described. Section 3 describes in detail the Graphics Processing Unit (GPU) and tools for MATLAB Computing with GPU. Section 4 contains the results of execution and benchmarking plots. The brief discussion about the novelty of the work done is explained in section 5 and section 6 concludes the work.

8.2. SPECIAL MATHEMATICAL FUNCTIONS USED IN FRACTIONAL CALCULUS

Numerical methods for the special mathematical functions have been mentioned in the Handbook of Mathematical Functions with Formulas, Graphs, and Mathematical Tables [8], by Milton Abramowitz and Irene Stegun published in 1964. It is the first source of information about the properties of special functions. These days the Handbook is being updated as NIST Digital Library of Mathematical Functions (DLMF) [9] and is also freely accessible in a Web version. This handbook includes several special functions that have appeared in applied mathematics, physical sciences and engineering. The Book on Numerical methods for special functions by Amparo Gil [2] was also helpful for the study of special mathematical functions. These sources give many properties of special functions which can be used for their numerical evaluation. Using these numerical methods, we have developed MATLAB codes for the implementation and acceleration using GPU.

The Special functions used in this work are described below:

8.2.1. Gamma Function

The major definitions of fractional derivatives *i.e* Grnwald-Letnikov, Riemann-Liouville and Caputo involve Gamma function. The Gamma function is a modification of the factorial function. The common method for determining the value of $z!$ is found by multiplying,

$$1 * 2 * 3 * 4 \ldots \ldots * (z - 2) * (z - 1) * z. \tag{8.1}$$

The Gamma function gives values that are similar to factorials of non-integer numbers. It is implemented using the following infinite product definition, given by Karl Weierstrass as mentioned in 8.2 [11].

$$\Gamma(Z) = \frac{e^{-\gamma z}}{z} \prod_{k=1}^{N} \left(1 + \frac{z}{k}\right)^{-1} e^{\frac{z}{k}}, \tag{8.2}$$

where, γ =0.5772156649, is the Euler-Mascheroni constant, N is the upper limit of summation (ideally should be infinity). It is valid for all complex numbers Z, except 0 and the negative integers [8, 11]. Mathematicians and geometers have discovered and developed many applications of Gamma function [12]. For the positive integer z relation between Gamma and factorial function is,

$$\Gamma(Z) = (Z - 1)!. \tag{8.3}$$

8.2.2. Mittag-Leffler Function

This function is closely related to the fractional calculus and problems arising due to its application. The 1-parameter Mittag-Leffler function (MLF) can be defined by the following series denoted by 4 when the real part of α is strictly positive,

$$E_\alpha(z) = \sum_{k=0}^{\infty} \frac{z^k}{\Gamma(\alpha k + 1)} \tag{8.4}$$

where, z is complex variable and $\Gamma(.)$ is Gamma function and $\alpha > 0$ [13, 14]. For $\alpha = 1$ the Mittag-Leffler function reduces to exponential function. For $0 < \alpha < 1$, it interpolates between the pure exponential function and hypergeometric function [14].

The 2-parameter Mittag-Leffler function is the generalization of 1-parameter version. It was proposed by Wiman in 1905. Later this function was extensively studied by Humbert and Agarwal [16]. It's series expansion is given by,

$$E_{\alpha,\beta}(z) = \sum_{k=0}^{\infty} \frac{z^k}{\Gamma(\alpha k + \beta)}, \tag{8.5}$$

where, $Re(\alpha) > 0$, $Re(\beta) > 0$. The Mittag-Leffler function of 3-parameters was given by Prabhakar. Its defined as 6 [18, 22].

$$E_{\alpha,\beta}^{\gamma}(z) = \sum_{k=0}^{\infty} \frac{(\gamma)_k}{k!\,\Gamma(\alpha k + \beta)} z^k, for\ \alpha > 0, \beta > 0, \gamma > 0. \tag{8.6}$$

Substituting $\gamma = 1$ reduces it to 2-parameter MLF, and $\gamma = \beta = 1$ gives the 1-parameter MLF.

A generalized Mittag-Leffler function of 4-parameters was given by Shukla and Prajapati [18, 19, 23] and is defined as :

$$E_{\alpha,\beta}^{\gamma,q}(z) = \sum_{k=0}^{\infty} \frac{(\gamma)_{qk} z^k}{k!\,\Gamma(\alpha k + \beta)}, for\ \alpha > 0, \beta > 0, \gamma > 0, q > 0. \tag{8.7}$$

The MLF can be represented in terms of other special functions as [13],

$$E_{1,1}(z) = e^z, \tag{8.8}$$

$$E_{2,1}(z^2) = cosh(z), \tag{8.9}$$

$$E_{2,1}(-z^2) = cos(z), \tag{8.10}$$

$$E_{1/2,1}(\pm z^{1/2}) = e^z erfc(\mp z^{1/2}). \tag{8.11}$$

The MLF is an eigen function of fractional differential equation and hence the solutions of fractional differential equations are expressed in terms of Mittage-Leffler function. This function can also be used in step or impulse response calculation of fractional-order systems. Mainardi [24] proved that the Mittag-

Leffler function is present whenever derivatives of fractional-order are introduced into the constitutive equations of a linear viscoelastic body [21]. The Mittag-Leffler function appears in the solution of the fractional master equation. This characterizes the renewal process with reward modeling by the random walk model known as Continuous Time Random Walk (CTRW). In that, the waiting time is assumed to be a continuous random variable. The basic role of the Mittag-Leffler waiting time probability density in time fractional continuous time random walk became well known since the fundamental paper by Hilfer (1995) [17]. The deviations of physical phenomena from exponential behavior can be governed by physical laws through Mittag-Leffler functions.

8.2.3. Pochhammer Symbol

The Pochhammer symbol was introduced by Leo August Pochhammer, and the notation is $(x)_n$, where n is a non-negative integer. Depending on the context, the Pochhammer symbol may be represent either by the rising factorial or by the falling factorial [31]. Pochhammer symbol $(x)_n$ is used to represent the falling factorial,

$$(x)_n = \begin{cases} 1 & n = 0; \\ x(x-1)(x-2)\ldots(x-n+1) & n > 0. \end{cases} \qquad (8.12)$$

When x is a non-negative integer, The symbol $(x)^n$ is used to represent the rising factorial as,

$$(x)^n = \begin{cases} 1 & n = 0; \\ x(x+1)\ldots(x+n-1) & n > 0. \end{cases} \qquad (8.13)$$

Using the gamma function, rising factorial can be represented for real values of n provided x and $x+n$ are not negative integers as,

$$(x)^n = \frac{\Gamma(x+n)}{\Gamma(x)}. \qquad (8.14)$$

and so can the Falling factorial as,

$$(x)_n = \frac{\Gamma(x+1)}{\Gamma(x-n+1)}. \qquad (8.15)$$

Pochhammer symbol is required is required in the definition of Gauss Hypergeometric function.

8.2.4. Gauss Hypergeometric Function

Gauss Hypergeometric function (GHF) is a solution of a second-order linear ordinary differential equation (ODE). Every second-order linear ODE with three regular singular points can be transformed into this equation. Computation of Gauss hypergeometric function is an extremely difficult task in practice [30]. The famous GHF is present everywhere in mathematical physics as many well-known partial differential equations may be reduced to Gauss equation *via* separation of variables. In mathematics, the Gaussian or ordinary hypergeometric function $_2F_1(a, b; c; z)$ is a special function represented by the hypergeometric series [33].

This hypergeometric function is defined for $|z| < 1$ by the power series as [32, 34],

$$_2F_1(a, b; c; z) = \sum_{n=0}^{\infty} \frac{(a)^n (b)^n z^n}{(c)^n n!}, \tag{8.16}$$

where $a, b, c \in \mathbb{R}$ and n is a non-negative integer. It is undefined if c is equal to non-positive integer. Here $(q)^n$ is the rising Pochhammer symbol, which is defined in 8.13.

8.2.5. Confluent Hypergeometric Function

The Confluent hypergeometric function is a special case of gauss hypergeometric function. It is also known as Kummer's function. In the name of Confluent hypergeometric function the term "Confluent" is used because, two regular singularities of gauss hypergeometric function at 1 and ∞, get merged into an irregular one at ∞. And it's regular singularity at point 0, remains unchanged [25, 33]. This function $M(a, b, x)$, arises naturally in both statistics and physics.

Kummer's equation may be written as,

$$z \frac{d^2 w}{dz^2} + (b - z) \frac{dw}{dz} - aw = 0. \tag{8.17}$$

It has a regular singularity at 0 and an irregular singularity at ∞. It has usually two linearly independent solutions, first is Kummer's confluent hypergeometric function $M(a, b, z)$ and second is Tricomi confluent hypergeometric function $U(a, b, z)$ [26]. Kummer's function is given by,

$$M(a, b, z) = \sum_{n=0}^{\infty} \frac{a^{(n)} z^n}{b^{(n)} n!} =_1 F_1(a; b; z), \tag{8.18}$$

where, $a^{(0)} = 1$, $a^{(n)} = a(a + 1)(a + 2) \cdots (a + n - 1)$, is the rising factorial.

The hypergeometric function has wide range of applications for which it is frequently referred. For example, it arises in the study of photon scattering from atoms, networks, Coulomb wave functions, binary stars, finance and many others [28, 29]. It is also required in study of certain random variables. The function also contains other functions as special case, including many that are widely used in mathematical physics. Its special cases include the Bessel function, the incomplete gamma function, Laguerre polynomials, Hermite polynomials, Coulomb wave function and Parabolic cylinder function [30].

When $\alpha = 1$ the Prabhakar function $E_{1,\beta}^{\gamma}(z)$ coincides with the Kummer confluent hypergeometric function $\phi(\gamma; \beta; z)$ apart from the constant factor $(\Gamma(\beta))^{-1}$

$$E_{\alpha,\beta}^{\gamma}(z) = \frac{1}{\Gamma(\beta)} \phi(\gamma; \beta; z). \tag{8.19}$$

8.2.6. Dawson's Function

Dawson fuction is named after H. G. Dawson [35], the origin it has the series expansion given by,

$$F(x) = x \sum_{k=0}^{\infty} \frac{(-2x^2)^k}{(2k+1)!!}, \tag{8.20}$$

where,

$$(2k + 1)!! = (2k + 1)(2k - 1)(2k - 3)\ldots..$$

where n!! is the double factorial defined as:

$$(n)!! = \begin{cases} n.(n-2)\ldots.5.3.1 & n > 0, Odd \\ n.(n-2)\ldots.6.4.2 & n > 0, Even \\ 1 & n = -1,0. \end{cases} \tag{8.21}$$

8.2.7. Bessel Wright Function

In mathematics, the Bessel– Maitland function, or Wright generalized Bessel function, is a generalization of the Bessel function, introduced by Edward Maitland Wright (1934) [35]. It is defined as 8.22,

$$J_\vartheta^\mu = \phi(\mu, \vartheta + 1; -z) = \sum_{k=0}^{\infty} \frac{1}{\Gamma(\mu k + \vartheta + 1)} \frac{(-z)^k}{k!}. \tag{8.22}$$

where, $k \in$ integer, $z \in \mathbb{C}$ $\mu \in \mathbb{R}$. The Bessel Wright function is a special case of generalized Bessel Wright function $J_{\vartheta, \lambda}^\mu$ which can be derived from 4-parameter Mittag-Leffler function [13]. In the next section, we present the MATLAB implementation of these functions using CPU and GPU environments.

8.3. GRAPHICS PROCESSING UNIT (GPU)

The Graphics Processing Unit (GPU) is widely used for realtime graphics rendering in applications like high definition gaming, computer aided design softwares, *etc*. Rendering is the process of generating an image from a 2D or 3D model by means of computer programs. This process demands huge number of calculations to be performed in fraction of seconds. The GPUs were introduced in 1990 for the same purpose [38]. GPU is becoming popular as cheap parallel supercomputer whose processing power can be utilized. The architecture of GPU is different from the traditional CPU. The CPU provides more resources to make single instruction execute faster, whereas the GPU makes use of hundreds of individual processing elements (cores) to execute single instruction on multiple data elements (elements in array) in parallel as shown in Fig. (**8.1**). But the requirement is, each element should be independent from one another that is, in the base programming the elements cannot communicate with each other. Hence a GPU program should be structured appropriately, such that it operates on many parallel elements which are executed simultaneously.

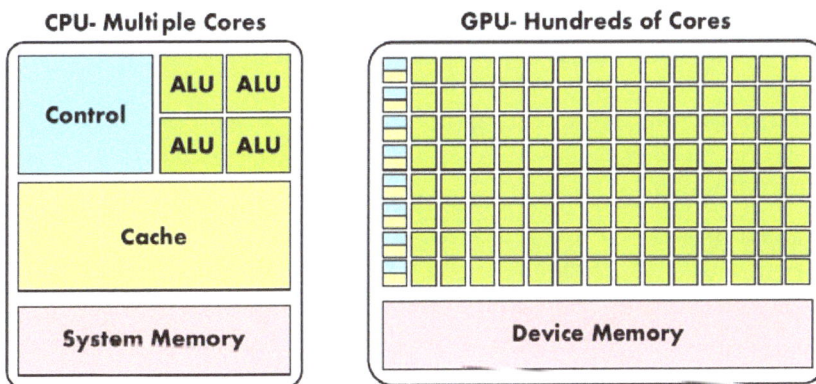

Fig. (8.1). Pictorial representation of CPU and GPU structures [36].

So, we can say that programmable units of the GPU follow Single Program Multiple Data (SPMD) model of programming. GPU architecture can be more energy efficient than CPU architecture for certain workloads. For GPU implementation, the algorithms must be throughput intensive, which means it processes lots of data elements so that there will be more operations to be performed in parallel [36-39].

Nowadays, various applications like weather forecasting, molecular dynamics, computational fluid dynamics, computational finance, medical imaging, ocean modelling and space sciences make use of parallel processing power of GPU. With the evolution of GPU hardware, the software solutions and Application Program Interface (APIs) for GPGPU technology are also developed. Some toolboxes of MATLAB are also available for GPU computing using which we can make use of benefits of MATLAB for GPU computing [40].

8.3.1. GPU Computing with MATLAB

Matrix Laboratory (MATLAB) is a high-performance language of technical computing [42]. MATLAB allows matrix manipulations, plotting of functions and data, implementation of algorithms, and creation of user interfaces, *etc*. Furthermore, MATLAB is a modern programming language environment: it has built-in editing and debugging tools, and supports object-oriented programming. These factors make MATLAB an excellent tool for teaching and research. The computational power of GPU can be easily accessed using toolboxes of MATLAB which are available for GPU computing. Using these toolboxes, the computationally intensive MATLAB code can be ported to execute on GPU with minor changes. There are three toolboxes that are extensively in use: Jacket, GPUmat, and Parallel Computing Toolbox (PCT) of MATLAB [37].

Parallel Computing Toolbox enables the user to program MATLAB to use GPU. In many cases, execution in the GPU is faster than in the CPU, so this feature offers improved performance. The PCT toolbox requires the NVIDIA CUDA enabled device with compute capability of 1.3 or greater. PCT provides a straightforward way to speed up MATLAB code by executing it on GPU. By changing the data type to gpuArray() function, selected MATLAB commands that have been overloaded for gpuArray() function. PCT provides more than 90 GPU enabled functions. A complete list of built-in MATLAB functions is available in the PCT documentation [42]. The few important functions which are used in GPU computing are listed in Table **8.1**.

Following steps are followed for converting the MATLAB code for GPU execution:

Step 1. Start with the code that gives correct results.

Step 2. Perform Vectorization of the code.

Step 3. Profile the code using Run & Time Function.

Step 4. Analyze and Convert the section that takes more time to execute on GPU.

Step 5. Transfer the variables used in the section described in step #4 to GPU memory using gpuArray() function.

Step 6. After GPU execution, transfer the output variables back to CPU memory using gather function.

Step 7. After receiving the gathered data from GPU, rest of the execution is performed on CPU.

Table 8.1. GPU overloaded functions in MATLAB.

Sr. No.	Function	Syntax	Description
1	gpuArray	$G = gpuArray(X);$	Transfers the specified array from MATLAB workspace on GPU memory.
2	arrayFun	$A = arrayfun(FUN,B)$	Evaluation of the function specified by arrayfun on GPU.
3	gather	$Gc_{pu} = gather(G);$	Transfers the array from GPU memory back to MATLAB workspace.
4	bsxfun	$C = bsxfun(FUN, A,B)$	Apply element-by-element operation to two arrays for built-in functions.

While converting MATLAB code to run on the GPU, it is best to start with a correct and optimized MATLAB code. The code should give the desired results correctly. For CPU programming, vectorization is considered to be a good practice, whereas for GPU, vectorization is essential for achieving high performance. Hence it is necessary to vectorize the code by replacing looped scalar operations with MATLAB matrix and vector operations. Profiling gives accurate information about the time taken by every line or section of the code.

The Run & Time feature of MATLAB is an effective tool which profiles the code while executing it [41]. Figs. (**8.2** and **8.3**) show the screenshots of the profiling windows for CPU and GPU executions of Gauss Hypergeometric function.

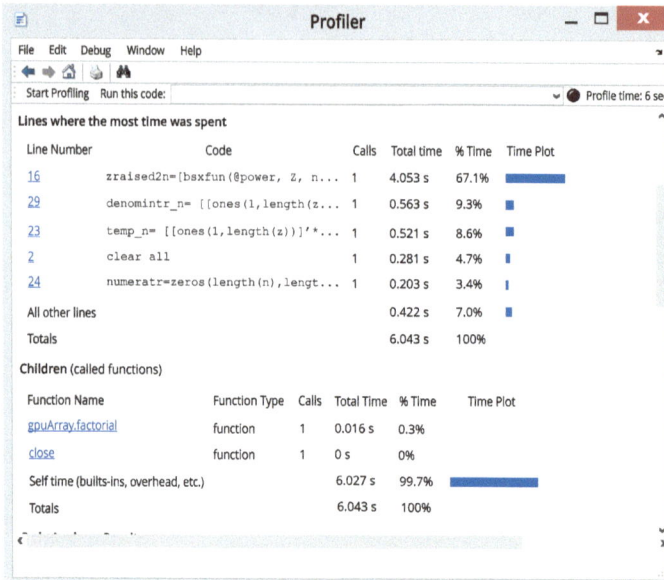

Fig. (8.2). Profiling of Gauss Hypergeometric function when executed on CPU.

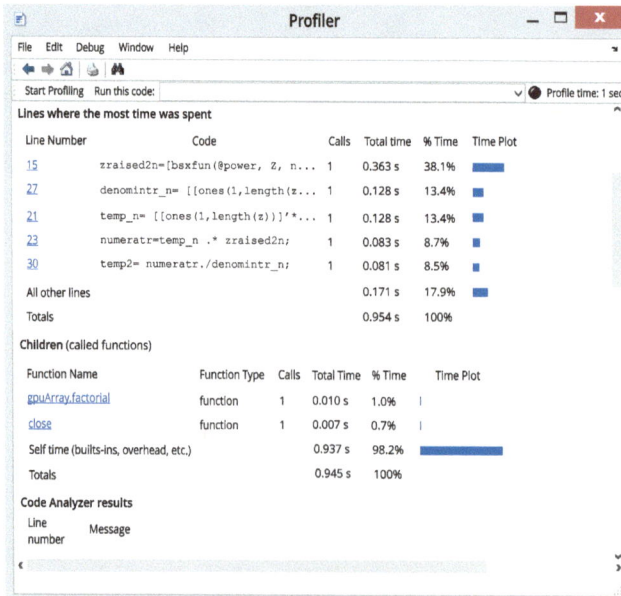

Fig. (8.3). Profiling of Gauss Hypergeometric function when executed on GPU.

From Figs. (**8.2** and **8.3**) we can clearly see how the profiler gives the time required for each line of code. If we compare these two windows, we can notice the reduction in execution time from 6.043 seconds on CPU to 0.954 seconds on GPU. So, using profiler we can identify the time consuming section of code. The part of the code highlighted by the profiler takes longer time for execution. This part is considered for possible parallelization and its evaluation using GPU. Writing the GPU code using the functions in MATLAB which are overloaded to run on GPU is advantageous. The variables or data required for execution on GPU device is first transferred to device memory using gpuArray() function. The functions with these variables as arguments are directly executed on GPU. After the processing on GPU, the results are copied back using gather() function. The gathered variables are then used for further operations [41, 42].

8.4. RESULTS

This section presents the results of GPU execution. Developing a parallel version of code enables us to run its application on large data (*e.g.* more pixels, a bigger physical model of system) in less amount of time. The result of successful parallelization is measured in terms of speedup. The speedup is the ratio of serial execution time to parallel execution time. The speedup can be calculated by using formula as: [43-45]

$$\text{Speedup} = \frac{\text{Program execution time on CPU}}{\text{Program execution time on GPU}}. \tag{8.23}$$

For example, if the serial version of an application executes in 150 seconds and the corresponding parallel version of the same application runs in 15 seconds, the speedup for the parallel application is 10x.

We have tested the computation of special functions on CPU and GPU of two stand-alone systems and compared the performance. The specifications of GPU and CPU of these systems are listed in Tables **8.2-8.5** respectively. The first system on which the codes were executed the codes is a computer with CUDA enabled NVIDIA GPU with 384 cores. The second system used is a server type system at the GPU Center of Excellence (GCoE) at IDP in Systems and Control Laboratory, Indian Institute of Technology, Bombay, India. This system is equipped with one CPU and a cluster of three GPUs.

Table 8.2. NVIDIA Geforce 940M GPU Specifications.

Model	NVIDIA Geforce 940M
Total Graphics Memory	2048 MB
No. of cores	384
Clock Rate	1071 MHz

Table 8.3. Intel i5 CPU Specifications.

Model	Intel i5-5200U
RAM	8 GB
No. of cores	2
Clock Rate	2.2 GHz

Table 8.4. Leopard Cluster GPU Specifications.

Model	Tesla K40 GPU (2 Nos.)	Tesla K20 GPU (1 Nos.)
Total Graphics Memory	12 GB	5 GB
No. of cores	2880	2496
Clock Rate	745 MHz	706 MHz

Table 8.5. IITB CPU Specifications.

Model	Intel(R) Xeon(R)CPU E5-2620
RAM	32 GB
No. of cores	24 cores
Clock Rate	2.00GHz

The programs were executed serially on CPU and in parallel on GPUs. We have used two different GPUs: NVIDIA Geforce 940M and Leopard cluster. The special functions described in section 2 were executed in these CPU and GPU environments.

In Tables **8.6-8.12**, for some values of t, it can be observed that the system gives Out of Memory Error (OME) for NVIDIA 940M GPU. For example, in Table **8.6** for arguments $\alpha = 0.1, \beta = 0.1, \gamma = 0.1$ and $t = 0: 1e^{-6}: 0.3$, OME occurs because the size of vector t is increased to a large value. Because of this increase in size of t, the number of calculations for respective function is increased. The GPU memory of NVIDIA 940M device is insufficient to carry out these calculation on this huge data. This OME limitation is overcomed by using the GPU with larger device memory.

Table 8.6. Results for Mittag-Leffler function of 3-parameters.

Sr.No.	Arguments					NVIDIA 940M			Leopard Cluster		
	α	β	γ	N	t	T_{CPU}(s)	T_{GPU}(s)	Speedup	T_{CPU}(s)	T_{GPU}(s)	Speedup
1	0.1	0.1	0.1	170	$0: 1e^{-6}: 0.1$	2.36	0.44	5.31	0.88	0.034	25.56
					$0: 1e^{-6}: 0.2$	5.34	0.87	6.17	1.75	0.058	29.99
					$0: 1e^{-6}: 0.3$	-	OME*	-	2.62	0.083	31.63
					$0: 1e^{-6}: 1.5$	-	OME*	-	12.32	0.37	33.18
2	0.5	0.5	0.5	170	$0: 1e^{-6}: 0.1$	2.79	0.46	6.07	0.89	0.038	23.26
					$0: 1e^{-6}: 0.2$	5.24	0.87	6.00	1.87	0.06	30.19
					$0: 1e^{-6}: 0.3$	-	OME*	-	2.84	0.08	33.06
					$0: 1e^{-6}: 1.5$	-	OME*	-	12.67	0.38	33.21
3	0.9	0.9	0.9	170	$0: 1e^{-6}: 0.1$	2.76	0.45	6.15	0.92	0.04	25.40
					$0: 1e^{-6}: 0.2$	5.11	0.86	5.97	1.77	0.05	29.58
					$0: 1e^{-6}: 0.3$	-	OME*	-	2.60	0.08	30.87
					$0: 1e^{-6}: 1.5$	-	OME*	-	12.37	0.37	33.32
4	0.5	0.1	0.1	170	$0: 1e^{-6}: 0.1$	2.46	0.45	5.42	0.86	0.035	24.13
					$0: 1e^{-6}: 0.2$	5.22	0.86	6.08	1.70	0.060	28.32
					$0: 1e^{-6}: 0.3$	-	OME*	-	2.55	0.083	30.54
					$0: 1e^{-6}: 1.5$	-	OME*	-	12.75	0.37	34.08

Note : *OME Stands for Out-of Memory error for GPU Device.

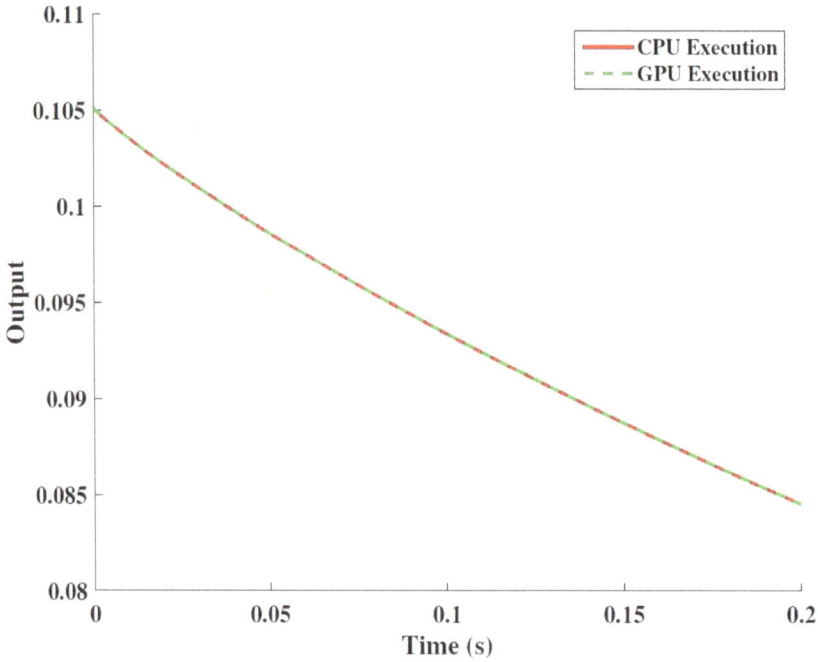

Fig. (8.4). Output benchmarking for Mittag-Leffler function of 3-parameters $\alpha = 0.9$, $\beta = 0.1, \gamma = 0.1, N = 170$.

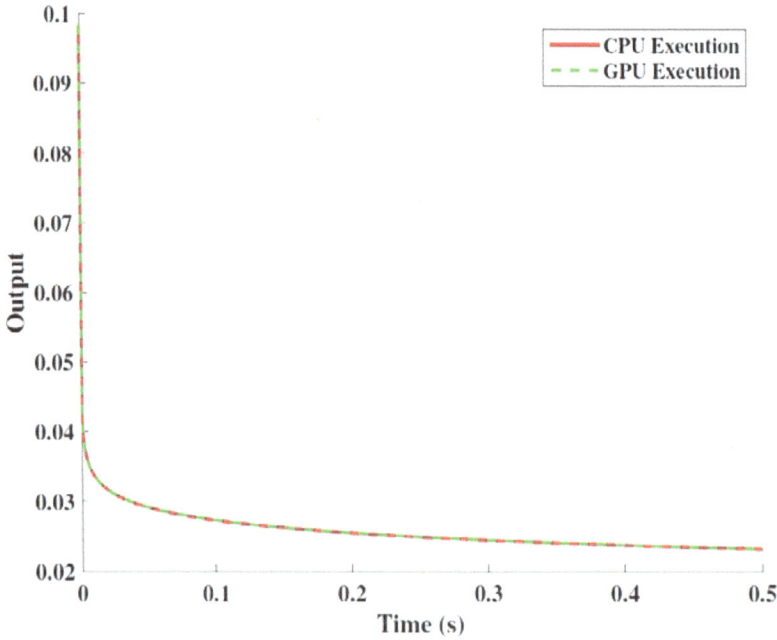

Fig. (8.5). Output benchmarking for Mittag-Leffler function of 4-parameters $a = 0.1$, $b = 0.9, p = 0.1, q = 0.1, N = 200$.

Table 8.7. Results for Mittag-Leffler function of 4-parameters.

Sr. No.	Arguments						NVIDIA 940M			Leopard Cluster		
	α	β	p	q	N	t	T_{CPU}(s)	T_{GPU}(s)	Speedup	T_{CPU}(s)	T_{GPU}(s)	Speedup
1	0.1	0.1	0.1	0.1	200	$0:1e^{-6}:0.5$	11.07	1.35	8.21	2.52	0.067	37.32
						$0:1e^{-6}:0.6$	14.35	1.61	8.89	3.04	0.09	33.56
						$0:1e^{-6}:1.5$	-	OME	-	8.88	0.22	39.29
						$0:1e^{-6}:2$	-	OME	-	11.79	0.29	39.50
2	0.9	0.9	0.9	0.9	200	$0:1e^{-6}:0.5$	12.29	1.36	9.03	2.58	0.07	33.31
						$0:1e^{-6}:0.6$	15.00	1.63	9.18	3.04	0.09	33.10
						$0:1e^{-6}:1.5$	-	OME	-	8.85	0.22	38.90
						$0:1e^{-6}:2$	-	OME	-	11.90	0.30	39.09
3	0.1	0.9	0.1	0.1	200	$0:1e^{-6}:0.5$	12.37	1.34	9.23	2.73	0.078	35.07
						$0:1e^{-6}:0.6$	14.85	1.61	9.24	3.35	0.092	36.22
						$0:1e^{-6}:1.5$	-	OME	-	8.69	0.21	39.59
						$0:1e^{-6}:2$	-	OME	-	11.82	0.29	39.83
4	0.9	0.1	0.1	0.1	200	$0:1e^{-6}:0.5$	12.38	1.36	9.09	3.12	0.07	39.53
						$0:1e^{-6}:0.6$	15.80	1.62	9.75	3.64	0.09	37.25
						$0:1e^{-6}:1.5$	-	OME	-	9.03	0.22	40.62
						$0:1e^{-6}:2$	-	OME	-	12.24	0.32	38.00

Table 8.8. Results for Confluent Hypergeometric Function.

Sr. No.	Arguments				NVIDIA 940M			Leopard Cluster		
	a	b	N	t	T_{CPU}(s)	T_{GPU}(s)	Speedup	T_{CPU}(s)	T_{GPU}(s)	Speedup
1	0.1	0.1	90	$0:1e^{-6}:0.1$	1.10	0.22	4.98	0.35	0.035	9.96
				$0:1e^{-6}:0.5$	5.50	1.01	5.47	1.54	0.08	18.47
				$0:1e^{-6}:1.5$	-	OME	-	4.44	0.21	21.08
				$0:1e^{-6}:2$	-	OME	-	6.03	0.27	21.91
2	0.5	0.5	90	$0:1e^{-6}:0.1$	1 14	0.23	4.84	0.35	0.03	9.75
				$0:1e^{-6}:0.5$	5.57	1.01	5.53	1.54	0.08	18.21
				$0:1e^{-6}:1.5$	-	OME	-	4.48	0.21	21.30
				$0:1e^{-6}:2$	-	OME	-	6.00	0.27	22.06

Table 8.8 contd....

Sr. No.	Arguments				NVIDIA 940M			Leopard Cluster		
	a	b	N	t	T_{CPU}(s)	T_{GPU}(s)	Speedup	T_{CPU}(s)	T_{GPU}(s)	Speedup
3	0.9	0.9	90	$0:1e^{-6}:0.1$	1.1	0.23	4.76	0.34	0.04	8.63
				$0:1e^{-6}:0.5$	5.53	1.22	4.54	1.51	0.08	18.10
				$0:1e^{-6}:1.5$	-	OME	-	4.46	0.21	20.98
				$0:1e^{-6}:2$	-	OME	-	5.97	0.27	21.98
4	0.1	0.9	90	$0:1e^{-6}:0.1$	1.12	0.23	4.91	0.37	0.03	10.11
				$0:1e^{-6}:0.5$	5.67	1.01	5.62	1.57	0.08	18.33
				$0:1e^{-6}:1.5$	-	OME	-	4.68	0.22	21.03
				$0:1e^{-6}:2$	-	OME	-	5.97	0.27	21.78

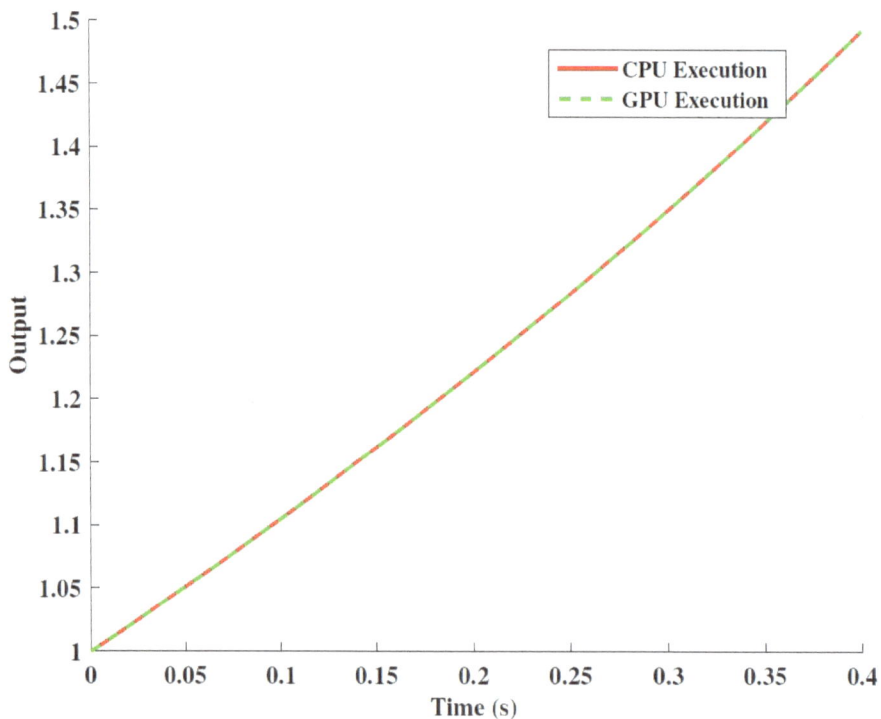

Fig. (8.6). Output benchmarking for Confluent Hypergeometric Function $a = 0.1$, $b = 0.1$, $N = 90$.

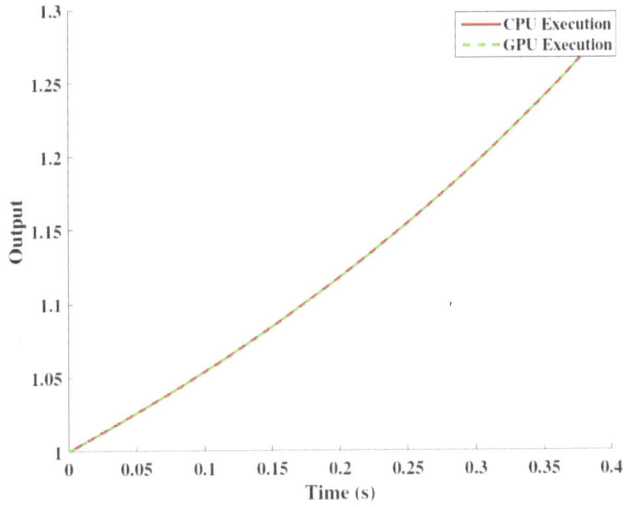

Fig. (8.7). Output benchmarking for Gauss hypergeometric function $a = 0.1$, $b = 0.5$, $c = 0.1$, $N = 90$.

Table 8.9. Results for Gauss hypergeometric function.

Sr. No.	Arguments					NVIDIA 940M			Leopard Cluster		
	a	b	c	N	t	T_{CPU}(s)	T_{GPU}(s)	Speedup	T_{CPU}(s)	T_{GPU}(s)	Speedup
1	0.1	0.1	0.1	90	$0:1e^{-6}:0.1$	1.10	0.22	4.95	0.38	0.03	10.71
					$0:1e^{-6}:0.5$	6.01	1.03	5.86	1.72	0.08	19.66
					$0:1e^{-6}:1.5$	-	OME	-	4.89	0.21	22.25
					$0:1e^{-6}:2$	-	OME	-	6.39	0.28	22.29
2	0.5	0.5	0.5	90	$0:1e^{-6}:0.1$	1.19	0.24	5.06	0.37	0.04	9.25
					$0:1e^{-6}:0.5$	6.19	1.21	5.11	1.69	0.08	19.65
					$0:1e^{-6}:1.5$	-	OME	-	4.96	0.21	23.38
					$0:1e^{-6}:2$	-	OME	-	6.51	0.27	23.69
3	0.9	0.9	0.9	90	$0:1e^{-6}:0.1$	1.20	0.22	4.95	0.39	0.04	9.47
					$0:1e^{-6}:0.5$	6.43	1.04	6.20	1.68	0.08	19.21
					$0:1e^{-6}:1.5$	-	OME	-	4.91	0.21	22.46
					$0:1e^{-6}:2$	-	OME	-	6.44	0.27	23.37
4	0.1	0.9	0.1	90	$0:1e^{-6}:0.1$	1.19	0.23	4.81	0.38	0.04	9.04
					$0:1e^{-6}:0.5$	6.13	1.02	6.03	1.65	0.08	18.95
					$0:1e^{-6}:1.5$	-	OME	-	4.85	0.21	22.93
					$0:1e^{-6}:2$	-	OME	-	6.43	0.27	23.35

Table 8.10. Results for $_2F_2$ Hypergeometric Function.

Sr. No.	Arguments						NVIDIA 940M			Leopard Cluster		
	a	b	p	q	N	t	T_{CPU}(s)	T_{GPU}(s)	Speedup	T_{CPU}(s)	T_{GPU}(s)	Speedup
1	0.1	0.1	0.1	0.1	90	$0:1e^{-6}:0.1$	1.16	0.22	5.16	0.39	0.04	9.41
						$0:1e^{-6}:0.5$	5.75	1.01	5.68	1.70	0.08	19.61
						$0:1e^{-6}:1.5$	-	OME	-	4.86	0.21	22.45
						$0:1e^{-6}:2$	-	OME	-	6.43	0.27	23.36
2	0.5	0.5	0.5	0.5	90	$0:1e^{-6}.1$	1.11	0.22	4.95	0.38	0.03	9.76
						$0:1e^{-6}:0.5$	5.81	1.01	5.76	1.68	0.09	18.53
						$0:1e^{-6}:1.5$	-	OME	-	4.90	0.21	22.49
						$0:1e^{-6}:2$	-	OME	-	6.56	0.27	23.56
3	0.9	0.9	0.9	0.9	90	$0:1e^{-6}:0.1$	1.13	0.23	4.77	0.37	0.03	9.53
						$0:1e^{-6}:0.5$	5.67	1.01	5.61	1.69	0.08	19.23
						$0:1e^{-6}:1.5$	-	OME	-	4.97	0.21	22.98
						$0:1e^{-6}:2$	-	OME	-	6.64	0.27	24.19
4	0.1	0.1	0.9	0.1	90	$0:1e^{-6}:0.1$	1.11	0.24	4.72	0.37	0.04	8.74
						$0:1e^{-6}:0.5$	5.97	1.02	5.84	1.69	0.08	18.94
						$0:1e^{-6}:1.5$	-	OME	-	4.88	0.21	22.57
						$0:1e^{-6}:2$	-	OME	-	6.45	0.27	23.18

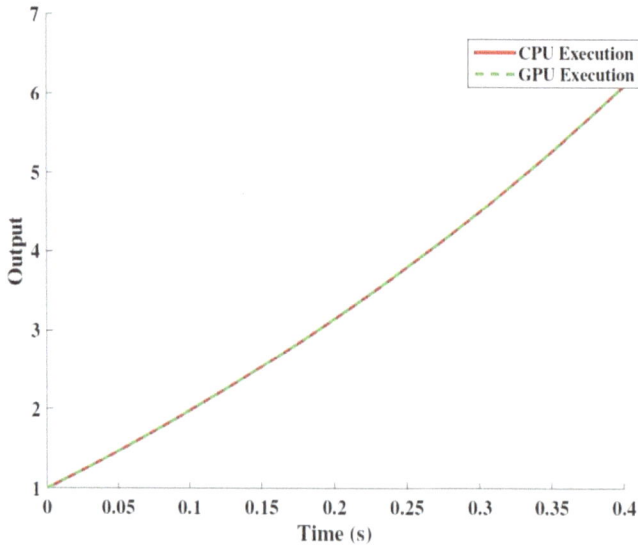

Fig. (8.8). Output benchmarking for $_2F_2$ Hypergeometric Function $a = 0.1$, $b = 0.9$, $c = 0.1$, $d = 0.1$, $N = 90$.

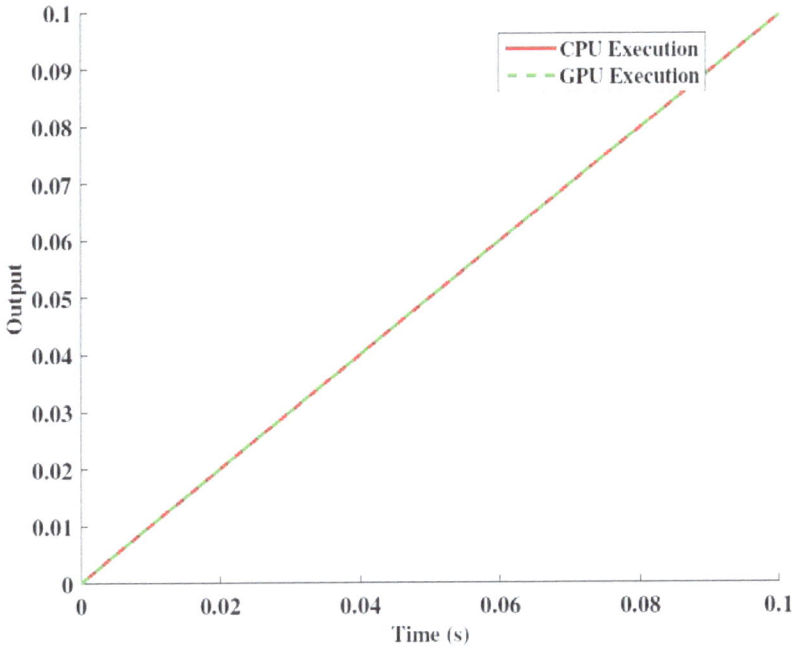

Fig. (8.9). Output benchmarking for Dawson's Function $N = 100$.

Table 8.11. Results for Dawson's Function.

Sr. No.	Arguments		NVIDIA 940M			Leopard Cluster		
	N	x	T_{CPU}(s)	T_{GPU}(s)	Speedup	T_{CPU}(s)	T_{GPU}(s)	Speedup
1	100	$0:1e^{-6}:0.1$	1.37	0.32	4.28	0.46	0.05	8.36
2	100	$0:1e^{-6}:0.2$	2.68	0.62	4.36	0.87	0.09	9.48
3	100	$0:1e^{-6}:0.3$	4.59	0.95	4.81	1.31	0.12	10.20
4	100	$0:1e^{-6}:0.4$	6.10	1.28	4.77	1.84	0.16	11.51
5	100	$0:1e^{-6}:0.5$	-	OME	-	2.15	0.20	10.71
6	100	$0:1e^{-6}:1$	-	OME	-	3.90	0.38	10.04
7	100	$0:1e^{-6}:2$	-	OME	-	7.75	0.75	10.28

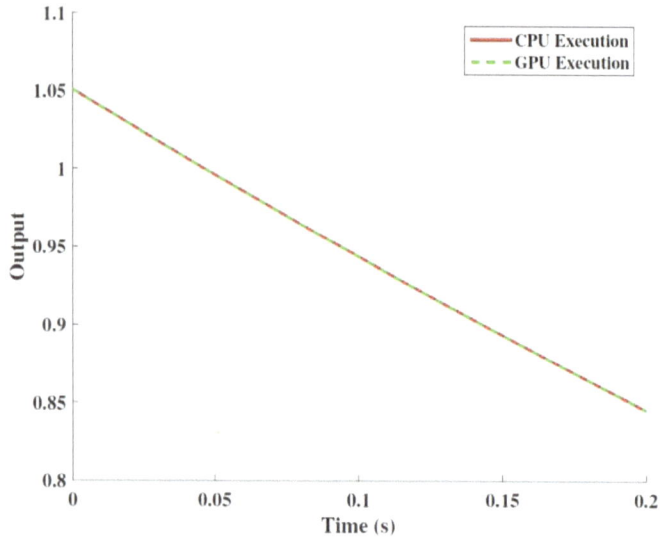

Fig. (8.10). Output benchmarking for Bessel Wright Function $\vartheta = 0.1$, $\mu = 0.5$, $N = 200$.

Table 8.12. Results for Bessel Wright Function.

Sr. No.	Arguments				NVIDIA 940M			Leopard Cluster		
	ϑ	μ	N	t	T_{CPU}(s)	T_{GPU}(s)	Speedup	T_{CPU}(s)	T_{GPU}(s)	Speedup
1	0.1	0.1	200	$0:1e^{-6}:0.1$	2.70	0.40	6.80	1.12	0.03	32.61
				$0:1e^{-6}:0.5$	-	OME	-	5.18	0.11	46.37
				$0:1e^{-6}:1$	-	OME	-	10.41	0.22	46.58
				$0:1e^{-6}:1.5$	-	OME	-	16.36	0.31	51.91
2	0.5	0.5	200	$0:1e^{-6}:0.1$	2.75	0.41	6.67	1.11	0.03	29.78
				$0:1e^{-6}:0.5$	-	OME	-	5.24	0.11	45.00
				$0:1e^{-6}:1$	-	OME	-	10.25	0.20	48.89
				$0:1e^{-6}:1.5$	-	OME	-	16.42	0.31	52.25
3	0.9	0.9	200	$0:1e^{-6}:0.1$	2.63	0.40	6.59	1.12	0.03	31.82
				$0:1e^{-6}:0.5$	-	OME	-	5.35	0.11	46.32
				$0:1e^{-6}:1$	-	OME	-	10.60	0.20	51.34
				$0:1e^{-6}:1.5$	-	OME	-	16.02	0.30	52.15
4	0.1	0.5	200	$0:1e^{-6}:0.1$	2.72	0.38	7.06	1.17	0.04	28.32
				$0:1e^{-6}:0.5$	-	OME	-	5.28	0.11	47.30
				$0:1e^{-6}:1$	-	OME	-	10.35	0.20	49.83
				$0:1e^{-6}:1.5$	-	OME	-	16.36	0.30	53.12

8.5. DISCUSSION

From the results obtained in the previous section, it is clear that the parallel implementation on GPU results in 4x to 50x speedup compared to execution on CPU.

1. The results of Mittag-Leffler function of 3-parameters are presented in Table **8.6**, the upper limit of summation is kept as 170 and execution time is recorded for different sets of parameters $(\alpha, \beta, \gamma, t)$. From the above results it can be concluded that, with change in the parameters $(\alpha, \beta, \gamma, t)$ there is no significant change in speedup but as with increase in the size of t, considerable speedup is obtained. For example, when $\alpha = 0.1, \beta = 0.1, \gamma = 0.1$ and $t = 0: 1e^{-6}: 0.1$ the speedup obtained is 25.56x and for the same values of α, β, γ with $t = 0: 1e^{-6}: 1.5$ speedup obtained is 33.18x. This highlights the fact that GPU is suitable for data intensive application, because it is noticed that speedup increases with decrease in the step size and increase in time vector size. Similarly for the Mittag-Leffler function of 4-parameters, from Table **8.7** it can be concluded that speedup varies from 30x to 40x as the size of time vector increases. For example the second case in Table **8.7** for parameters $\alpha = \beta = p = q = 0.9$, the upper limit of summation is 200 and $t = 0: 1e^{-6}: 0.5$ the execution is 33.31 times faster. For the same parameters by increasing the step size $t = 0: 1e^{-6}: 2$ speedup is 39.09x.

2. From Tables **8.8-8.10** it is clear that for various types of hypergeometric functions speedup is in the range of 9x to 24x. As in for the case of gauss hypergeometric function for the set of parameters $a = b = c = 0.5$ with upper limit of summation 90 and step size $t = 0: 1e^{-6}: 2$ speedup is 23.69x while execution on Leopard cluster. But for the same set of parameters when execution is carried out on NVIDIA 940M device Out of Memory error occurs. This is because the device memory is not sufficient for handling the huge amount of calculations for $t = 0: 1e^{-6}: 2$.

3. Bessel Wright function is given in Table **8.12**. The value of k is kept constant at 200. When different sets of arguments values ϑ and μ are used, it is noted that effect on execution time or speedup is not significant. But when different values of time (t) vector were used, it resulted in different

values of speedup. For first set of arguments with $t = 0: 1e^{-6}: 0.1$ the speedup is 32.61x and with $t = 0: 1e^{-6}: 1.5$ it is 51.91x.

4. Table **8.11** shows the results for Dawson's Function. In this case the upper limit of summation is 100 and the time vector is varied from $x = 0: 1e^{-6}: 0.1$ to $x = 0: 1e^{-6}: 2$. Upto $x = 0: 1e^{-6}: 0.4$ execution on NVIDIA 940M device dose not give OME but for the sets of parameters after that OME occurs. The value of speedup is in the range of 8x to 11x. It is observed that with increase in size of time vector, speedup is not increasing significantly as the number of calculations involved is less as compared to other functions.

Similarly, for all the functions it can be interpreted that with k as constant the speedup is dependent on value of t. As the number of elements in the time vector t increases speedup also increases. Hence, it can be stated that the speedup increases when the amount of data or number of iterations of an operation to be performed on that data increases. All the results are executed on two different GPUs and as per the specifications of the GPU, more speedup is obtained for Lepord Clusture GPU. By observing the results of different functions it can be concluded that as the computation in the function increases the speedup also increases. For NVIDIA 940M GPU, because of the limited GPU memory, out of memory error occurs and this problem gets resolved by using Leopard cluster GPU which has high GPU memory. Speedup mentioned here can be increased by optimizing the MATLAB codes and the use of more powerful platforms like CUDA. For all the functions, the output benchmarking plots are obtained as shown in Figs. (**8.4-8.10**). By observing these plots, it is concluded that the output of CPU and GPU execution exactly overlaps with each other for given set of parameters.

8.6. CONCLUSION

In this work, the acceleration of computation for different special mathematical functions used in fractional calculus using Graphical processing unit (GPU) is carried out. This acceleration is achieved using multiple processing cores of graphics processor. Special codes for these functions were developed. Then using Parallel Computing Toolbox of MATLAB vectorized version of these codes are developed. The vectorization is performed for the section of sequential code which consumes majority of execution time. Thus we can observe that the special functions of fractional calculus can be successfully implemented on GPU. This allows faster execution of these functions. The speedups obtained are in the range of 4x to 10x for systems with NVIDIA 940M GPU and 10x to 50x for leopard cluster GPU. From the results it may be concluded that speedup depends on the amount of data and number of iterations of an operation. To extend this work further, the MATLAB codes can be optimized and more powerful platforms like NVIDIA Compute Unified Device Architecture (CUDA) can be utilized.

CONSENT FOR PUBLICATION

Not applicable.

CONFLICT OF INTEREST

The authors declare no conflict of interest, financial or otherwise.

ACKNOWLEDGEMENTS

Declared none.

APPENDIX

Sample MATLAB Codes

I. MATLAB code for 1-parameter Mittag Leffler Function

Main Program:

```
1.  clc
2.  clear all
3.  close all
4.
5.  %% Input Parameters
6.  alpha_in=0.9;
7.  t_in = 0:1e-6:0.6;
8.  bta_in=1;
9.  N_in=200;
```

10.

11. %% Function Calls

12. tic;

13. E4_f=mlf_vectorized_cpu_func(alpha_in,bta_in,t_in,N_in);

14. time_mlf_vectorized_cpu=toc;

15.

16. tic;

17. E5_f=mlf_vectorized_gpu_func(alpha_in,bta_in,t_in,N_in);

18. time_mlf_vectorized_gpu=toc;

19.

20. %% Speedup Calculation

21. speedup=time_mlf_vectorized_cpu/time_mlf_vectorized_gpu;

22.

23. %% Output Benchmarking Plot

24. plot(t_in,E4_f,'r',t_in,E5_f,'g–','LineWidth',0.5);

25. h=legend('Output of CPU Execution','Output of GPU Execution');

26. set(h,'FontSize',12);

27. grid on;

28. xlabel('Time (s)','fontsize',12);

29. ylabel('Output','fontsize',12);

Function for execution on CPU:

1.   function [E4_f] = mlf_vectorized_cpu_func(alpha_arg,bta_arg,t_arg,N_arg)

2.

3.   alpha = alpha_arg;

4.   bta = bta_arg;

5.   t = t_arg;

6.   N = N_arg;

7.

8.   %% Initialization

9.   E4= zeros(length(t),(N+1));

10. E4_f=zeros(1,length(t));

11. arr1= zeros(length(t),(N+1));

12. arr2= zeros(length(t),(N+1));

13. z = 1:length(t);

14.

15. %% Calculate Numerator

16. arr1a= (-t(z).^alpha);

17. arr1b=[0:200];

18. arr1=[bsxfun(@power, arr1a, arr1b')]';

19.

20. %% Calculate Denominator

21. arr2 = gamma(ones(1,N+1) + alpha*[0:N]);

22.

23. %% Perform Division

24. E4=[bsxfun(@rdivide, arr1, arr2)];

25.

26. %% Perform Summation from k=0 to N

27. E4_f=sum(E4,2)'; % Adds all the elements of rows of 2d matrix

Function for execution on GPU:

1.

2. function [E5_f] = mlf_vectorized_gpu_func(alpha_arg,bta_arg,t_arg,N_arg)

3.

4. alpha = alpha_arg;

5. bta = bta_arg;

6. t = t_arg;

7. N = N_arg;

8.

9. %Initialization

10. E5= zeros(length(t),(N+1));

11. E5_f=zeros(1,length(t));

12. arr1= zeros(length(t),(N+1));

13. arr2= zeros(length(t),(N+1));

14. z = 1:length(t);

15.

16. %Calculate arr1

17. arr1a=gpuArray((-t(z).^alpha));

18. arr1b=gpuArray([0:200]);

19. arr1=[bsxfun(@power, arr1a, arr1b')]';

20.

21. %Calculate arr2

22. arr2 =gpuArray(gamma(ones(1,N+1) + alpha*[0:N]));

23.

24. %Perform Division

25. E5=[bsxfun(@rdivide, arr1, arr2)];

26.

27. %Perform Summation from k=0 to N

28. E5_f=sum(E5,2)'; % Adds all the elements of rows of 2d matrix

29. E5_f=gather(E5_f);

II. MATLAB code for Dawson's Function

Main Program:

1. clc

2. clear all

3. close all

4.

5. x_in = 0:1e-6:0.1;

6. n_in =0:1:100;

7.

8. tic;

9. op_cpu = dawson_vectorized_cpu_func(x_in,n_in);

10. time_2F2_vectorized_cpu=toc;

11.

12. tic;

13. op_gpu = dawson_vectorized_gpu_func(x_in,n_in);

14. time_2F2_vectorized_gpu=toc;

15.

16. speedup=time_2F2_vectorized_cpu/time_2F2_vectorized_gpu;

17.

18. figure,

19. plot(x_in,op_cpu,'r',x_in,op_gpu,'g--','LineWidth',0.5);

20. h=legend('Output of CPU Execution','Output of GPU Execution');

21. set(h,'FontSize',12);

22. grid on;

23.

24. xlabel('Time (s)','fontsize',12);

25. ylabel('Output','fontsize',12);

Function for execution on CPU:

1. function [vectorized_op_gpu] = dawson_vectorized_gpu_func(x_arg,N_arg)

2. x=gpuArray(x_arg);

3. n=gpuArray(N_arg);

4. numeratr=gpuArray(zeros(length(n),length(x)));

5.

6. %% Calculation of Numerator

7. temp= (-2.*(x.^2));

8.   temp_n= [[ones(1,length(n))]'* temp];

9.   n_n= [[ones(1,length(x))]'* n]';

10. numeratr=temp_n.^n_n;

11.

12. %% Calculation of Denominator

13. denomintr= [factorial(2.*n+1)] ./ [(2.^n) .* factorial(n)];

14. denomintr_n= [[ones(1,length(x))]'* denomintr]';

15.

16. %% Calculate term2

17. term2=numeratr./denomintr_n;

18. sum_term2=sum(term2,1);

19.

20. %% Final Output

21.

22. vectorized_op_gpu=gather(x .* sum_term2);

23. end

Function for execution on GPU:

1.   function [vectorized_op_cpu] = dawson_vectorized_cpu_func(x_arg,N_arg)

2.

3.   x=x_arg;

4.   n=N_arg;

5.   numeratr=zeros(length(n),length(x));

6.

7. %% Calculation of Numerator

8. temp= (-2.*(x.^2));

9. temp_n= [[ones(1,length(n))]'* temp]; % Convert temp to 2D from 1D.

10. n_n= [[ones(1,length(x))]'* n]'; %Convert n to 2D from 1D.

11. numeratr=temp_n.^n_n;

12.

13. %% Calculation of Denominator

14. denomintr= [factorial(2.*n+1)] ./ [(2.^n) .* factorial(n)];

15. denomintr_n= [[ones(1,length(x))]'* denomintr]';

16.

17. %% Calculate term2

18. term2=numeratr./denomintr_n;

19. sum_term2=sum(term2,1);

20.

21. %% Final Output

22. vectorized_op_cpu=x .* sum_term2;

23. end

REFERENCES

[1]   V. Kiryakova, "The special functions of fractional calculus as generalized fractional calculus operators of some basic functions," *Comput. Math. Appl.*, vol. 59, no. 3, pp. 1128–1141, Feb. 2010.

[2]   A. Gil, J. Segura, and N.M. Temme, *Numerical Methods for Special Functions*. Society for Industrial and Applied Mathematics: Philadelphia, USA, Jan. 2007.

[3]   R.L. Magin, "Fractional calculus in bioengineering," In: 13th International Carpathian Control Conference (ICCC), 2012.

[4]   S. Rogosin, and H.M. Srivastava, "The role of the Mittag-Leffler function in fractional modeling," *Mathematics*, vol. 3, no. 2, pp. 368–381, May 2015.

[5]   C.A. Monje, Y. Chen, B.M. Vinagre, D. Xue, V. Feliu-Batlle, *Fractional-Order Systems and Controls: Fundamentals and Applications*. Springer Verlag: London, 2010.

[6]   H. Buchholz, *The Confluent Hypergeometric Function: With Special Emphasis on its Applications*. Springer Science & Business Media: London, 2013.

[7]   J.B. Seaborn, *Hypergeometric Functions and Their Applications*. Springer Science & Business Media: London, 2013.

[8]   M. Abramowitz, and I.A. Stegun, *Handbook of Mathematical Functions: With Formulas, Graphs, and Mathematical Tables*. Courier Corporation, 1964.

[9]   D.W. Lozier, "NIST digital library of mathematical functions," *Ann. Math. Artif. Intell.*, vol. 38, no. 1, pp. 105–119, May 2003.

[10]   P. Sebah, and X. Gourdon, "Introduction to the gamma function," 2002. [Online]: Available: http://www.csie.ntu.edu.tw/ b89089/link/gammaFunction.pdf

[11]   W. Wang, "Some properties of k-gamma and k-beta functions," *ITM Web of Conferences*, vol. 7, pp. 07003, EDP Sciences, 2016.

[12]   D.W. Lozier, and F.W.J. Olver, "Numerical evaluation of special functions," *AMS Proceedings of Symposia in Applied Mathematics*, vol. 48, pp. 79-125, 1994.

[13]   R. Gorenflo, A. Kilbas, F. Mainardi, and S. Rogosin, *Mittag-Leffler Functions, Related Topics and Applications*. Springer: Berlin, 2014.

[14]   F. Mainardi, G. Rudolf, "On Mittag-Leffler-type functions in fractional evolution processes," *J. Comput. Appl. Math.*, vol. 118, pp. 283–299, Jun. 2000.

[15]   R. Garrappa, "Numerical evaluation of two and three parameters Mittag-Leffler functions," *SIAM J. Numer. Anal.*, vol. 53, pp. 1350–1369, May 2015.

[16]   S. Rogosin, "The role of the Mittag-Leffler function in fractional modeling," *Mathematics*, vol. 3, pp. 368–381, May 2015.

[17]   H.J. Seybold, and R. Hilfer, "Numerical results for the generalized Mittag-Leffler function," *Fract. Calc. Appl. Anal.*, vol. 8, pp. 127–139, 2005.

[18]   J.C. Prajapati and A.K. Shukla, "Decomposition of generalized Mittag-Leffler function and its properties," *Adv. Pure Math.*, vol. 2, no. 1, pp. 8–14, Jan. 2012.

[19]   A.K. Shukla and J.C. Prajapati, "On a generalization of Mittag-leffler function and its properties," *J. Math. Anal. Appl.*, vol. 336, no. 2, pp. 797–811, Dec. 2007.

[20]   M. Garg, A. Sharma and P. Manohar, "A generalized Mittag-Leffler type function with four parameters," *Thai J. Math.*, vol. 14, no. 3, pp. 637–649, Nov. 2016.

[21]   Y. Li, Y. Chen, and I. Podlubny, "Mittag-Leffler stability of fractional order nonlinear dynamic systems," *Automatica*, vol. 45, no. 8, pp. 1965–1969, Aug. 2009.

[22]   R. Garra, and R. Garrappa, "The Prabhakar or three parameter Mittag-Leffler function: theory and application," *Commun. Nonlinear Sci. Numer. Simul.*, vol. 56, pp. 314–329, Mar. 2018.

[23]   M.D. Ortigueira, A.M. Lopes, R. Garrappa, and J.A. Tenreiro, "On the computation of the multidimensional Mittag-Leffler function," *Commun. Nonlinear Sci. Numer. Simul.*, vol. 53, pp. 278–287, Dec. 2017.

[24]   F. Mainardi, *Fractional Calculus and Waves in Linear Viscoelasticity*. Imperial College Press and World Scientific Co. Pte. Ltd.: Singapore, 2010.

[25]   E.W. Weisstein, "Euler's Hypergeometric Transformations," MathWorld-A Wolfram Web Resource. [Online]. Available: http://mathworld.wolfram.com/HypergeometricFunction.html [Accessed: 24th May 2017].

[26]   M. Nardin, W.F. Perger, and A. Bhalla, "Numerical evaluation of the confluent hypergeometric function for complex arguments of large magnitudes," *J. Comput. Appl. Math.*, Elsevier, vol. 39, no. 2, pp. 193-200, 1992.

[27]   J.W. Pearson, S. Olver, and M.A. Porter, "Numerical Methods for the computation of the confluent and Gauss hypergeometric functions," *Numer. Algo.*, vol. 74, no. 3, pp. 821–866, Mar. 2017.

[28]   A. Deano, and N.M. Temme, "On modified asymptotic series involving confluent hypergeometric functions,"*Electron, Trans. Numer. Anal.*, vol. 35, pp. 88–103, 2009.

[29]   A.D. MacDonald, "Properties of the confluent hypergeometric function," *Stud. Appl. Math.*, vol. 28, no. 1–4, pp. 183–191, Apr. 1949.

[30]   K.E. Muller, "Computing the confluent hypergeometric function, M (a,b,x)," *Numer. Math.*, vol. 90, no. 1, pp. 179–196, Nov. 2001.

[31]   R. Daz, and E. Pariguan, "On hypergeometric functions and pochhammer k-symbol," *Divulgaciones Matemtcas*, vol. 15, no. 2, pp. 179–192, 2007.

[32]   J. Doornik, "Numerical evaluation of the Gauss hypergeometric function by power summations," *Math. Comput.*, vol. 84, no. 294, pp. 1813–1833, Jul. 2015.

[33]   J.W. Pearson, S. Olver, and M.A. Porter, "Numerical methods for the computation of the confluent and Gauss hypergeometric functions," *Numer. Algo.*, Springer, vol. 74, no. 3, pp. 821-866, 2017.

[34]   J.W. Pearson, *"Computation of hypergeometric functions,"* PhD thesis, University of Oxford, UK, 2009.

[35]   Digital Library of Mathematical Functions, National Institute of Standards and Technology [Online]. Available: http://dlmf.nist.gov [Accessed: 6th June 2017].

[36]   B. Zhang, S. Xu, F. Zhang, Y. Bi, and L. Huang , "Accelerating MATLAB code using GPU: A review of tools and strategies", In: *Artificial Intelligence, Management Science and Electronic Commerce(AIMSEC): 2nd International Conference,* Dengleng, China, 2011, pp. 1875–1878.

[37]   "MATLAB Parallel Computing Toolbox Users Guide," Ver. 6.7, The MathWorks, Inc.,Natick, Massachusetts, United States, Sep. 2015.

[38]   Nvidia, CUDA C Programming Guide , Ver. 7.5, Sep. 2015. [Online]. Available: http://docs.nvidia.com [Accessed: 6th June 2017].

[39]   Jacket- The GPU Acceleration Engine for MATLAB [Online]. Available: http://www.omatrix.com/jacket.html [Accessed 8th June 2017].

[40]   P. Messmer, P.J. Mullowney, and B. Granger, "GPULib: GPU computing in high-level languages," *Comput. Sci. Eng.*, vol. 10, no. 5, pp. 70–73, Sep. 2008.

[41]   "Best Practices for MATLAB GPU Coding," *Cornell University Center for Advanced Computing.* [Online]. Avalilable: https://www.cac.cornell.edu/matlab [Accessed 8th June 2017].

[42]   "MATLAB Documenatation 2017," *The MathWorks, Inc.*, Natick, Massachusetts, United States. [Online]. Available: https://in.mathworks.com/help/matlab/ [Accessed 10th June 2017].

[43]   "Predicting and Measuring Parallel Performance," February 1, 2012. [Online]. Available: https://software.intel.com [Accessed 10th June 2017].

[44]   O. Muller, A. Baghdadi, and M. Jézéquel, "Parallelism efficiency in convolutional turbo decoding," *EURASIP J. Adv. Signal Process.*, vol. 2010, no. 1, pp. 927–920, Dec. 1, 2010.

[45]   "Parallel Computing," *Wolfram Mathworld.* [Online]. Available: http://mathworld.wolfram.com/ParallelComputing.html [Accessed 5th June 2017].

[46]   V.E. Tarasov, *Fractional Dynamics: Applications of Fractional Calculus to Dynamics of Particles, Fields and Media*. Springer Science & Business Media: USA, 2010.

[47]   D. Baleanu, K. Diethelm, E. Scalas, and J.J. Trujillo, *Fractional Calculus: Models and Numerical Methods*, World Scientific Publishing: Singapore, 2016.

[48]   K.B. Oldham, and J. Spanier, *The Fractional Calculus: Theory and Applications of Differentiation and Integration to Arbitrary Order*. Dover Publications: Mineola, NY, 2006.

[49]   A.A. Kilbas, H.M. Srivastava, and J.J. Trujillo, *Theory and Applications of Fractional Differential Equations*. Elsevier: Netherlands, 2006.

[50]   J. Liu, C. Gong, W. Bao, G. Tang, and Y. Jiang, "Solving the Caputo fractional reaction-diffusion equation on GPU," *Discrete Dyn. Nat. Soc.*, vol. 2014, Jun. 2014.

[51]   N.E. Banks, *"Insights from the parallel implementation of efficient algorithms for the fractional calculus,"* PhD thesis, University of Chester, United Kingdom, 2015.

[52]   S. Wei, W. Chen, and Y. Hon, "Implicit local radial basis function method for solving two-dimensional time fractional diffusion equations," *Therm. Sci.*, vol. 19, no. 1, pp. 59–67, 2015.

[53]   Z. Fu, W. Chen, and H. Yang, "Boundary particle method for Laplace transformed time fractional diffusion equations," *J. Comput. Phys.*, vol. 235, pp. 52–66, Feb. 2013.

[54]   Z. Fu, W. Chen, and L. Ling, "Method of approximate particular solutions for constant-and variable-order fractional diffusion models," *Engineering Analysis with Boundary Elements*, vol. 57, pp. 37–46, Aug. 2015.

[55]   G. Pang, W. Chen, and Z. Fu, "Space-fractional advection–dispersion equations by the Kansa method," *J. Comput. Phys.*, vol. 293, pp. 280–296, Jul. 2015.

New Iterative Method: A Review

Varsha Daftardar-Gejji[1,*] and Manoj Kumar[1,2]

[1]Department of Mathematics, Savitribai Phule Pune University, Pune - 411007, India;
[2]National Defence Academy, Khadakwasala Pune- 411023, India

Abstract: In this article we provide a comprehensive review of new iterative method (NIM) proposed by Daftardar-Gejji and Jafari [1]. This method is useful for solving functional equation of the form $u = f + N(u)$, which encompasses a wide class of nonlinear equations. This method has received considerable attention and has been successfully employed to solve numerous ordinary, partial, integral and fractional differential equations.

Keywords: New iterative method, convergence, algebraic equations, partial differential equations, fractional differential equations, integral equations.

9.1. INTRODUCTION

In 1980, G. Adomian proposed a decomposition method, known as Adomian Decomposition Method (ADM) to solve the functional equation

$$u = f + N(u), \tag{9.1}$$

which encompasses a wide class of equations viz integral, differential, delay differential, integro-differential, partial differential, fractional differential equations [2-6]. This method gives a solution in terms of rapidly convergent series. First few terms of the series give approximate solution to the problem. There are certain advantages of ADM as it works without linearization, perturbation or discretisation [7]. One of the limitations of this method is that it involves tedious calculations of Adomians polynomials.

In pursuance to this, Daftardar-Gejji and Jafari in 2006 [1], proposed a new decomposition method, termed as New Iterative method (NIM). The algorithm of

*Corresponding author Varsha Daftardar-Gejji: Department of Mathematics, Savitribai Phule Pune University, Pune - 411007, India; Tel/Fax: 91-25601356/57; E-mails: vsgejji@unipune.ac.in, vsgejji@gmail.com

NIM is simple, easy to understand and implement. There are many cases in which we obtain the exact solutions, or else two-three terms are enough to get good approximate solutions. NIM is economical in terms of computer power/memory and does not involve tedious calculations such as Adomian polynomials in case of ADM [7, 8] or construction of homotopy in Homotopy perturbation method (HPM) [9, 10], or obtaining Lagrange's multipliers in variational iteration method (VIM) [11]. NIM can be easily employed using computer algebra packages such Mathematica, Maple, Matlab and requires much less computational work as compared to ADM, HPM or VIM. This method has been employed for solving various linear/ non-linear ordinary, integral, partial differential equations of integer and fractional order. In this article we provide comprehensive review of NIM and its applications.

This article is organized as follows. NIM is described briefly in section 9.2. NIM and its relation to Taylor series expansion is discussed in section 9.3. In section 9.4, applications of NIM existing in the literature are given. NIM's applicability to various problems such as solution of nonlinear algebraic equations, initial value problems, boundary value problems, integro-differential equations is described in sections 9.5, 9.6, 9.7 and 9.8 respectively. In section 9.9, various methods associated with NIM such as Three step iterative method, Iterative Laplace Transform method, New predictor-corrector method and new numerical methods for ordinary differential equations are briefly discussed. Finally the conclusions are drawn.

9.2. NEW ITERATIVE METHOD

To illustrate the basic idea of the NIM, we consider the following general functional equation [1]:

$$u = f + L(u) + N(u), \tag{9.2}$$

where N is a nonlinear operator, L a linear operator and f a known function. We are looking for a solution u of (9.2) having the series form

$$u = \sum_{i=0}^{\infty} u_i. \tag{9.3}$$

Since L is a linear operator, therefore

$$L(\sum_{i=0}^{\infty} u_i) = \sum_{i=0}^{\infty} L(u_i). \tag{9.4}$$

The non-linear operator N can be decomposed as

$$N\left(\sum_{i=0}^{\infty} u_i\right) = N(u_0) + \sum_{i=1}^{\infty}\left(N\left(\sum_{j=0}^{i} u_j\right) - N\left(\sum_{j=0}^{i-1} u_j\right)\right). \tag{9.5}$$

Now using the above equations (9.3), (9.4) and (9.5) in (9.2)

$$\sum_{i=0}^{\infty} u_i = f + \sum_{i=0}^{\infty} L(u_i) + N(u_0) + \sum_{i=1}^{\infty}\left(N\left(\sum_{j=0}^{i} u_j\right) - N\left(\sum_{j=0}^{i-1} u_j\right)\right). \tag{9.6}$$

Define the recurrence relation in the following way:

$$u_0 = f,$$

$$u_1 = L(u_0) + N(u_0),$$

$$u_2 = L(u_1) + N(u_0 + u_1) - N(u_0),$$

$$u_{n+1} = L(u_n) + N(u_0 + u_1 + \ldots + u_n) - N(u_0 + u_1 + \ldots + u_{n-1}),$$

$$n = 1,2,\ldots \tag{9.7}$$

Then,

$$(u_1 + u_2 + \ldots + u_{n+1}) = L(u_0 + u_1 + \ldots + u_n) + N(u_0 + u_1 + \ldots + u_n), n = 1,2,\ldots \tag{9.8}$$

and

$$\sum_{i=0}^{\infty} u_i = f + \sum_{i=0}^{\infty} L(u_i) + N\left(\sum_{j=0}^{\infty} u_j\right). \tag{9.9}$$

The k-term approximate solution of (9.2) is given by $u \approx \sum_{i=0}^{k-1} u_i$.

9.3. TAYLOR SERIES AND NIM

Define

$$G_1 = N(u_0 + u_1) - N(u_0)$$

$$= N(u_0) + N'(u_0)u_1 + N''(u_0)\frac{u_1^2}{2!} + \cdots - N(u_0) \tag{9.10}$$

$$= \sum_{k=1}^{\infty} N^{(k)}(u_0)\frac{u_1^k}{k!},$$

$$G_2 = N(u_0 + u_1 + u_2) - N(u_0 + u_1) = N'(u_0 + u_1)u_2 + N''(u_0 + u_1)\frac{u_2^2}{2!}\cdots$$

$$= \sum_{j=1}^{\infty} \left[\sum_{i=0}^{\infty} N^{(i+j)}(u_0)\frac{u_1^i}{i!} \right] \frac{u_2^j}{j!}, \tag{9.11}$$

$$G_3 = \sum_{i_3=1}^{\infty} \sum_{i_2=0}^{\infty} \sum_{i_1=0}^{\infty} N^{(i_1+i_2+i_3)}(u_0)\frac{u_3^{i_3}}{i_3!}\frac{u_2^{i_2}}{i_2!}\frac{u_1^{i_1}}{i_1!} \tag{9.12}$$

In general,

$$G_n = \sum_{i_n=1}^{\infty} \sum_{i_{n-1}=0}^{\infty} \cdots \sum_{i_1=0}^{\infty} \left[N^{(\sum_{k=1}^{n} i_k)}(u_0)(\prod_{j=1}^{n} \frac{u_j^{i_j}}{i_j!}) \right]. \tag{9.13}$$

Hence,

$$N(u) = G_0 + G_1 + G_2 + G_3 + \cdots$$

$$= N(u_0) + \sum_{k=1}^{\infty} N^{(k)}(u_0)\frac{u_1^k}{k!} + \sum_{j=1}^{\infty} \left[\sum_{i=0}^{\infty} N^{(i+j)}(u_0)\frac{u_1^i}{i!} \right] \frac{u_2^j}{j!}$$

$$+ \sum_{i_3=1}^{\infty} \sum_{i_2=0}^{\infty} \sum_{i_1=0}^{\infty} N^{(i_1+i_2+i_3)}(u_0)\frac{u_3^{i_3}}{i_3!}\frac{u_2^{i_2}}{i_2!}\frac{u_1^{i_1}}{i_1!}\cdots$$

$$= N(u_0) + N'(u_0)[u_1 + u_2 + u_3 + \cdots]$$

$$+ N''(u_0)\left[\frac{u_1^2}{2!} + (u_1 u_2 + \frac{u_2^2}{2!}) + (u_3 u_2 + u_3 u_1 + \frac{u_3^2}{2!}) + \cdots\right]$$

$$+ N^{(3)}(u_0)\left[\frac{u_1^3}{3!} + (u_2\frac{u_1^2}{2!} + u_1\frac{u_2^2}{2!} + \frac{u_2^3}{3!}) + \cdots\right] + \cdots. \tag{9.14}$$

$$N(u) = N(u_0) + [u_1 + u_2 + u_3 + \cdots]N'(u_0)$$

$$+ [u_1 + u_2 + u_3 + \cdots]^2\frac{N''(u_0)}{2!}$$

$$+ [u_1 + u_2 + u_3 + \cdots]^3\frac{N^{(3)}(u_0)}{3!}\cdots.$$

$$= N(u_0) + (u - u_0)N'(u_0) + (u - u_0)^2\frac{N^{(2)}(u_0)}{2!}$$

$$+ (u - u_0)^3\frac{N^{(3)}(u_0)}{3!} + \cdots. \tag{9.15}$$

Eq. (9.15) is a Taylor series of $N(u)$ arround u_0.

9.3.1. Convergence of New Iterative Method

Here we discuss the criteria for the convergence of NIM [12].

Theorem 9.1. *If N is a analytic in a neighborhood of u_0 and $||N^n(u_0)|| = Sup\{N^n(u_0)(h_1, h_2, \cdots, h_n): ||h_i|| \leq 1, 1 \leq i \leq n\} \leq L$, for any n and for some real $L > 0$ and $||u_i|| \leq M < \frac{1}{e}, i = 1, 2, \cdots$, then the series $\sum_{n=0}^{\infty} G_n$ is absolutely convergent and moreover,*

$$||G_n|| \leq LM^n e^{n-1}(e-1), n = 1, 2, \cdots.$$

   Proof. In view of (9.13) we get,

$$||G_n|| \leq LM^n \sum_{i_n=1}^{\infty} \sum_{i_{n-1}=0}^{\infty} \cdots \sum_{i_1=0}^{\infty} \left(\prod_{j=1}^{n} \frac{1}{i_j!}\right) = LM^n e^{n-1}(e-1). \quad \textbf{(9.16)}$$

Thus, the series $\sum_{n=1}^{\infty} ||G_n||$ is dominated by the following convergent series

$$LM(e-1) \sum_{n=1}^{\infty} (Me)^{n-1},$$

where $M < 1/e$. Hence, using comparison test $\sum_{n=0}^{\infty} G_n$ is absolutely convergent.

Note: For the boundedness of $||u_i||, \forall i$ a more useful result is proved in the following theorem.

Theorem 9.2. *Is N is analytic and $||N^{(n)}|| \leq M \leq e^{-1}, \forall n$, then the series $\sum_{n=0}^{\infty} G_n$ is absolutely convergent.*

9.4. APPLICATIONS OF NEW ITERATIVE METHOD

Most physical phenomena are modelled by non-linear equations. For deep insights and physical interpretations, one needs their exact solutions, which is not an easy task. So, one has to rely on numerical methods which give best approximate solution. NIM gives solutions to a wide class of nonlinear equations in terms of rapidly convergent series. Because of its simplicity and correctness, this method has been widely used by many researchers to solve linear/ nonlinear differential equations of integer and fractional orders.

   Daftardar-Gejji and Bhalekar have applied NIM for solving partial differential equations [13], evolution equations [14], and fractional boundary value problems

[15, 16]. Usman *et al.* [17] have employed it for several equations such as telegraph equation, Helmoltz equation, coupled Burger equations. In addition, they have solved system of coupled PDEs and conclude that NIM is more powerful and effective than other existing methods. Loghmani *et al.* [18] have applied it for solving nonlinear sequential fractional differential equations. A new mathematical model proposed by Srivastava and Rai [19] for oxygen delivery through a capillary to tissues, using multi-term fractional diffusion equation, has been solved using NIM. Saeed and Aziz [20] have derived a new two-step and three-step iterative methods using NIM for solving nonlinear algebraic equations. Yun [21] also proposed a new three-step iterative method for solving nonlinear equations, which was a significant improvement of the method proposed by Noor and Noor [22]. They showed that this iterative method has fourth-order convergence.

NIM also has been employed to solve various models. Ghori *et al.* [23] have used it for predictive microbial growth model. M. A. AL-Jawary [24] has applied NIM for epidemic model and the prey and predator models. Further, he has used it for Cauchy problems [25], linear and non-linear Volterra integro-differential equations and systems of linear and non-linear Volterra integro-differential equations [26]. Hemeda has used NIM for solving integro-differential equations [27], fractional partial differential equations in Fluid Mechanics [28], fuzzy integro-differential equations [29], fractional differential equations [30] and gas dynamic equation [31].

NIM has been used extensively for many other equations such as biharmonic equations [32], linear and non-linear diffusion equations [33], singular boundary value problems [34], fifth-order boundary value problems [35], time-fractional Schrödinger equations [36], Fornberg-Whitham equation [37], Telegraph equation [38] and fractional Davey-Stewartson equations [39], the Newell-Whitehead-Segel equation [40]. The numerical study of the nonlinear Baranayi and Robert's model [23] which characterizes the microbial growth in static as well as dynamic environmental conditions, has been done using NIM. NIM is also used to obtain exact solutions of non-linear fractional equations modeling biological population [41], Laplace equation [42], the stiff system of equations [43]. In the following sections we provide various types of illustrative examples solved by NIM.

9.5. SOLVING NON-LINEAR ALGEBRAIC EQUATIONS

In this section, we apply NIM to solve non-linear algebraic equations.

Example 9.1. *Consider the nonlinear algebraic equation*

$$x^5 - 2x^4 + 4x^3 - x^2 - 7x + 5 = 0. \tag{9.17}$$

Eq. (9.17) can be written as

$$x = \frac{5}{7} + \frac{1}{7}[x^5 - 2x^4 + 4x^3 - x^2] = x_0 + N(x), \tag{9.18}$$

where $x_0 = \frac{5}{7}$ and $N(x) = \frac{1}{7}[x^5 - 2x^4 + 4x^3 - x^2]$.

Now employing NIM algorithm given in (9.7), we get

$$x_1 = N(x_0) = 0.0875486,$$

$$x_2 = N(x_0 + x_1) - N(x_0) = 0.0444366,$$

$$x_3 = N(x_0 + x_1 + x_2) - N(x_0 + x_1) = 0.0274979,$$

$$\vdots \tag{9.19}$$

The n-term NIM solution $\sum_{i=0}^{n} x_i$ converges to the exact solution/root $x = 1$, when n is very large.

Example 9.2. *Consider the system of following nonlinear algebraic equations [44]*

$$x^2 - 10x + y^2 + 8 = 0,$$

$$xy^2 + x - 10y + 8 = 0. \tag{9.20}$$

The above system can be written as:

$$x = \frac{8}{10} + \frac{1}{10}x^2 + \frac{1}{10}y^2 = x_0 + N_1(x, y),$$

$$y = \frac{8}{10} + \frac{1}{10}x + \frac{1}{10}xy^2 = y_0 + N_2(x, y). \tag{9.21}$$

Take $x_0 = \frac{8}{10}$, $y_0 = \frac{8}{10}$ and $N_1(x, y) = \frac{1}{10}x^2 + \frac{1}{10}y^2$, $N_2(x, y) = \frac{1}{10}x + \frac{1}{10}xy^2$.

$$x_1 = N(x_0) = 0.12800000,$$

$$y_1 = N(y_0) = 0.13120000,$$

$$x_2 = N(x_0 + x_1) - N(x_0) = 0.044831744,$$

$$y_2 = N(y_0 + y_1) - N(y_0) = 0.042069983,$$

$$\vdots \qquad\qquad\qquad\qquad\qquad\qquad\qquad \textbf{(9.22)}$$

Hence, the six-term NIM solution is

$$x = x_0 + x_1 + \ldots + x_5 = 0.99831880,$$

$$y = y_0 + y_1 + \ldots + y_5 = 0.99832056.$$

This converges to the exact solution $x = 1, y = 1$ as the number of iterations become very large.

9.6. SOLVING INITIAL VALUE PROBLEMS

In this section we present various integer and fractional order initial value problems solved in the literature using NIM.

9.6.1. Notations and Preliminaries

The basic definitions of fractional integral/derivatives [45] are given below:

Definition 9.1. Mittag-Leffler function of order α is defined as

$$E_\alpha(z) = \sum_{k=0}^{\infty} \frac{z^k}{\Gamma(\alpha k + 1)}, Re(\alpha) > 0. \qquad \textbf{(9.23)}$$

Definition 9.2. Riemann-Liouville integral formula of order $\alpha \in R$, is defined as

$$I_a^\alpha f(t) = \frac{1}{\Gamma(\alpha)} \int_a^t f(\tau)(t - \tau)^{\alpha-1} d\tau, t > a. \qquad \textbf{(9.24)}$$

Definition 9.3. Let $f \in C^n[a, b]$. Then Caputo fractional derivative of order $\alpha > 0$ is defined as

$$^cD_a^\alpha f(t) := \begin{cases} \frac{1}{\Gamma(n-\alpha)} \int_a^t \frac{f^n(\tau)}{(t-\tau)^{\alpha-n+1}} d\tau, & n - 1 < \alpha < n, n \in N, \\ \frac{d^n}{dt^n} f(t), & \alpha = n \in N. \end{cases} \qquad \textbf{(9.25)}$$

Theorem 9.3. *Let $f \in C^n[a,b]$ and $n - 1 < \alpha < n, n \in N$ then*

$$I_a^\alpha {}^c D_a^\alpha f(t) = f(t) - \sum_{k=0}^{n-1} \frac{f^k(a)}{k!}(t-a)^k, t > a \qquad (9.26)$$

Notation: We denote $I_t f(t) = \int_0^t f(\tau)d\tau$.

9.6.2. Non-linear Partial Differential Equations

Here we present some examples of non-linear partial differential equations solved by NIM.

Example 9.3. *Consider the following KdV equation [46]:*

$$u_t - 3(u^2)_x + u_{xxx} = 0, u(x,0) = 6x. \qquad (9.27)$$

Eq. (9.27) is equivalent to the following integral equation

$$u = 6x + I_t[3(u^2)_x - u_{xxx}] = u_0 + N(u). \qquad (9.28)$$

Using the recurrence relation (9.7), we get $u_1 = N(u_0) = 6x(36t)$, $u_2 = 6x(1 + 36t)$ so on. NIM solution converge to the exact solution $u(x,t) = \frac{6x}{1-36t}, |36t| < 1$.

Example 9.4. *Consider the cubic Boussinesq equation [46]:*

$$u_{tt} - u_{xx} + 2(u^3)_{xx} - u_{xxxx} = 0, u(x,0) = \frac{1}{x}, u_t(x,0) = -\frac{1}{x^2}. \qquad (9.29)$$

Eq. (9.29) is equivalent to the following integral equation

$$u(x,t) = \frac{1}{x} - \frac{t}{x^2} + \int_0^t \int_0^t [u_{xx} - 2(u^3)_{xx} + u_{xxx}]dtdt. \qquad (9.30)$$

Taking $u_0 = \frac{1}{x} - \frac{t}{x^2}$ and $N(u) = \int_0^t \int_0^t [u_{xx} - 2(u^3)_{xx} + u_{xxx}]dtdt$,

following the NIM algorithm (9.7) we obtain

$$u(x,t) = \frac{1}{x+t}, \qquad (9.31)$$

as an exact solution.

Example 9.5 Consider the following non-linear Perona-Malik equation :

$$u_t = \frac{1}{u_x^2+u_y^2}(u_y^2 u_{xx} - 2u_x u_y u_{xy} + u_x^2 u_{yy}),\qquad (9.32)$$

along with the initial condition:

$$u(x,y,0) = \sqrt{x^2 + y^2} - 1.\qquad (9.33)$$

The initial value problem (9.32-9.33) is equivalent to the following integral equation:

$$u(x,y,t) = \sqrt{x^2 + y^2} - 1 + I_t[\frac{1}{u_x^2+u_y^2}(u_y^2 u_{xx} - 2u_x u_y u_{xy} + u_x^2 u_{yy})].\quad (9.34)$$

Taking $u_0 = \sqrt{x^2 + y^2} - 1$ and $N(u) = I_t[\frac{1}{u_x^2+u_y^2}(u_y^2 u_{xx} - 2u_x u_y u_{xy} + u_x^2 u_{yy})]$
and using the recurrence relation (9.7), we get

$$u_1(x,y,t) = N(u_0) = \frac{t}{\sqrt{x^2+y^2}},$$

$$u_2(x,y,t) = N(u_0 + u_1) - N(u_0) = -\frac{t^2}{2(x^2+y^2)^{3/2}},$$

$$u_3(x,y,t) = N(u_0 + u_1 + u_2) - N(u_0 + u_1) = \frac{t^3}{2(x^2+y^2)^{5/2}},$$

$$u_4(x,y,t) = N(u_0 + u_1 + u_2 + u_3) - N(u_0 + u_1 + u_2) = -\frac{5t^4}{8(x^2+y^2)^{7/2}},$$

$$\vdots \qquad\qquad\qquad\qquad\qquad\qquad (9.35)$$

Five-term approximate solution is given by:

$$u(x,y,t) = -1 + (x^2 + y^2)^{1/2} + \frac{t}{(x^2+y^2)^{1/2}} - \frac{t^2}{2(x^2+y^2)^{3/2}}$$

$$+ \frac{t^3}{2(x^2+y^2)^{5/2}} - \frac{5t^4}{8(x^2+y^2)^{7/2}}.\qquad (9.36)$$

The same solution has been obtained using HPM in [47].

Example 9.6. *Consider the fifth order KdV equation [48]*

$$u_t + 45u^2 u_x + 15u_x u_{2x} + 15uu_{3x} + u_{5x} = 0,\qquad (9.37)$$

along with the following initial condition

$$u(x, 0) = 2k^2 \text{sech}^2(k(x - x_0)). \tag{9.38}$$

The initial value problem (9.37-9.38) is equivalent to

$$u(x, t) = u(x, 0) + I_t[-(45u^2 u_x + 15u_x u_{2x} + 15uu_{3x} + u_{5x})],$$

$$= u_0 + N(u). \tag{9.39}$$

Now using the recurrence relation (9.7), we get

$$u_1 = N(u_0) = 64k^7 t \tanh(k(x - x_0))\text{sech}^2(k(x - x_0)),$$

$$u_2 = N(u_0 + u_1) - N(u_0)$$

$$= 8k^{12}t^2 \text{sech}^{10}(k(x - x_0))[-2(276480k^{10}t^2 + 17) \times$$

$$\cosh(2k(x - x_0)) + (92160k^{10}t^2 - 8)\cosh(4k(x - x_0)) +$$

$$460800k^{10}t^2 + 29440k^5 t \sinh(2k(x - x_0)) -$$

$$14080k^5 t \sinh(4k(x - x_0)) + 1280k^5 t \sinh(6k(x - x_0)) +$$

$$2\cosh(6k(x - x_0)) + \cosh(8k(x - x_0)) - 25]. \tag{9.40}$$

Therefore, 3-term NIM solution of (9.37) with initial condition (9.38) is $u(x, t) \approx u_0 + u_1 + u_2$. The numerical comparison for Eq. (9.37) between HPM and NIM is given in Table **9.1**.

Example 9.7. *Consider the following seventh order KdV equation [48]*

$$u_t + 140u^3 u_x + 70u_x^3 + 280uu_x u_{2x} + 70u^2 u_{3x} + 70u_{2x}u_{3x} +$$

$$42u_x u_{4x} + 14uu_{5x} + u_{7x} = 0, \tag{9.41}$$

with the following initial condition

$$u(x, 0) = 2k^2 \text{sech}^2(kx). \tag{9.42}$$

The exact solution of (9.41-9.42) is given by

$$u(x, t) = 2k^2 \text{sech}^2(k(x - 64k^6 t)). \tag{9.43}$$

Table 9.1. Error comparison between 3-term NIM solution and 10-term HPM solution of (9.37) for $k = 0.01$.

x	t	*Exact solution*	$10 - termHPM Abs. err.$	$3 - termNIM Abs. err.$
0.1	0.1	1.999998×10^{-4}	4.800×10^{-16}	2.71051×10^{-20}
	0.2	1.999998×10^{-4}	9.600×10^{-16}	0
	0.3	1.999998×10^{-4}	1.440×10^{-15}	8.13152×10^{-20}
	0.4	1.999998×10^{-4}	1.920×10^{-15}	2.71051×10^{-20}
	0.5	1.999998×10^{-4}	2.400×10^{-15}	2.71051×10^{-20}
0.3	0.1	1.999982×10^{-4}	1.440×10^{-15}	2.71051×10^{-20}
	0.2	1.999982×10^{-4}	2.880×10^{-15}	5.42101×10^{-20}
	0.3	1.999982×10^{-4}	4.320×10^{-15}	0
	0.4	1.999982×10^{-4}	5.760×10^{-15}	0
	0.5	1.999982×10^{-4}	7.200×10^{-15}	5.42101×10^{-20}
0.5	0.1	1.999950×10^{-4}	2.400×10^{-14}	8.13152×10^{-20}
	0.2	1.999950×10^{-4}	4.800×10^{-14}	5.42101×10^{-20}
	0.3	1.999950×10^{-4}	7.200×10^{-14}	5.42101×10^{-20}
	0.4	1.999950×10^{-4}	9.600×10^{-14}	5.42101×10^{-20}
	0.5	1.999950×10^{-4}	1.200×10^{-14}	5.42101×10^{-20}

The Eq. (9.41) along with the initial condition (9.42) can be written as

$$u(x,t) = u(x,0) + I_t[-(140u^3u_x + 70u_x^3 + 280uu_xu_{2x} +$$

$$70u^2u_{3x} + 70u_{2x}u_{3x} + 42u_xu_{4x} + 14uu_{5x} + u_{7x})], \qquad (9.44)$$

Taking

$$N(u) = I_t[-(140u^3u_x + 70u_x^3 + 280uu_xu_{2x} + 70u^2u_{3x} + 70u_{2x}u_{3x} +$$

$$42u_xu_{4x} + 14uu_{5x} + u_{7x})]. \qquad (9.45)$$

Now using the NIM algorithm (9.7) in Eq. (9.45)

$$u_1 = N(u_0) = 256k^9 t\tanh(kx)\text{sech}^2(kx). \qquad (9.46)$$

$$u_2 = N(u_0 + u_1) - N(u_0)$$

$$= \frac{16}{3} k^{16} t^2 \text{sech}^{13}(kx)[45097156608k^{21}t^3 \sinh(kx) -$$

$$19730006016k^{21}t^3 \sinh(3kx) + 2818572288k^{21}t^3 \sinh(5kx) -$$

$$564264960k^{14}t^2 \cosh(5kx) + 41287680k^{14}t^2 \cosh(7kx) -$$

$$882(1638400k^{14}t^2 + 1)\cosh(kx) + 6(318832640k^{14}t^2 - 87) \times$$

$$\cosh(3kx) - 30994432k^7 t\sinh(kx) - 15009792k^7 t\sinh(3kx) +$$

$$13834240k^7 t\sinh(5kx) - 2085888k^7 t\sinh(7kx) +$$

$$64512k^7 t\sinh(9kx) - 153\cosh(5kx) + 3\cosh(7kx) +$$

$$15\cosh(9kx) + 3\cosh(11kx)]. \tag{9.47}$$

Therefore, the three-term NIM solution of Eq. (9.41) along with initial condition (9.42) is given by $u(x,t) \approx u_0 + u_1 + u_2$.

Table **9.2**. gives the numerical comparison between ADM and NIM for Eq.(9.41)

Table 9.2. Error comparison between 3-term NIM solution and 5-term ADM solution of (9.41) for $k = 0.1$.

x	t	*Exact solution*	$5 - termADM\ Abs.err.$	$3 - termNIM\ Abs.err.$
	0.1	1.99980×10^{-2}	1.31265×10^{-8}	3.46945×10^{-18}
	0.2	1.99980×10^{-2}	2.46564×10^{-8}	0
0.1	0.3	1.99980×10^{-2}	3.60575×10^{-8}	0
	0.4	1.99980×10^{-2}	4.72723×10^{-8}	0
	0.5	1.99980×10^{-2}	5.82447×10^{-8}	3.46945×10^{-18}
	0.1	1.99820×10^{-2}	3.50406×10^{-8}	0
	0.2	1.99820×10^{-2}	6.95085×10^{-8}	3.46945×10^{-18}
0.3	0.3	1.99820×10^{-2}	1.03998×10^{-7}	3.46945×10^{-18}
	0.4	1.99820×10^{-2}	1.38303×10^{-7}	3.46945×10^{-18}
	0.5	1.99820×10^{-2}	1.72205×10^{-7}	6.93889×10^{-18}
	0.1	1.99500×10^{-5}	1.36124×10^{-8}	0
	0.2	1.99500×10^{-5}	4.60129×10^{-8}	3.46945×10^{-18}
0.5	0.3	1.99500×10^{-5}	1.08644×10^{-7}	6.93889×10^{-18}
	0.4	1.99500×10^{-5}	1.73664×10^{-7}	6.93889×10^{-18}

9.6.3. Non-linear Fractional Differential Equations

Here we present various examples of fractional differential equations in which NIM is used for getting their solutions.

Example 9.8. *Consider the following ZKE(3,3,3) [49] of the form:*

$$D_t^\alpha u + (u^3)_x + 2(u^3)_{xxx} + 2(u^3)_{xyy} = 0, \tag{9.48}$$

where $0 < \alpha \leq 1$.

For $\alpha = 1$, the exact solution of (9.48) with respect to the following initial condition [50]

$$u(x, y, 0) = \frac{3}{2} p \sinh\left(\frac{x+y}{6}\right), \tag{9.49}$$

where ρ is an arbitrary constant, is defined as

$$u(x, y, t) = \frac{3}{2} p \sinh\left(\frac{1}{6}(x + y - pt)\right). \tag{9.50}$$

Eq. (9.48) is equivalent to the following integral equation

$$u(x, y, t) = u(x, y, 0) + I_t^\alpha[-(u^3)_x - 2(u^3)_{xxx} - 2(u^3)_{xyy}]. \tag{9.51}$$

Let $u_0 = \frac{3}{2} p \sinh\left(\frac{x+y}{6}\right)]$ and $N(u) = I_t^\alpha[-(u^3)_x - 2(u^3)_{xxx} - 2(u^3)_{xyy}]$.

In view of (9.7), we get

$$u_1 = N(u_0)$$

$$= 3p^3 \left(5\cosh\left(\frac{x+y}{6}\right) - 9\cosh\left(\frac{x+y}{2}\right)\right)\frac{t^\alpha}{32\Gamma(\alpha+1)}, \tag{9.52}$$

$$u_2 = N(u_0 + u_1) - N(u_0)$$

$$= 768p^5\left[-621\sinh\left(\frac{x+y}{2}\right) + 70\sinh\left(\frac{x+y}{6}\right) + \right.$$

$$\left. 765 \sinh\left(\frac{5(x+y)}{6}\right)\right]\frac{t^{2\alpha}}{131072\Gamma[2\alpha+1]} -$$

$$3p^7C\left[1385\cosh\left(\frac{x+y}{6}\right) + 9\left(75\cosh\left(\frac{x+y}{2}\right) - \right.\right.$$

$$1615 \cosh\left(\frac{5(x+y)}{6}\right) + 1827\cosh\left(\frac{7(x+y)}{6}\right)\Big)\Big] \frac{t^{3\alpha}}{8192B^2\Gamma(3\alpha+1)}$$

$$-3p^9 t^{4\alpha} \Big[3550\sinh\left(\frac{x+y}{6}\right) - 9) - 3412\sinh\left(\frac{x+y}{2}\right) -$$

$$10935\sinh\left(\frac{3(x+y)}{2}\right) + 1700\sinh\left(\frac{5(x+y)}{6}\right) +$$

$$9135\sinh\left(\frac{7(x+y)}{6}\right)\Big)\Big] \frac{t^{4\alpha}}{131072B^3\Gamma(4\alpha+1)}, \tag{9.53}$$

where $B = \Gamma(\alpha + 1), C = \Gamma(2\alpha + 1)$.

Hence, the three-term NIM solution of Eq. (9.48) is given as

$$u(x, y, t) = u_0 + u_1 + u_2. \tag{9.54}$$

In Table **9.3**, three-term NIM solution is compared with exact and HPM solutions for Eq. (9.48).

Table 9.3. Three-term NIM solution compared with exact and HPM solutions, when $\alpha = 1$ and $p = 0.001$.

x	y	t	*Exact solution*	*NIM Solution*	*HPM Solution*
0.1	0.1	0.2	$4.995923204 \times 10^{-5}$	$5.000918398 \times 10^{-5}$	$5.000895773 \times 10^{-5}$
		0.3	$4.993421817 \times 10^{-5}$	$5.000914609 \times 10^{-5}$	$5.000880670 \times 10^{-5}$
		0.4	$4.990920434 \times 10^{-5}$	$5.000910819 \times 10^{-5}$	$5.000865568 \times 10^{-5}$
0.6	0.6	0.2	$3.019530008 \times 10^{-4}$	$3.020038994 \times 10^{-4}$	$3.020036280 \times 10^{-4}$
		0.3	$3.019274992 \times 10^{-4}$	$3.020038472 \times 10^{-4}$	$3.020034401 \times 10^{-4}$
		0.4	$3.019019978 \times 10^{-4}$	$3.020037950 \times 10^{-4}$	$3.020032522 \times 10^{-4}$
0.9	0.9	0.2	$4.567281735 \times 10^{-4}$	$4.567802963 \times 10^{-4}$	$4.567799629 \times 10^{-4}$
		0.3	$4.567020404 \times 10^{-4}$	$4.567802244 \times 10^{-4}$	$4.567797243 \times 10^{-4}$
		0.4	$4.566759074 \times 10^{-4}$	$4.567801525 \times 10^{-4}$	$4.567794858 \times 10^{-4}$

Example 9.9. *Consider the following nonlinear time-fractional gas dynamics equation [51]:*

$$D_t^\alpha u(x, t) + \frac{1}{2}(u^2)_x - u(1 - u) = 0, t > 0, 0 < \alpha \le 1, \tag{9.55}$$

along with initial condition

$$u(x,0) = e^{-x}. \tag{9.56}$$

The initial value problem (9.55-9.56) is equivalent to the following integral equation:

$$u(x,t) = e^{-x} - I_t^\alpha(\tfrac{1}{2}(u^2)_x - u(1-u)). \tag{9.57}$$

Taking $u_0 = e^{-x}$ and $N(u) = -I_t^\alpha(\tfrac{1}{2}(u^2)_x - u(1-u))$ and using the recurrence relation (9.7), we get

$$u_1(x,t) = N(u_0) = e^{-x}(\frac{t^\alpha}{\Gamma(\alpha+1)}),$$

$$u_2(x,t) = N(u_0 + u_1) - N(u_0) = e^{-x}(\frac{t^{2\alpha}}{\Gamma(2\alpha+1)}),$$

$$u_3(x,t) = N(u_0 + u_1 + u_2) - N(u_0 + u_1) = e^{-x}(\frac{t^{3\alpha}}{\Gamma(3\alpha+1)}),$$

$$\vdots \tag{9.58}$$

NIM gives closed form solution as

$$u(x,t) = \sum_{i=0}^\infty u_i(x,t) = e^{-x}E_\alpha(t^\alpha), \tag{9.59}$$

where $E_\alpha(t^\alpha)$ is a Mittag-Leffler function.

Example 9.10. *Consider the following time-fractional coupled Burgers equations in (1+1)- dimension [51]:*

$$D_t^\alpha u(x,t) - u_{xx} - 2uu_x + (uv)_x = 0,$$

$$D_t^\alpha v(x,t) - v_{xx} - 2uu_x + (uv)_x = 0, \tag{9.60}$$

with the initial conditions:

$$u(x,0) = e^x, v(x,0) = e^x. \tag{9.61}$$

The initial value problem (9.60-9.61) is equivalent to the following integral equation:

$$u(x,t) = e^x + I_t^\alpha(u_{xx} + 2uu_x - (uv)_x),$$

$$v(x,t) = e^x + I_t^\alpha(v_{xx} + 2uu_x - (uv)_x). \tag{9.62}$$

Taking $u_0 = e^x, v_0 = e^x$ and

$$N(u) = I_t^\alpha(u_{xx} + 2uu_x - (uv)_x),$$

$$N(v) = I_t^\alpha(v_{xx} + 2uu_x - (uv)_x). \tag{9.63}$$

Now using the recurrence relation (9.7), we get

$$u_1(x,t) = N(u_0) = e^x(\frac{t^\alpha}{\Gamma(\alpha+1)}),$$

$$v_1(x,t) = N(v_0) = e^x(\frac{t^\alpha}{\Gamma(\alpha+1)}),$$

$$u_2(x,t) = N(u_0 + u_1) = e^x(\frac{t^{2\alpha}}{\Gamma(2\alpha+1)}),$$

$$v_2(x,t) = N(v_0 + v_1) = e^x(\frac{t^{2\alpha}}{\Gamma(2\alpha+1)}),$$

$$\vdots \tag{9.64}$$

NIM gives closed form solution as

$$u(x,t) = \sum_{i=0}^\infty u_i(x,t) = e^x E_\alpha(t^\alpha),$$

$$v(x,t) = \sum_{i=0}^\infty v_i(x,t) = e^x E_\alpha(t^\alpha). \tag{9.65}$$

Example 9.11. *Consider the following time-fractional coupled Burgers equations in (2+1)- dimension [51]:*

$$D_t^\alpha u(x,y,t) - \nabla^2 u - 2u\nabla u + (uv)_x + v(uv)_y = 0,$$
$$D_t^\alpha v(x,y,t) - \nabla^2 v - 2v\nabla v + (uv)_x + v(uv)_y = 0, 0 < \alpha \le 1, \tag{9.66}$$

with the initial conditions:

$$u(x,y,0) = e^{x+y}, v(x,y,0) = e^{x+y}. \tag{9.67}$$

The initial value problem (9.66-9.67) is equivalent to the following integral equation:

$$u(x,y,t) = e^{x+y} + I_t^\alpha[\nabla^2 u + 2u\nabla u - (uv)_x - v(uv)_y],$$

$$v(x, y, t) = e^{x+y} + I_t^\alpha [\nabla^2 v + 2v\nabla v - (uv)_x - v(uv)_y].$$

Taking $u_0 = e^{x+y}, v_0 = e^{x+y}$ and

$$N(u) = I_t^\alpha [\nabla^2 u + 2u\nabla u - (uv)_x - v(uv)_y],$$

$$N(v) = I_t^\alpha [\nabla^2 v + 2v\nabla v - (uv)_x - v(uv)_y].$$

In view of (9.7), we get

$$u_n(x, y, t) = e^{x+y} \left(\frac{(2t^\alpha)^n}{\Gamma(n\alpha+1)} \right),$$

$$v_n(x, y, t) = e^{x+y} \left(\frac{(2t^\alpha)^n}{\Gamma(n\alpha+1)} \right),$$

$$\vdots \tag{9.68}$$

NIM gives the closed from solution as

$$
\begin{aligned}
u(x, y, t) &= \sum_{n=0}^\infty u_n(x, y, t) = e^{x+y} E_\alpha(2t^\alpha), \\
v(x, y, t) &= \sum_{n=0}^\infty v_n(x, y, t) = e^{x+y} E_\alpha(2t^\alpha).
\end{aligned}
\tag{9.69}
$$

Example 9.12. *Consider the following time-fractional biological population equation with initial condition [41]:*

$$D_t^\alpha u = (u^2)_{xx} + (u^2)_{yy} + hu, \ t > 0, 0 < \alpha \le 1,$$

$$u(x, y, 0) = \sqrt{xy}. \tag{9.70}$$

Eq. (9.70) is equivalent to the following integral equation:

$$u(x, y, t) = \sqrt{xy} + I_t^\alpha [(u^2)_{xx} + (u^2)_{yy} + hu]. \tag{9.71}$$

NIM algorithm (9.7) gives us the exact solution defined as

$$u(x, y, t) = \sqrt{xy} E_\alpha(ht^\alpha). \tag{9.72}$$

Example 9.13. *Consider the following time-fractional biological population equation with initial condition [41]:*

$$D_t^\alpha u = (u^2)_{xx} + (u^2)_{yy} + hu(1 - ru), t > 0, 0 < \alpha \le 1, \qquad (9.73)$$

$$u(x, y, 0) = e^{\sqrt{hr/8}(x+y)}. \qquad (9.74)$$

NIM solution turns to the exact solution:

$$u(x, y, t) = e^{\sqrt{hr/8}(x+y)} \sum_{k=0}^{\infty} \frac{(ht)^k}{k!} = e^{\sqrt{hr/8}(x+y)+ht}. \qquad (9.75)$$

Example 9.14. *Consider the following system of three nonlinear time-fractional partial differential equations [41] $t > 0, 0 < \alpha \le 1$,*

$$D_t^\alpha u = -v_x w_y + v_y w_x - u, \qquad (9.76)$$

$$D_t^\alpha v = -w_x u_y - w_y u_x + v, \qquad (9.77)$$

$$D_t^\alpha w = -u_x v_y - u_y v_x + w, \qquad (9.78)$$

with initial conditions

$$u(x, y, 0) = e^{x+y}, v(x, y, 0) = e^{x-y}, w(x, y, 0) = e^{-x+y}. \qquad (9.79)$$

NIM solution turns to the exact solution

$$u(x, y, t) = \sum_{i=0}^{\infty} u_i = e^{x+y} \sum_{k=0}^{\infty} \frac{(-t^\alpha)^k}{\Gamma(k\alpha+1)} = e^{x+y} E_\alpha(-t^\alpha), \qquad (9.80)$$

$$v(x, y, t) = \sum_{i=0}^{\infty} v_i = e^{x-y} \sum_{k=0}^{\infty} \frac{(t^\alpha)^k}{\Gamma(k\alpha+1)} = e^{x-y} E_\alpha(t^\alpha), \qquad (9.81)$$

$$w(x, y, t) = \sum_{i=0}^{\infty} w_i = e^{-x+y} \sum_{k=0}^{\infty} \frac{(t^\alpha)^k}{\Gamma(k\alpha+1)} = e^{-x+y} E_\alpha(t^\alpha). \qquad (9.82)$$

Example 9.15. *Consider two dimensional IBVP [52]*

$$D_t^\alpha u = \frac{1}{2}(y^2 u_{xx} + x^2 u_{yy}), 0 < x, y < 1, 0 < \alpha \le 1, t > 0, \qquad (9.83)$$

subject to the Neumann boundary conditions

$$u_x(0, y, t) = 0, u_x(1, y, t) = 2\sinh t,$$

$$u_y(x, 0, t) = 0, u_y(x, 1, t) = 2\cosh t, \qquad (9.84)$$

and the initial condition

$$u(x, y, 0) = y^2. \tag{9.85}$$

The integral equation corresponding to Eqs. (9.83, 9.85) is

$$u(x, y, t) = \sum_{k=0}^{m-1} \frac{\partial^k}{\partial t^k} u(x, y, 0) \frac{t^k}{k!} + \frac{1}{2} I_t^\alpha (y^2 u_{xx} + x^2 u_{yy}).$$

$$= u_0 + N(u). \tag{9.86}$$

Now using NIM algorithm, we get

$$u_n(x, y, t) = [x^2 \frac{1-(-1)^n}{2} + y^2 \frac{1+(-1)^n}{2}] \frac{\Gamma(1)}{\Gamma(n\alpha+1)} t^{n\alpha}, n \in N. \tag{9.87}$$

For $\alpha = 1$, NIM gives the closed form solution as

$$u(x, y, t) = x^2 \sinh t + y^2 \cosh t. \tag{9.88}$$

Example 9.16. *Consider the two-dimensional IBVP [52]*

$$D_t^\alpha u = \frac{1}{12} (x^2 u_{xx} + y^2 u_{yy}), 0 < x, y < 1, 1 < \alpha \leq 2, t > 0, \tag{9.89}$$

subject to the Neumann conditions

$$u_x(0, y, t) = 0, u_x(1, y, t) = 4\cosh t,$$

$$u_y(0, y, t) = 0, u_y(1, y, t) = 4\sinh t, \tag{9.90}$$

and the initial conditions

$$u(x, y, 0) = x^4, u_t(x, y, 0) = y^4. \tag{9.91}$$

The integral equation corresponding to Eqs. (9.89, 9.91) is

$$u(x, y, t) = \sum_{k=0}^{m-1} \frac{\partial^k}{\partial t^k} u(x, y, 0) \frac{t^k}{k!} + \frac{1}{12} I_t^\alpha (x^2 u_{xx} + y^2 u_{yy}).$$

$$= u_0 + N(u). \tag{9.92}$$

Using NIM algorithm (9.7), we get

$$u_n(x, y, t) = x^4 \frac{\Gamma(2)}{\Gamma(n\alpha+1)} t^{n\alpha} + y^4 \frac{\Gamma(2)}{\Gamma(n\alpha+2)} t^{n\alpha+1}, n \in N. \tag{9.93}$$

Thus, the NIM solution in series form is given by

$$u(x, y, t) = x^4 \left(1 + \frac{t^\alpha}{\Gamma(\alpha+1)} + \frac{t^{2\alpha}}{\Gamma(2\alpha+1)} + \ldots + \frac{t^{n\alpha}}{\Gamma(n\alpha+1)} + \ldots\right)$$

$$+ y^4 \left(t + \frac{t^\alpha+1}{\Gamma(\alpha+2)} + \frac{t^{2\alpha}+1}{\Gamma(2\alpha+2)} + \ldots + \frac{t^{n\alpha}+1}{\Gamma(n\alpha+2)} + \ldots\right). \tag{9.94}$$

For $\alpha = 2$, the series (9.94) has closed form solution:

$$u(x, y, t) = x^4 \cosh t + y^4 \sinh t, \tag{9.95}$$

which is the exact solution of IVP (9.89-9.91).

9.7. SOLVING BOUNDARY VALUE PROBLEMS

NIM can also be used for solving boundary value problems.

Consider the following fractional boundary value problem (BVP) [16]:

$$D_t^\alpha u = u_{xx} + A(u), \ t > 0, \ 0 < x < l, \tag{9.96}$$

where $m - 1 < \alpha \le m, m = 1,2$ and $A(u)$ is a given nonlinear function of u together with the following conditions

$$\frac{\partial^k u}{\partial t^k}(x, 0) = p_k(x), k = 0,1,\ldots,(m - 1), \tag{9.97}$$

$$u(0, t) = f_0(t), u(l, t) = f_l(t). \tag{9.98}$$

Let T be the inverse operator of $\partial^2 / \partial x^2$ which is defined as [53, 54]

$$T = \int_0^x \int_0^{x_1} dx_2 dx_1 - \frac{x}{l} \int_0^l \int_0^{x_1} dx_2 dx_1. \tag{9.99}$$

Now applying I_t^α on both sides of Eq. (9.96) and also using the initial conditions (9.97), we get

$$u = \sum_{k=0}^{m-1} \frac{t^k p_k}{k!} + I_t^\alpha u_{xx} + I_t^\alpha A(u). \tag{9.100}$$

Now applying T on both sides of Eq. (9.96) and using the boundary conditions (9.98), we get

$$u = f_0 + \frac{x}{l}(f_l - f_0) + T(D_t^\alpha u) - T(A(u)). \tag{9.101}$$

From (9.100) and (9.101), we get

$$u = \frac{1}{2}\left[\left[\sum_{k=0}^{m-1}\frac{t^k p_k}{k!} + f_0 + \frac{x}{l}(f_l - f_0)\right] + [I_t^\alpha u_{xx} + T(D_t^\alpha u)] + [I_t^\alpha A(u) - T(A(u))]\right]. \tag{9.102}$$

Setting,

$$u_0 = \frac{1}{2}\left[\sum_{k=0}^{m-1}\frac{t^k p_k}{k!} + f_0 + \frac{x}{l}(f_l - f_0)\right], \tag{9.103}$$

$$L(u) = [I_t^\alpha u_{xx} + T(D_t^\alpha u)], \tag{9.104}$$

$$N(u) = [I_t^\alpha A(u) - T(A(u))]. \tag{9.105}$$

We get,

$$u = u_0 + L(u) + N(u). \tag{9.106}$$

Now the above equation can be solved by NIM.

Example 9.17. *Consider the following linear fractional diffusion equation along with initial and boundary conditions [16]:*

$$D_t^\alpha u = u_{xx}, 0 < x < l, t > 0, 0 < \alpha \le 1, \tag{9.107}$$

$$u(x,0) = x^2 + 2x + 1, \tag{9.108}$$

$$u(0,t) = \frac{2t^\alpha}{\Gamma(\alpha+1)} + 1, \tag{9.109}$$

$$u(l,t) = l^2 + 2l + 1 + \frac{2t^\alpha}{\Gamma(\alpha+1)}. \tag{9.110}$$

In view of (9.103), we get

$$u_0 = \frac{1}{2}\left[x^2 + 4x + 2 + \frac{2t^\alpha}{\Gamma(\alpha+1)} + xl\right], \tag{9.111}$$

and also using the recursive formulas (9.7) and (9.104), we get

$$u_n = \frac{1}{2^{n+1}}\left[x^2 + \frac{2t^\alpha}{\Gamma(\alpha+1)} - xl\right], n = 1,2,\ldots \qquad \textbf{(9.112)}$$

Hence the solution of (9.107)

$$u(x,t) = \sum_{i=0}^{\infty} u_i = x^2 + 2x + \frac{2t^\alpha}{\Gamma(\alpha+1)} + 1. \qquad \textbf{(9.113)}$$

Example 9.18. *Consider the nonlinear wave equation of fractional order:*

$$D_t^\alpha u = u_{xx} + xu^2, 0 < x < 1, t > 0, 1 < \alpha \le 2,$$

$$u(x,0) = x, u_t(x,0) = 1,$$

$$u(0,t) = t, u(1,t) = 2t. \qquad \textbf{(9.114)}$$

Using NIM, we get

$$u_0 = \frac{1}{2}[x(1+t) + 2t],$$

$$u_1 = \frac{x}{480}[3 + 26t + 63t^2 - 40t^2x^2 - 20tx^2 - 20t^2x^3$$

$$-3x^4 - 6tx^4 - 3t^2x^4 + \frac{120t^{\alpha+2}(x+2)^2}{\Gamma(\alpha+3)}].$$

Hence, two term NIM solution is:

$$u = u_0 + u_1. \qquad \textbf{(9.115)}$$

9.8. SOLVING INTEGRO-DIFFERENTIAL EQUATIONS

In this section we solve integro-differential equations using NIM.

Consider the general nth order linear nonlinear integro-differential equation [27]:

$$y^{(n)}(x) + f(x)y(x) + \int_a^b w(x,t)y^{(q)}(t)y^{(m)}(t)dt = g(x), \qquad \textbf{(9.116)}$$

along with the initial conditions

$$y(a) = \alpha_0, y'(a) = \alpha_1, y''(a) = \alpha_2, \cdots, y^{(n-1)}(a) = \alpha_{n-1}, \quad (9.117)$$

where $\alpha_i, i = 0,1,2, \cdots, n-1$, are real constants, q, m and n are integers with $q \le m \le n$. In Eq. (9.116) the functions f, g and w are known and the solution y is to be determined.

In view of the NIM algorithm, nth-order integro-differential equation (9.116-9.117) is equivalent to the following integral equation:

$$y(x) = y_0(x) - I_x^n[\int_a^b w(x,t)y^{(q)}(t)y^{(m)}(t)dt] = y_0(x) - N(y), \quad (9.118)$$

where $y_0(x)$ is the solution of the nth-order differential equation:

$$\frac{d^n y_0}{dx^n} = g(x) - f(x)y(x), \frac{d^k y(a)}{dx^k} = \alpha_k, k = 0,1,2,\ldots,n-1. \quad (9.119)$$

Some particular cases of Eq. (9.116) are considered below.

Example 9.19. *Consider the following first order integro-differential equation [27]:*

$$y'(x) = 1 - \frac{x}{3} + \int_0^1 xty(t)dt, y(0) = 0. \quad (9.120)$$

This is equivalent to the following equation

$$y(x) = x - \frac{x^2}{6} + I_x[\int_0^1 xty(t)dt]. \quad (9.121)$$

Take $y_0 = x - \frac{x^2}{6}$ and $N(y) = I_x[\int_0^1 xty(t)dt]$. Now applying the recurrence relation (9.7)

$$y_1(x) = N[y_0(x)] = \frac{7x^2}{48}, \quad (9.122)$$

$$y_2(x) = N[y_0(x) + y_1(x)] - N[y_0(x)] = x - \frac{x^2}{384},$$

$$\vdots$$

and so on. NIM solution converges to the exact solution $y = x$.

Example 9.20. *Consider the following second-order integro-differential equation* [27]:

$$y''(x) = e^x - x + \int_0^1 xty(t)dt, y(0) = 1, y'(0) = 1. \qquad \textbf{(9.123)}$$

Eq. (9.123) is equivalent to the following integral equation:

$$y(x) = e^x - \frac{x^3}{6} + I_x^2[\int_0^1 xty(t)dt]. \qquad \textbf{(9.124)}$$

Let $y_0(x) = e^x - \frac{x^3}{6}$ and $N(y) = I_x^2[\int_0^1 xty(t)dt]$. A few iterations of NIM are

$$y_1(x) = N[y_0(x)] = \frac{29x^3}{180},$$

$$y_2(x) = N[y_0(x) + y_1(x)] - N[y_0(x)] = \frac{29x^3}{5400},$$

$$y_3(x) = N[y_0(x) + y_1(x) + y_2(x)] - N[y_0(x) + y_1(x)] = \frac{29x^3}{162000},$$

$$\vdots \qquad \qquad \textbf{(9.125)}$$

NIM solution converges to $y = e^x$, which is the exact solution.

Example 9.21. *Consider the third-order nonlinear integro-differntial equation* [27]:

$$y'''(x) = e^x - \frac{x^4}{4}(e^x + 1) + \int_0^1 xty''^2(t)dt, \qquad \textbf{(9.126)}$$

$$y(0) = 0, y'(0) = y''(0) = 1.$$

Eq. (9.126) is equivalent to the following integral equation

$$y(x) = e^x - \frac{e^2+1}{96}x^4 + I_x^3[\int_0^1 xty''^2(t)dt]. \qquad \textbf{(9.127)}$$

Take $y_0(x) = e^x - \frac{e^2+1}{96}x^4$ and $N(y) = I_x^3[\int_0^1 xty''^2(t)dt]$.

Using NIM algorithm (9.7), we get

$$y_1(x) = N[y_0(x)] = 0.0457858,$$

$$y_2(x) = N[y_0(x) + y_1(x)] - N[y_0(x)] = 0.0198916,$$

$$y_3(x) = N[y_0(x) + y_1(x) + y_2(x)] - N[y_0(x) + y_1(x)] = 0.0099483,$$

$$\vdots \tag{9.128}$$

and so on. NIM solution converges to the exact solution $y = e^x$.

9.9. METHODS BASED ON NIM

In this section we review some methods developed by NIM.

9.9.1. Three-step Iterative Method

In 2006, Noor and Noor [22] have suggested a new three-step iterative method for solving nonlinear algebraic equations. They proved that this method has third order convergence and does not involve the higher order differentials of the function as compared to methods developed by Adomian decomposition method [55]. Later on in 2008, Jae Heon Yun has improved this method and shown that it has fourth-order convergence, not third-order convergence. The algorithm of this method is as given below [21].

Consider the following nonlinear equation:

$$f(x) = 0. \tag{9.129}$$

Let x_0 be the initial guessed solution of Eq. (9.129). Then the approximate solution x_{n+1} using this method is given as

1. Predictor:

$$y_n = x_n - \frac{f(x_n)}{f'(x_n)}, f'(x_n) \neq 0, \tag{9.130}$$

$$z_n = -\frac{f(y_n)}{f'(x_n)}. \tag{9.131}$$

2. Corrector:

$$x_{n+1} = y_n + z_n - \frac{f(y_n+z_n)}{f'(x_n)}, n = 0,1,2,\dots \tag{9.132}$$

This algorithm is called as the three-step iterative method for solving the nonlinear Eq. (9.129). Its convergence has been discussed in [21].

9.9.2. Iterative Laplace Transform Method

This method has been developed by Jafari *et al.* [56] for solving systems of linear and nonlinear fractional partial differential equations. This method is a combination of Laplace transform and NIM, known as Iterative Laplace Transform Method (ILTM). Yan has used it for solving Fractional Fokker-Planck Equations [57]. Najafia *et al.* [58] have used this method for getting exact solution of fractional gas dynamics equation. The basic idea of this method is illustrated here.

Consider the following system of fractional partial differential equations:

$$^cD_t^{\alpha_i}u_i(\bar{x},t) = A_i(u_1(\bar{x},t),\ldots,u_n(\bar{x},t)), m_{i-1} < \alpha_i \leq m_i, i = 1,2,\ldots,n, \quad \textbf{(9.133)}$$

$$\frac{\partial^{(k_i)}u_i(\bar{x},0)}{\partial t^{(k_i)}} = h_{i,k_i}(\bar{x}), k_i = 0,1,\ldots,m_{i-1}, m_i \in N,$$

where A_i are nonlinear operators and $u_i(\bar{x},t)$ are unknown functions. Taking the Laplace transform(denoted by L) on both sides of Eq. (9.133), we get

$$s^{\alpha_i}L[u_i(\bar{x},t)] - \sum_{k=0}^{m_i-1} s^{\alpha_i-k-1}u_i^{(k)}(\bar{x},0) = L[A_i((u_1(\bar{x},t),\ldots,u_n(\bar{x},t)))], \quad \textbf{(9.134)}$$
$$i = 1,2,\ldots,n.$$

Laplace transform of the Caputo fractional derivative is defined as

$$L[^cD^\alpha f(t)] = s^\alpha - \sum_{k=0}^{n-1} s^{(\alpha-k-1)}f^{(k)}(0), n - 1 < \alpha \leq n.$$

By taking the inverse Laplace (L^{-1}) on both sides of Eq. (9.134),we get

$$u_i(\bar{x},t) = L^{-1}[\sum_{k=0}^{m_i-1} s^{-k-1}u_i^{(k)}(\bar{x},0)]$$

$$+L^{-1}[s^{-\alpha_i}L[A_i(u_1(\bar{x},t),\ldots,u_n(\bar{x},t))]], \quad \textbf{(9.135)}$$

which can be written as

$$u_i(\bar{x},t) = f_i + N_i(u_i(\bar{x},t),\ldots,u_n(\bar{x},t)), i = 1,2,\ldots,n, \quad \textbf{(9.136)}$$

where

$$f_i = L^{-1}[\sum_{k=0}^{m_i-1} s^{-k-1}u_i^{(k)}(\bar{x},0)],$$

$$N_i(u_1(\bar{x},t),\ldots,u_n(\bar{x},t)) = L^{-1}[s^{-\alpha_i}L[A_i(u_1(\bar{x},t),\ldots,u_n(\bar{x},t))]]. \quad (9.137)$$

Following the NIM algorithm (9.7) in Eq. (9.135), we get the following k-term solution of Eq. (9.133) as

$$u_i(\bar{x},t) \approx u_{i1}(\bar{x},t) + \ldots + u_{ik}(\bar{x},t), i = 1,2\ldots,n. \quad (9.138)$$

Convergence of this method is same as described in [59].

Example 9.22. *Consider the following system of fractional partial differential equation [56]:*

$$D_t^\alpha u(x,t) - v_x(x,t) + u(x,t) + v(x,t) = 0,$$

$$D_t^\beta u(x,t) - u_x(x,t) + u(x,t) + v(x,t) = 0, (0 < \alpha, \beta \le 1), \quad (9.139)$$

along with initial conditions

$$u(x,0) = \sinh(x), v(x,0) = \cosh(x). \quad (9.140)$$

By taking the Laplace transform of Eq. (9.139), we get

$$s^\alpha L[u(x,t)] - s^{\alpha-1}u(x,0) + L[-v_x(x,t) + u(x,t) + v(x,t)] = 0,$$

$$s^\beta L[u(x,t)] - s^{\beta-1}u(x,0) + L[-u_x(x,t) + u(x,t) + v(x,t)] = 0. \quad (9.141)$$

Now, by taking the inverse Laplace transform of Eq. (9.141), we get

$$u(x,t) = L^{-1}[s^{-1}u(x,0)] + L^{-1}[s^{-\alpha}L[v_x(x,t) - u(x,t) - v(x,t)]],$$

$$v(x,t) = L^{-1}[s^{-1}v(x,0)] + L^{-1}[s^{-\beta}L[u_x(x,t) - u(x,t) - v(x,t)]]. \quad (9.142)$$

Taking $N(u) = L^{-1}[s^{-\alpha}L[v_x(x,t) - u(x,t) - v(x,t)]]$ and

$N(v) = L^{-1}[s^{-\beta}L[u_x(x,t) - u(x,t) - v(x,t)]]$. Using the NIM algorithm (9.7), we get,

$$u_0(x,t) = L^{-1}[s^{-1}u(x,0)] = L^{-1}[s^{-1}\sinh(x)] = \sinh(x),$$

$$v_0(x,t) = L^{-1}[s^{-1}v(x,0)] = L^{-1}[s^{-1}\cosh(x)] = \cosh(x), \quad (9.143)$$

$$u_1(x,t) = L^{-1}[s^{-\alpha}L[v_{0_x}(x,t) - u_0(x,t) - v_0(x,t)]] = -\frac{\cosh(x)t^{\alpha}}{\Gamma(\alpha+1)},$$

$$v_1(x,t) = L^{-1}[s^{-\beta}L[u_{0_x}(x,t) - u_0(x,t) - v_0(x,t)]] = -\frac{\sinh(x)t^{\beta}}{\Gamma(\beta+1)}, \quad (9.144)$$

$$\vdots$$

so on.

Hence the solution has the following series form.

$$u(x,t) = \sinh(x)(1 + \frac{t^{\alpha+\beta}}{\Gamma(\alpha+\beta+1)} + \ldots)$$

$$-\cosh(x)(\frac{t^{\alpha}}{\Gamma(\alpha+1)} + \frac{t^{\alpha+\beta}}{\Gamma(\alpha+\beta+1)} - \frac{t^{2\alpha}}{\Gamma(2\alpha+1)} + \ldots),$$

$$v(x,t) = \cosh(x)(1 + \frac{t^{\alpha+\beta}}{\Gamma(\alpha+\beta+1)} + \ldots) \quad\quad (9.145)$$

$$-\sinh(x)(\frac{t^{\beta}}{\Gamma(\beta+1)} + \frac{t^{\alpha+\beta}}{\Gamma(\alpha+\beta+1)} - \frac{t^{2\beta}}{\Gamma(2\beta+1)} + \ldots). \quad\quad (9.146)$$

For $\alpha = \beta = 1$, we get the following closed form solution of (9.139).

$$u(x,t) = \sinh(x)(1 + \frac{t^2}{2!} + \frac{t^4}{4!} + \ldots) - \cosh(x)(t + \frac{t^3}{3!} + \frac{t^5}{5!} + \ldots)$$

$$= \sinh(x - t),$$

$$v(x,t) = \cosh(x)(1 + \frac{t^2}{2!} + \frac{t^4}{4!} + \ldots) - \sinh(x)(t + \frac{t^3}{3!} + \frac{t^5}{5!} + \ldots)$$

$$= \cosh(x - t). \quad\quad (9.147)$$

9.9.3. New Predictor-corrector Method

Daftardar-Gejji *et al.* [60] have recently proposed a new predictor-corrector method to solve the non-linear fractional differential equations. This method is a combination of fractional Adams method [61, 62] and NIM [1], termed as New Predictor- Corrector method (NPCM). It is observed that the new predictor corrector method is more accurate and takes less time as compared to the fractional Adams method (FAM). Further, NPCM has better stability properties than FAM. The algorithm of this method is as follows.

Consider the initial value problem

$$^cD_0^\alpha x(t) = f(t, x(t)), x^{(k)}(0) = x_0^{(k)}, k = 0,1,\ldots,\lceil\alpha\rceil - 1. \qquad \textbf{(9.148)}$$

The new predictor-corrector formula for solving Eq. (9.148) is given as:

$$y_{n+1}^p = \sum_{k=0}^{\lceil\alpha\rceil-1} x_0^{(k)} \frac{t_{n+1}^k}{k!} + \frac{h^\alpha}{\Gamma(\alpha+2)} \sum_{j=0}^n a_{j,n+1} f(t_j, x_j), \qquad \textbf{(9.149)}$$

$$z_{n+1}^p = \frac{h^\alpha}{\Gamma(\alpha+2)} f(t_{n+1}, y_{n+1}^p), \qquad \textbf{(9.150)}$$

$$x_{n+1}^c = y_{n+1}^p + \frac{h^\alpha}{\Gamma(\alpha+2)} f(t_{n+1}, y_{n+1}^p + z_{n+1}^p), \qquad \textbf{(9.151)}$$

where y_{n+1}^p and z_{n+1}^p are predictors and x_{n+1}^c corrector, where $a_{j,n+1}$ are defined as

$$a_{j,n+1} = \begin{cases} n^{\alpha+1} - (n-\alpha)(n+1)^\alpha & \text{if } j = 0, \\ (n-j+2)^{\alpha+1} + (n-j)^{\alpha+1} - 2(n-j+1)^{\alpha+1} & \text{if } 1 \le j \le n, \\ 1 & \text{if } j = n+1. \end{cases} \qquad \textbf{(9.152)}$$

This method is known as three-step NPCM for solving non-linear equation of fractional order (9.148). NPCM is further extended for solving fractional differential equations (FDEs) involving delay. [63].

9.9.4. New Numerical Methods

Recently, Sukale and Daftardar Gejji [64] have developed new numerical methods (NNMs) for solving ordinary differential equations with and without delay terms. They have used NIM to improve the existing methods such as trapezoidal rule and Adams Moulton methods (2-step and 3-step). Further, they have also studied consistency of the method and performed error and stability analysis. Several examples have been solved to demonstrate efficiency of this method. This new numerical methods yields more accurate results than the existing methods.

Patade and Bhalekar [40] have derived a new numerical method for ordinary differential equations by applying NIM to implicit trapezium formula. They have also discussed the error, stability and convergence analysis for this method and also applied it for various problems. They have developed a software package for this method. Similar idea has been used to solve Volterra integro-differential equations [65] .

Sukale and Daftardar-Gejji [64] have considered the initial value problem (IVP) for the delay differential equation:

$$\frac{dy}{dt} = f(t, y(t), y(t - \tau)),$$

$$y(t) = \phi(t), -\tau \le t \le 0. \tag{9.153}$$

The existence and uniqueness of the solution of Eq. (9.153) is defined in [66] and the delay term $y(t_n - \tau)$ is denoted by v_n and defined in [67].

For solving Eq. (9.153) over the interval $[0, T]$, divide the interval into l subintervals. Let $h = \frac{T}{l}, t_n = nh; n = 0,1,2,\ldots, l \in Z^+$. After integrating Eq. (9.153) from t_n to t_{n+1}, we get

$$y_{n+1} = y(t_{n+1}) = y(t_n) + \int_{t_n}^{t_{n+1}} f(t, y(t), y(t - \tau)) dt. \tag{9.154}$$

9.9.4.1. Modified Trapezoidal Rule [64]

The definite integral in Eq. (9.154) is approximated by trapezoidal rule as

$$y_{n+1} = y_n + \frac{h}{2} [f(t_n, y_n, v_n) + f(t_{n+1}, y_{n+1}, v_{n+1})]. \tag{9.155}$$

Eq. (9.155) is of the form $y_{n+1} = g + N(y_{n+1})$, where

$$g = y_n + \frac{h}{2} f(t_n, y_n, v_n), \tag{9.156}$$

and

$$N(y_{n+1}) = \frac{h}{2} f(t_{n+1}, y_{n+1}, v_{n+1}). \tag{9.157}$$

The three term approximation of NIM scheme gives the following two-step predictor-corrector formula

$$u_0 = y_{n+1}^p = y_n + \frac{h}{2} f(t_n, y_n, v_n), \tag{9.158}$$

$$u_1 = N(u_0) = z_{n+1}^p = \frac{h}{2} f(t_{n+1}, y_{n+1}^p, v_{n+1}). \tag{9.159}$$

The three-term approximate solution of Eq. (9.155) is $u = u_0 + u_1 + u_2 = u_0 + N(u_0 + u_1)$ *i.e.*

$$y_{n+1}^c = y_{n+1}^p + \frac{h}{2} f(t_{n+1}, y_{n+1}^p + z_{n+1}^p, v_{n+1}), \tag{9.160}$$

Eqs. (9.158)-(9.160) constitute a modified trapezoidal rule. The same idea has been continued for solving a system of delay differential equations.

9.9.4.2. Modified 2-Step Adams Moulton Method [64]

Using the 2-step Adams Moulton implicit rule we approximate the definite integral of Eq. (9.154) as

$$y_{n+1} = y_n + \frac{h}{12} [5f(t_{n+1}, y_{n+1}, v_{n+1}) + 8f(t_n, y_n, v_n) - f(t_{n-1}, y_{n-1}, v_{n-1})]. \tag{9.161}$$

The Eq. (9.161) is of the form $y_{n+1} = g + N(y_{n+1})$, where

$$g = y_n + \frac{8h}{12} f(t_n, y_n, v_n) - \frac{h}{12} f(t_{n-1}, y_{n-1}, v_{n-1}), \tag{9.162}$$

and

$$N(y_{n+1}) = \frac{5h}{12} f(t_{n+1}, y_{n+1}, v_{n+1}). \tag{9.163}$$

Using three term NIM solution of DDEs (9.161), we get the following 2-step modified Adam Moulton method, where the predictors y_{n+1}^p, z_{n+1}^p are defined as

$$y_{n+1}^p = y_n + \frac{8h}{12} f(t_n, y_n, v_n) - \frac{h}{12} f(t_{n-1}, y_{n-1}, v_{n-1}), \tag{9.164}$$

$$z_{n+1}^p = \frac{5h}{12} f(t_{n+1}, y_{n+1}^p, v_{n+1}), \tag{9.165}$$

and the corrector y_{n+1}^c is defined as

$$y_{n+1}^c = y_{n+1}^p + \frac{5h}{12} f(t_{n+1}, y_{n+1}^p + z_{n+1}^p, v_{n+1}). \tag{9.166}$$

On similar lines Modified 3-step Adams Moulton Method has been developed [64].

Note that the above mentioned modified methods are also hold for solving ordinary differential equation

$$\frac{dy}{dt} = f(t, y(t)), y(0) = y_0,$$

(9.167)

when the delay term is put to zero.

9.10. CONCLUSIONS

It is a proven fact that NIM is a powerful method for solving functional equations of the form $u = f + N(u)$, which encompasses a wide class of equations *viz.* integral equations, differential equations, delay differential equations, integro-differential equations, partial differential equations, fractional differential equations. This method does not involve linearizion, perturbation or discretization and gives rapidly converging series solution. The algorithm of NIM is simple and easily employable using computer algebra packages such as Mathematica, Maple, Matlab and requires much less computational work as compared to ADM, HPM or VIM. The method has attracted attention of many researchers for solving variety of problems. In many cases NIM gives closed form solutions or else very good approximate numerical solutions.

CONSENT FOR PUBLICATION

Not applicable.

CONFLICT OF INTEREST

The authors declare no conflict of interest, financial or otherwise.

ACKNOWLEDGEMENTS

Declared none

REFERENCES

[1]   V. Daftardar-Gejji and H. Jafari, "An iterative method for solving nonlinear functional equations," *J. Math. Anal. Appl.*, vol. 316, pp. 753-763, Apr. 2006.
[2]   J. Biazar, E. Babolian, and R. Islam, "Solution of the system of ordinary differential equations by Adomian decomposition method," *Appl. Math. Comput.*, vol. 147, pp. 713-719, Jan. 2004.
[3]   D. J. Evans and K. Raslan, "The Adomian decomposition method for solving delay differential equation," *Int. J. Comput. Math.*, vol. 82, pp. 49-54, Jan. 2005.
[4]   I. Hashim, "Adomian decomposition method for solving bvps for fourth-order integro-differential equations," *J. Comput. Appl. Math.*, vol. 193, pp. 658-664, Sep. 2006.

[5]   A.-M. Wazwaz, "The decomposition method applied to systems of partial differential equations and to the reaction-diffusion Brusselator model," *Appl. Math. Comput.*, vol. 110, pp. 251-264, Apr. 2000.

[6]   V. Daftardar-Gejji and H. Jafari, "Adomian decomposition: a tool for solving a system of fractional differential equations," *J. Math. Anal. Appl.*, vol. 301, pp. 508-518, Jan. 2005.

[7]   G. Adomian, *Solving frontier problems of physics: the decomposition method*. Kluwer: Boston, 1994.

[8]   S. M. El-Sayed and D. Kaya, "An application of the adm to seven-order Sawada-Kotera equations," *Appl. Math. Comput.*, vol. 157, pp. 93-101, Sep. 2004.

[9]   J.-H. He, "Homotopy perturbation technique," *Comput. Methods Appl. Mech. Engrg.*, vol. 178, pp. 257-262, Aug. 1999.

[10]   M. Ghasemi, A. Azizi, and M. Fardi, "Numerical solution of seven-order Sawada-Kotera equations by homotopy perturbation method," *Math. Sci. J*, vol. 1, pp. 69-77, Jan. 2011.

[11]   J.-H. He, "Variational iteration method-a kind of non-linear analytical technique: some examples," *Internat. J. Non-Linear Mech.*, vol. 34, pp. 699-708, Jul. 1999.

[12]   S. Bhalekar and V. Daftardar-Gejji, "Convergence of the new iterative method," *Int. J. Differential Equations*, vol. 2011, pp. 1-10, Nov. 2011.

[13]   S. Bhalekar and V. Daftardar-Gejji, "New iterative method: application to partial differential equations," *Appl. Math. Comput.*, vol. 203, pp. 778-783, Sep. 2008.

[14]   S. Bhalekar and V. Daftardar-Gejji, "Solving evolution equations using a new iterative method," *Numer. Methods Partial Differential Equations*, vol. 26, pp. 906-916, Jul. 2010.

[15]   V. Daftardar-Gejji and S. Bhalekar, "Solving fractional diffusion-wave equations using a new iterative method," *Fract. Calc. Appl. Anal.*, vol. 11, no. 2, pp. 193p-202p, 2008.

[16]   V. Daftardar-Gejji and S. Bhalekar, "Solving fractional boundary value problems with Dirichlet boundary conditions using a new iterative method," *Comput. Math. Appl.*, vol. 59, pp. 1801-1809, Mar. 2010.

[17]   M. Usman, A. Yildirim, and S. T. Mohyud-Din, "A reliable algorithm for physical problems," *Int. J. Phys. Sci.*, vol. 6, pp. 146-153, Jan. 2011.

[18]   G. Loghmani and S. Javanmardi, "Numerical methods for sequential fractional differential equations for Caputo operator," *Bull. Malays. Math. Sci. Soc.(2)*, vol. 35, Apr. 2012.

[19]   V. Srivastava and K. Rai, "A multi-term fractional diffusion equation for oxygen delivery through a capillary to tissues," *Math. Comput. Modell.*, vol. 51, pp. 616-624, Mar. 2010.

[20]   R. K. Saeed and K. M. Aziz, "An iterative method with quartic convergence for solving nonlinear equations," *Appl. Math. Comput.*, vol. 202, pp. 435-440, Aug. 2008.

[21]   J. H. Yun, "A note on three-step iterative method for nonlinear equations," *Appl. Math. Comput.*, vol. 202, pp. 401-405, Aug. 2008.

[22]   M. A. Noor and K. I. Noor, "Three-step iterative methods for nonlinear equations," *Appl. Math. Comput.*, vol. 183, no. 1, pp. 322-327, 2006.

[23]   M. B. Ghori, M. Usman, and S. T. Mohyud-Din, "Numerical studies for solving a predictive microbial growth model using a new iterative method," *Int. J. Modern Biol. Med*, vol. 5, no. 1, pp. 33-39, 2014.

[24]   A.-J. Majeed, "A reliable iterative method for solving the epidemic model and the prey and predator problems," *Int. J. Basic Appl. Sci.*, vol. 3, pp. 441-450, Nov. 2014.

[25]   M. AL-Jawary, "A reliable iterative method for Cauchy problems," *Mathematical Theory and Modeling*, vol. 4, no. 13, pp. 148-153, 2014.

[26]   M. AL-Jawary and H. AL-Qaissy, "A reliable iterative method for solving Volterra integro-differential equations and some applications for the Lane-Emden equations of the first kind," *Mon. Notices Royal Astron. Soc.*, vol. 448, pp. 3093-3104, Mar. 2015.

[27]   A. Hemeda, "New iterative method: application to nth-order integro-differential equations," *International Mathematical Forum*, vol. 7, no. 47, pp. 2317-2332, 2012.

[28]   A. Hemeda, "Solution of fractional partial differential equations in fluid mechanics by extension of some iterative method," *Abstr. Appl. Anal.*, vol. 2013, Dec. 2013.

[29]   A. Hemeda, "Formulation and solution of nth-order derivative fuzzy integro-differential equation using new iterative method with a reliable algorithm," *J. Appl. Mathe.*, vol. 2012, Oct. 2012.

[30]   A. Hemeda, "New iterative method: an application for solving fractional physical differential equations," *Abstr. Appl. Anal.*, vol. 2013, May. 2013.

[31]   A. Hemeda, "New iterative method for solving gas dynamic equation," *Int. J. Appl. Math. Research*, vol. 3, pp. 190-195, May. 2014.

[32]   O. A. Taiwo and O. S. Odetunde, "Biharmonic equations by an iterative decomposition method," *J. Math. Sci.*, vol. 20, no. 1, pp. 37-44, 2009.

[33]   M. Sari, A. Gunay, and G. Gurarslan, "Approximate solutions of linear and non-linear diffusion equations by using Daftardar-Gejji-Jafari's method," *Int. J. Mathematical Modelling and Numerical Optimisation*, vol. 2, pp. 376-386, Jan. 2011.

[34]   H. Jafari, M. Ahmadi, and S. Sadeghi, "Solving singular boundary value problems using daftardar-jafari method," *Appl. Appl. Math.*, vol. 7, pp. 357-364, Jun. 2012.

[35]   S. T. Mohyud-Din, A. Yildirim, and M. Hosseini, "An iterative algorithm for fifth-order boundary value problems," *World Appl. Sci. J.*, vol. 8, no. 5, pp. 531-535, 2010.

[36]   A. Bibi, A. Kamran, U. Hayat, and S. T. Mohyud-Din, "New iterative method for time-fractional Schrödinger equations," *World J. Model. Simul.*, vol. 9, no. 2, pp. 89-95, 2013.

[37]   M. A. Ramadan and M. S. Al-luhaibi, "New iterative method for solving the Fornberg-Whitham equation and comparison with homotopy perturbation transform method," *Br. J. Math. Computer Sci.*, vol. 4, pp. 1213-1227, May 2014.

[38]   M. Sari, A. Gunay, and G. Gurarslan, "A solution to the telegraph equation by using DGJ method," *Int. J. Nonlinear Sci.*, vol. 17, no. 1, pp. 57-66, 2014.

[39]   H. Jafari, H. Tajadodi, A. Bolandtalat, and S. Johnston, "A decomposition method for solving the fractional Davey-Stewartson equations," *Int. J. Appl. Comput. Math.*, vol. 1, pp. 559-568, Dec. 2015.

[40]   J. Patade and S. Bhalekar, "Approximate analytical solutions of Newell-Whitehead-Segel equation using a new iterative method," *World J. Model. Simul.*, vol. 11, no. 2, pp. 94-103, 2015.

[41]   H. Kocak and A. Yldrm, "An efficient new iterative method for finding exact solutions of nonlinear time-fractional partial differential equations," *Nonlinear Anal. Model. Control*, vol. 16, pp. 403-414, Dec. 2011.

[42]   M. Yaseen, M. Samraiz, and S. Naheed, "Exact solutions of Laplace equation by DJ method," *Results in Physics*, vol. 3, pp. 38-40, Dec. 2013.

[43]   M. S. U. Rehman, M. J. Amir, and T. Kamran, "Using new iterative method to find the exact solution for a class of stiff systems of equations," *Int. J. Sci. Research*, 2014.

[44]   H. Jafari and S. Seyfi, "An iterative method for solving a system of nonlinear algebraic equations," *J. Appl. Math.*, Jan. 2008.

[45]   V. Daftardar-Gejji, *Fractional Calculus: Theory and its Applications (Ed)*. Narosa Publishing House, 2014.

[46]   M. Yaseen, M. Ahmad, and M. Samraiz, "Rational solutions of kdv, k(2,2), Burgers and cubic Boussinesq equations by using DJ method," *Nonlinear Analysis Forum*, no. 18, pp. 163-168, 2013.

[47]   K. Yahya, J. Biazar, H. Azari, and P. R. Fard, "Homotopy perturbation method for image restoration and denoising," *arXiv preprint arXiv:1008.2579*, Aug. 2010.

[48]   M. Kumar and A. S. Saxena, "New iterative method for solving higher order kdv equations," *Int. J. Advance Research Sci. Eng.*, 2016.

[49]   M. Kumar, *Decomposition methods for ordinary and fractional differential equations*. MPhil thesis, SPPU, 2015.

[50]   M. Inc, "Exact solutions with solitary patterns for the zakharov-kuznetsov equations with fully nonlinear dispersion," *Chaos Solitons Fractals*, vol. 33, pp. 1783-1790, Aug. 2007.

[51]   M. S. Al-luhaibi, "New iterative method for fractional gas dynamics and coupled Burger's equations," *The Scientific World Journal*, vol. 2015, Mar. 2015.

[52]   W. Al-Hayani, "Daftardar-Jafari method for fractional heat-like and wave-like equations with variable coefficients," *Appl. Math.*, vol. 8, p. 215, Feb. 2017.

[53]   D. Lesnic, "A computational algebraic investigation of the decomposition method for time-dependent problems," *Appl. Math. Comput.*, vol. 119, pp. 197-206, Apr. 2001.

[54]   D. Lesnic, "The Cauchy problem for the wave equation using the decomposition method," *Appl. Math. Lett.*, vol. 15, pp. 697-701, Aug. 2002.

[55] S. Abbasbandy, "Improving Newton-Raphson method for nonlinear equations by modified Adomian decomposition method," *Appl. Math. Comput.*, vol. 145, pp. 887-893, Dec. 2003.

[56] H. Jafari, M. Nazari, D. Baleanu, and C. Khalique, "A new approach for solving a system of fractional partial differential equations," *Comput. Math. Appl.*, vol. 66, pp. 838-843, Sep. 2013.

[57] L. Yan, "Numerical solutions of fractional Fokker-Planck equations using iterative Laplace transform method," *Abstr. Appl. Anal.*, vol. 2013, Dec. 2013.

[58] R. Najafi, G. D. Küçük, and E. Çelik, "Modified iteration method for solving fractional gas dynamics equation," *Math. Method Appl. Sci.*, vol. 40, pp. 939-946, Mar. 2017.

[59] V. Daftardar-Gejji and H. Jafari, "An iterative method for solving nonlinear functional equations," *J. Math. Anal. Appl.*, vol. 316, pp. 753-763, Apr. 2006.

[60] V. Daftardar-Gejji, Y. Sukale, and S. Bhalekar, "A new predictor-corrector method for fractional differential equations," *Appl. Math. Comput.*, vol. 244, pp. 158-182, Oct. 2014.

[61] K. Diethelm, N. J. Ford, and A. D. Freed, "A predictor-corrector approach for the numerical solution of fractional differential equations," *Nonlinear Dyn.*, vol. 29, pp. 3-22, Jul. 2002.

[62] K. Diethelm, N. J. Ford, and A. D. Freed, "Detailed error analysis for a fractional Adams method," *Numer. Algorithms*, vol. 36, pp. 31-52, May 2004.

[63] S. Bhalekar and V. Daftardar-Gejji, "Chaos in fractional order financial delay system," *Comput. Math. Appl.*, Mar. 2016.

[64] Y. Sukale and V. Daftardar-Gejji, "New numerical methods for solving differential equations," *Int. J. Appl. Comput. Math.*, pp. 1-22, 2016.

[65] S. Bhalekar and J. Patade, "A novel third order numerical method for solving Volterra integro-differential equations," *arXiv preprint arXiv:1604.08863*, 2016.

[66] M. Zennaro, "Asymptotic stability analysis of Runge-Kutta methods for nonlinear systems of delay differential equations," *Numerische Mathematik*, vol. 77, pp. 549-563, Oct. 1997.

[67] V. Daftardar-Gejji, Y. Sukale, and S. Bhalekar, "Solving fractional delay differential equations: A new approach," *Fract. Calc. Appl. Anal*, vol. 18, pp. 400-418, Mar. 2015.

CHAPTER 10

Derivatives with Non-Singular Kernels from the Caputo-Fabrizio Definition and Beyond: Appraising Analysis with Emphasis on Diffusion Models

Jordan Hristov*

Dept. Chem. Eng, UCTM, Sofia 1756, 8 Kl. Ohridsky Blvd, Bulgaria

> *You observe that in the life of the intellect there is also a law of inertia. Everything continues to move along its old rectilinear path, and every change, every transition to new and modern ways, meets strong resistance.*
>
> *Fleix Klein (1849-1925).*

Abstract: This chapter presents an attempt to collate existing data about fractional derivatives with non-singular kernels conceived by Caputo and Fabrizio in 2015. The idea attracted immediately the interest of the researcher and the text encompasses the consequent developments of the idea with new derivatives of Riemann-Liouville type and the generalization with kernels expressed by the Mittag-Leffler function. The chapter especially stresses the attention on diffusion equations where the Caputo-Fabrizio time-fractional derivative naturally appears as a relaxation term when the constitutive equation relating the flux and the gradient contains either Cattaneo exponential kernel or Jeffrey kernel. Four models are considered demonstrating the technology of diffusion model derivation. A special section is devoted to a spatial derivative of Caputo-type with exponential non-singular kernel for materials exhibiting spatial memory. Critical comments and suggestions are devoted to the formalistic fractionalization approach and the outcomes of this reasonless operation.

Keywords: Caputo-Fabrizio derivative, non-singular kernels, diffusion equation.

AMS Subject Classification: 26A33, 35K05, 40C10.

10.1. INTRODUCTION

The present chapter focuses on developments in a new branch of fractional calculus based on fractional operators with exponential, non-singular smooth kernel conceived in 2015 by Caputo and Fabrizio [1]. This is a hot topic in modelling of

---

*Corresponding author **Jordan Hristov:** Dept. Chem. Eng, UCTM, Sofia 1756, 8 Kl. Ohridsky Blvd, Bulgaria; Tel: +359 885 82 77 12; E-mail: jordan.hristov@mail.bg

Sachin Bhalekar (Ed.)

dissipative phenomena by fractional derivatives with non-singular kernels in contrast to the widely used fractional derivatives with weakly singular memories of Riemann-Liouville and Caputo type [2]. We will skip analyzes of such type derivatives and dissipative phenomena modeled by them and will concentrate the attention on the new trend in fractional modeling which began after the seminal work of Caputo and Fabrizio [1].

The new definition of a fractional derivative immediately attracted the interest of the scientific society and numerous articles were published for the last two years. This chapter makes an attempt to present briefly the properties of these new derivatives (fractional operators) (see section 1) as well as new results provoked by the idea conceived in [1]. We try to encompass the basic definitions, their properties and the relevant Laplace transformations. The second, important part of this chapter (section 2 and section 3) is devoted to diffusion problems and how the Caputo-Fabrizio derivative emerges in them. Solutions are developed for diffusion problems where the formulation starts from well-defined constitutive equations relating the flux and the gradient. Section 4 focuses the attention on the formalistic fractionalization approach with the Caputo-Fabrizio integral operator with memory (we use this term also in the chapter) and the emerging problems. Section 5 is devoted to an attempt to explain the origin of the Caputo-Fabrizio integral operator on the basis of the Kohlrausch stretched-exponentially relaxation function and its asymptotic behaviours at small and large times. Some real world examples are commented in section 5 as a support of the Caputo-Fabrizio integral operator with memory in modelling of dissipative phenomena.

10.1.1. Fractional Derivatives with Non-Singular Kernels

10.1.1.1. Basic Definition of Caputo-Fabrizio Time-fractional Derivative

The appearance of the new definition of fractional derivative with exponential kernel was an answer to the demand that many classical constitutive equations (see the comments in [1] and the references therein) cannot model adequately dissipative process of transfer of energy (heat), mass (diffusion) and stress behaviour in many new materials appearing in modern technologies.

The answer of Caputo and Fabrizio to the new call in modeling of relaxation process in dissipative phenomena was the following constitutive definition [1]

$$_{cf}D_t^\alpha f(t) = \frac{M(\alpha)}{1-\alpha} \int_0^t exp\left[-\frac{\alpha(t-s)}{1-\alpha}\right]\frac{df(s)}{dt}ds \qquad (10.1)$$

where in (10.1) $M(\alpha)$ is a normalization function such that $M(0) = M(1) = 1$.

This is a definition of Caputo-type since under the integral sign we have the derivative $df(t)/dt$. In some cases in this chapter we will denote it as $^c_{cf}D^\alpha_t$ to distinguish it from the so-called Riemann-Liouville type (see further in this chapter) denoted as $^R_{cf}D^\alpha_t$

In accordance with the explanations of Caputo and Fabrizio [1] the definition (10.1) can be easily developed if in the classical Caputo derivative [2]

$$_cD^\alpha_t = \frac{1}{\Gamma(1-\alpha)} \int_a^1 \frac{df(s)}{dt} \frac{1}{(t-s)^\alpha} ds \qquad (10.2)$$

with $\alpha \in [0,1]$ and $a \in [-\infty, t]$ and $f(t) \in H^1(a,b)$ for $b > a$, the kernel $(t-s)^{-\alpha}$ is replaced by the exponential function $exp[-(\alpha t/(1-\alpha)]$ and $1/\Gamma(1-\alpha)$ by $M(\alpha) = M(\alpha)/(1-\alpha)$. In this way we get the definition (10.1).

This is a formal explanation without any physics behind. However, it is possible to demonstrate, and this will be done in this chapter, that the definition (10.1) comes naturally from the constitutive equation relating the flux and gradient by exponential damping functions. Nevertheless before focusing the attention deeply on the physics invoking use of this new derivative we will present some basic mathematical properties of it.

10.1.1.2. Basic Properties of $^c_{cf}D^\alpha_t$

In accordance with the definition (10.1) it follows directly that if $f(t) = C = const.$, then $^c_{cf}D^\alpha_t C = 0$ as in the classical definition (10.2) but now the kernel is not singular. The definition (10.1) can be applied to functions which are not from $H^1(a,b)$ and then from (1) we have

$$^c_{cf}D^\alpha_t = \frac{\alpha M(\alpha)}{1-\alpha} \int_{-\infty}^t [f(t) - f(a)] exp\left[-\frac{\alpha}{1-\alpha}(t-s)\right] ds \qquad (10.3)$$

for $f \in L^1(-\infty, b)$ and for any $\alpha \in [0,1]$.

If we change the notation as

$$\sigma = \frac{1-\alpha}{\alpha} \in [0,\infty], \alpha = \frac{1}{1+\sigma} \in [0,1] \qquad (10.4)$$

then the basic definition (10.1) can be expressed as

$$\,^c_{cf}D^\alpha_t = \frac{N(\sigma)}{\sigma} \int_0^1 \frac{df(s)}{dt} exp\left[-\frac{(t-s)}{\sigma}\right] ds \qquad (10.5)$$

where $N(\sigma)$ is the normalization term of the function $M(\alpha)$ such that $N(0) = N(\infty) = 1$ as well as

$$lim_{\sigma\to0}\frac{1}{\sigma}exp\left[-\frac{t-s}{\sigma}\right] = \delta(t-s) \qquad (10.6)$$

Obviously for $\alpha \to 1$ we get $\sigma \to 0$. Consequently (10.1) can be re-formulated as [1]

$$\,^c_{cf}D^\alpha_t{}_{\alpha\to1} = lim_{\alpha\to1}\frac{M(\alpha)}{1-\alpha}\int_0^t exp\left[-\frac{\alpha(t-s)}{1-\alpha}\right]\frac{df(t)}{dt}ds =$$

$$= lim_{\sigma\to0}\frac{N(\sigma)}{\sigma}\int_0^t exp\left[-\frac{\alpha(t-s)}{1-\alpha}\right]\frac{df(t)}{dt}ds = \frac{df(t)}{dt} \qquad (10.7)$$

In the opposite case when $\alpha \to 0$ we have [1]

$$\,^c_{cf}D^\alpha_t{}_{\alpha\to0} = lim_{\alpha\to0}\frac{M(\alpha)}{1-\alpha}\int_0^t exp\left[-\frac{\alpha(t-s)}{1-\alpha}\right]\frac{df(t)}{dt}ds =$$

$$= (lim_{\sigma\to1}\frac{N(\sigma)}{\sigma}\int_0^t exp\left[-\frac{\alpha(t-s)}{1-\alpha}\right]\frac{df(t)}{dt}ds = f(t) - f(a) \qquad (10.8)$$

The definitions presented up to this moment directly say that the new derivative is a convolution of $f(t)$ and the convolution operator [3]

$$K = exp\left[-\frac{\alpha}{1-\alpha}(t-s)\right]\frac{d}{ds} \qquad (10.9)$$

Moreover, the integration by parts in (10.1) results in [3]

$$\,^c_{cf}D^\alpha_t = \frac{1}{1-\alpha}f(t) - \frac{\alpha}{1-\alpha}\int_a^t f(s)exp\left[-\frac{\alpha}{1-\alpha}\right]ds, t \geq a \qquad (10.10)$$

and when $\alpha > 0$ in an equivalent form [3]

$$\,^c_{cf}D^\alpha_t = \frac{\alpha}{(1-\alpha)^2}\int_a^t [f(t) - f_a(s)]\left[-\frac{\alpha(t-s)}{1-\alpha}\right]ds, t > a \qquad (10.11)$$

This definition requires $f_a(t) = f(t)$ for $a \leq t < \infty$ and $f_a(t) = 0$, that is $f(t)$ is a casual function. The second term of (10.10) mimics the well-known form of integral in the Riemann-Liouville derivative with singular kernel [2] and will be commented further in this chapter.

Moreover, it follows that

$$lim_{\alpha \to 1}\left[\,^{c}_{cf}D^{\alpha}_{t}\right] = \frac{d}{dt}f(t) \tag{10.12}$$

while

$$lim_{\alpha \to 0}\left[\,^{c}_{cf}D^{\alpha}_{t}\right] = f(t) - f(a) \tag{10.13}$$

Further, if $n > 0$ and $\alpha \in [0,1]$, then [1]

$$^{c}_{cf}D^{\alpha}_{t}f(t) = {}^{c}_{cf}D^{\alpha}_{t}\left(D^{(n)}_{t}f(t)\right) \tag{10.14}$$

For example when $n = 1$ it follows that

$$^{c}_{cf}D^{\alpha}_{t}\left(D^{(n)}_{t}f(t)\right) = \frac{M(\alpha)}{1-\alpha}\int_{a}^{t}\frac{d^2}{dt^2}f(t)exp\left[-\frac{\alpha(t-s)}{1-\alpha}\right]ds \tag{10.15}$$

From the integration by parts and with the assumption that $df(a)/dt = 0$ it follows from the definition (10.15) that

$$^{c}_{cf}D^{\alpha}_{t}\left(D^{(1)}_{t}f(t)\right) = \frac{M(\alpha)}{1-\alpha}\left[\int_{a}^{t}\frac{d}{dt}f(s)exp\left[-\frac{\alpha(t-s)}{1-\alpha}\right]ds\right] \tag{10.16}$$

In all these derivations Caputo and Fabrizio accepted that the normalization function is $M(\alpha) = 1$.

10.1.1.3. Normalization Function M(α), Associated Fractional Integral and an Alternative Definition of $^{c}_{cf}D^{\alpha}_{t}$

Losada and Nieto [4] considered the fractional equation

$$^{c}_{cf}D^{\alpha}_{t}f(t) = u(t), t > 0 \tag{10.17}$$

with a solution

$$f(t) = \frac{2(1-\alpha)}{(2-\alpha)M(\alpha)}u(t) + \frac{2(1-\alpha)}{(2-\alpha)M(\alpha)}\int_{0}^{t}u(s)ds + f(0), 0 \leq \alpha \leq 1 \tag{10.18}$$

and suggested the following associated fractional integral ${}^{cf}_{LN}I^\alpha f(t)$

$$ {}^{cf}_{LN}I^\alpha f(t) = \frac{2(1-\alpha)}{(2-\alpha)M(\alpha)}u(t) + \frac{2(1-\alpha)}{(2-\alpha)M(\alpha)}\int_0^t u(s)ds, 0 \le \alpha \le 1 \quad (10.19) $$

Therefore, from the definition (10.19) it follows that the associated fractional inte-gral ${}^{cf}_{LN}I^\alpha f(t)$ is an average between the function $f(t)$ and its integral of order one.

From the condition [4]

$$ \frac{2(1-\alpha)}{(2-\alpha)M(\alpha)} + \frac{2\alpha}{(2-\alpha)M(\alpha)} = 1 \quad (10.20) $$

we may define directly $M(\alpha)$ as

$$ M(\alpha) = \frac{2}{2-\alpha}, 0 \le \alpha \le 1 \quad (10.21) $$

The definition (10.21) differs from $M(\alpha) = 1$ used by Caputo and Fabrizio [1, 3]. Because of that, Losada and Nieto [4] suggested a new definition of the fractional derivative of Caputo-Fabrizio of order α of a function $f(t)$, namely

$$ {}_{cf}D_t^\alpha = \frac{1}{1-\alpha}\int_0^t exp\left[-\frac{\alpha(t-s)}{1-\alpha}\right]\frac{df(s)}{dt}ds \quad (10.22) $$

Hence, $M(\alpha)$ is accepted to equal 1.

Here we use the notation ${}^{cf}_{LN}I^\alpha f(t)$ to distinguish this fractional integral defined by Losada and Nieto [4] from that used by Caputo and Fabrizio [3] in constitutive relations leading to fractional diffusion equation (see section 2.1.1).

10.1.1.4. Relation to the Associated Ordinary Derivative

The result of the integer order differentiation of ${}^c_{cf}D_t^\alpha$ that is ${}^c_{cf}D_t^\alpha\left(D_t^{(1)}f(t)\right)$ was commented above. Recently Ciancio and Flora [5] formulated a very interesting problem, namely

If $g(t)$ is defined by

$$g(t) = \int_a^t f(s) exp\left[\frac{\gamma}{1-\gamma}(t-s)\right] ds, \gamma \in (0,1) \tag{10.23}$$

Then,

$$\frac{d}{dt}[g(t)] = D_t^{(1)}[g(t)] = \left[\frac{1-\gamma}{M(\gamma)}\right] D_t^{(\gamma)}[f(t)] \tag{10.24}$$

From the definition (10.1) it follows that

$$D_t^{(1)}[g(t)] = \int_a^t \frac{d}{ds}[f(s)] exp\left[\frac{\gamma}{1-\gamma}(t-s)\right] ds \tag{10.25}$$

Taking into account the basic definition (10.1), then the result (10.25) can be presented as

$$D_t^{(1)}[g(t)] = D_t^{(\gamma)}[f(t)] \tag{10.26}$$

10.1.1.5. Laplace Transform

The Laplace transform L_T of $_{cf}^{c}D_t^{\alpha}$ with a lower terminal $a = 0$ given with p variable [1] taking into account the general rule of Laplace transform of a convolution is

$$L_T\left[\,_{cf}^{c}D_t^{\alpha}\right] = \frac{1}{1-\alpha} L_T[f(t)]L_T\left[exp\left(-\frac{\alpha t}{1-\alpha}\right)\right] \tag{10.27}$$

Hence,

$$L_T\left[\,_{cf}^{c}D_t^{\alpha}\right] = \frac{pL_T[f(t)-f(0)]}{p+\alpha(1-p)} \tag{10.28}$$

Consequently,

$$L_T\left[\,_{cf}^{c}D_t^{(\alpha+1)}\right] = \frac{p^2 L_T[f(t)-pf(t)-f\prime(0)]}{p+\alpha(1-p)} \tag{10.29}$$

and generally,

$$L_T\left[\,_{cf}^{c}D_t^{(\alpha+n)}\right] = \frac{p^{n+1}L_T[f(t)-p^n f(0)-p^{n-1}f\prime(0)...f^{(n)}(0)]}{p+\alpha(1-p)} \tag{10.30}$$

10.1.1.6. Fractional Derivative of Elementary and Transcendental Functions

Linear function $f(t) = t$

If $f(t) = t$, then [1]

$$_{cf}^{c}D_t^\alpha(t) = \frac{M(\alpha)}{1-\alpha}\int_0^t exp\left[-\frac{\alpha}{1-\alpha}(t-s)\right]ds = \frac{M(\alpha)}{1-\alpha}\left[1 - exp\left[-\frac{\alpha}{1-\alpha}t\right]\right], 0 < \alpha < 1 \quad \textbf{(10.31)}$$

Power-law function $f(t) = t^\beta$

When $f(t) = t^\beta$, $\beta > 0$ the fractional derivative $_{cf}^{c}D_t^\alpha(t^\beta)$ [6] can be straightforwardly obtained through the Laplace transform, namely

$$L_T\left[_{cf}^{c}D_t^\alpha(t\beta)\right] = \frac{1}{1-\alpha}L_T[t^\beta]L_T\left[exp\left(-\frac{\alpha}{1-\alpha}t\right)\right] \quad \textbf{(10.32)}$$

With

$$L_T[t^\beta] = \frac{\Gamma(1+\beta)}{p^{1+\beta}} \quad \textbf{(10.33)}$$

The inverse Laplace transform, assuming $M(\alpha) = 1$, results in

$$_{CF}D_t^\alpha\left(t^\beta\right) = \frac{1}{\alpha}\left[1 - exp\left(-\frac{\alpha}{1-\alpha}t\right)\right]\beta t^{\beta-1} \quad \textbf{(10.34)}$$

For $\alpha = 1$ the expression (10.34) reduces to $\beta t^{\beta-1}$

Exponential function $f(t) = exp(\beta t)$ [6]

Hence, the Laplace transform $L_T[_{cf}^{c}D_t^\alpha f(t)] = L_T[_{cf}^{c}D_t^\alpha exp(\beta t)]$ for $\beta > 0$ is

$$L_T[_{cf}^{c}D_t^\alpha f(t) = exp(\beta t)] = \frac{1}{(1-\alpha)}\frac{\beta}{(p-\beta)}\frac{1}{(p+A)}, A = \frac{\alpha}{1-\alpha} \quad \textbf{(10.35)}$$

The inverse Laplace transform of (10.35) yields

$$\lim_{\alpha \to 1}[_{cf}^{c}D_t^\alpha exp(\beta t)] = \frac{\beta exp(\beta t) - exp(-At)}{\beta + \alpha(1-\beta)} \quad \textbf{(10.36)}$$

For $\alpha \to 1$ the second exponential term in the nominator of (10.36) goes to zero and therefore $lim_{\alpha \to 1}[_{cf}^{c}D_t^\alpha exp(\beta t)] \to \beta exp(\beta t)$.

For $\beta < 0$ the Laplace transform of $_{cf}^{c}D_t^\alpha exp(-\beta t)$ is [6]

$$L_T[^C_{cf}D^\alpha_t exp(-\beta t)] = \left(\frac{1}{1-\alpha}\right)\frac{-\beta}{(p+\beta)}\frac{1}{(p+A)} \tag{10.37}$$

Then, the inverse Laplace transform of (10.37) is

$$^c_{cf}D^\alpha_t exp(-\beta t) = \frac{-\beta exp(-\beta t)+exp(-At)}{A-\beta} \tag{10.38}$$

For $\alpha \to 1$ the second exponential term in the nominator of (10.38) goes to zero and therefore $lim_{\alpha\to 1}[^C_{cf}D^\alpha_t exp(-\beta t)] \to -\beta exp(-\beta t)$.

The more general expression is provided by Caputo and Fabrizio [1], namely (in original notations)

$$^C_{cf}D^\alpha_t[exp(\omega t)] = \frac{E(\alpha)\omega\left(1-exp\left[-\left(\omega+\frac{\alpha}{1-\alpha}\right)t\right]\right)}{\left(\frac{\alpha}{1-\alpha}+\omega\right)}exp(\omega t), E(\alpha) = \frac{M(\alpha)}{1-\alpha} \tag{10.39}$$

For $f(t) = sin(\omega t)$ we have

$$^c_{cf}D^\alpha_t[sin(\omega t)] = E(\alpha)\int_0^t \omega exp\left[-\frac{\alpha}{1-\alpha}(t-s)\right]cos(\omega s)ds, E(\alpha) = \frac{M(\alpha)}{1-\alpha} \tag{10.40}$$

From (10.40) it follows that

$$^c_{cf}D^\alpha_t[sin(\omega t)] = E(\alpha)cos[a_\omega sin(\omega t + a_\omega)] - sin\left[a_\omega exp\left(-\frac{\alpha t}{1-\alpha}\right)\right] \tag{10.41}$$

where a_ω is defined through the relationships

$$tan a_\omega = \frac{a\omega}{1-\alpha}, sin\{a_\omega\} = \frac{\frac{\alpha}{1-\alpha}}{\left[\left(\frac{\alpha}{1-\alpha}\right)^2+\omega^2\right]^{\frac{1}{2}}},$$

$$cos a_\omega = \frac{\omega}{\left[\left(\frac{\alpha}{1-\alpha}\right)^2+\omega^2\right]^{\frac{1}{2}}} \tag{10.42}$$

10.1.2. Fractional Derivative with Non-singular Kernel of Riemann-Liouville Type

From the definition (10.1) and with the background of fractional derivative with singular kernels such as that defined by (10.2) the natural question which immediately springs to minds is: *Is it possible to construct a fractional derivative without singular kernel which mimics the construction of the well-known Riemann-Liouville derivative* [2]?

$$_{RL}D_t^\alpha = \frac{1}{\Gamma(1-\alpha)} \frac{d}{dt} \int_a^t f(s) \frac{1}{(t-s)^\alpha} ds \qquad (10.43)$$

The clear answer was given by several authors. In the second article of Caputo and Fabrizio [3] it was briefly mentioned (section 2.1 of [3]) that after integration by parts in (10.1) and combination with the convolution kernel (10.9) the desired fractional derivative can be defined as

$$_{cf}^R D_t^\alpha f(t) = \frac{M(\alpha)}{1-\alpha} \frac{d}{dt} \int_a^t f(s) exp\left[-\frac{\alpha}{1-\alpha}(t-s)\right] ds \qquad (10.44)$$

The same definition was given by Goufo and Atangana [7].

With the definition of Losada and Nieto (10.22) is it possible to express (10.44) as

$$_{cf}^R D_t^\alpha f(t) = \frac{(2-\alpha)M(\alpha)}{2(1-\alpha)} \frac{d}{dt} \int_a^t f(s) exp\left[-\frac{\alpha}{1-\alpha}(t-s)\right] ds \qquad (10.45)$$

It is clear that for $\alpha \to 1$ we have

$$lim_{\alpha \to 1}[_{cf}^R D_t^\alpha f(t)] = \frac{d}{dt} f(t) \qquad (10.46)$$

In the opposite case when $\alpha \to 0$ we have

$$lim_{\alpha \to 0}[_{cf}^R D_t^\alpha f(t)] = f(t) \qquad (10.47)$$

which differs from (10.13)

In accordance with the definition (10.44) we have a derivative of convolution, namely

$$_{cf}^R D_t^\alpha f(t) = \frac{M(\alpha)}{1-\alpha} \frac{d}{dt} [f(t) * g(t)], g(t) = exp\left[-\frac{\alpha}{1-\alpha}t\right] \qquad (10.48)$$

where in the standard notation the convolution product is defined by the integral

$$f(t) * g(t) = \int_a^t f(t) g(t-s) ds \qquad (10.49)$$

The $(n + \alpha)$ derivative of $f(t)$ with $_{cf}^{C}D_t^{\alpha}f(t)$ if $n \in \mathbb{N}$ and $\alpha \in [0,1]$, is defined as

$$_{cf}^{R}D_t^{(\alpha+n)}f(t) = _{cf}^{R}D_t^{\alpha}\left[_{cf}^{R}D_t^{n}f(t)\right] \tag{10.50}$$

The Laplace transform of $_{cf}^{R}D_t^{\alpha}f(t)$, taking $a = 0$ and the definition through the derivative of convolution (see (10.48) and (10.49)) is

$$L_T\left[_{cf}^{R}D_t^{\alpha}f(t), p\right] = \frac{M(\alpha)}{1-\alpha} L_T\left[\frac{d}{dt}\int_0^t f(s)exp(-\frac{\alpha}{1-\alpha}(t-s))\,ds\right] =$$

$$= \frac{M(\alpha)}{1-\alpha} L_T\left[\frac{d}{dt}(f(t) * g(t))\right] = \frac{M(\alpha)}{1-\alpha}[L_Tf(t),p][L_Tg(t),p] \tag{10.51}$$

Therefore,

$$L_T\left[_{cf}^{R}D_t^{\alpha}f(t), p\right] = \frac{pM(\alpha)}{p+\alpha(1-p)} L_T[f(t),p] \tag{10.52}$$

As in the case with the classic fractional derivatives with singular kernels, either $_{cf}^{C}D_t^{\alpha}$ or $_{cf}^{R}D_t^{\alpha}$ do not obey the chain rule in differentiation [7], namely

$$_{cf}^{C}D_t^{\alpha}f[g(t)] \neq R_{cf}D_x^{\alpha}f(x)\frac{dx(t)}{dx}, and \; _{cf}^{C}D_t^{\alpha}f[g(t)] \neq \frac{df(x)^C}{dt}\,_{cf}D_x^{\alpha} \tag{10.53}$$

Goufo [8] used this new definition of fractional derivative in analysis of Korteveg-de Vries-Burgers equation and its numerical solution.

At this point, in order to be correct with respect to the scientific society and mainly with respect to the scientists who invented new tools in fractional calculus, we have to mention that the definition (10.44) has a space-dependent replica (proposed in December 2015 and published in the middle of 2016) [9] (with respect to space coordinate x), namely (in original notations)

$$_{cf}^{Y}D_x^{\alpha}f(x) = \frac{R(v)}{1-v}\frac{d}{dx}\int_a^x f(\lambda)exp\left[-\frac{v}{q-v}(x-\lambda)\right]d\lambda, a \leq x, v \in [0,1] \tag{10.54}$$

This definition, was commented earlier in the book of Yang, Baleanu and Srivastava [10].

Further, $^Y_{cf}D^\alpha_x$, named as Yang-Srivastava-Machado space derivative, was used to solve simple problems related to steady-state heat conduction [11, 12] and flow problems [13]. This derivative and the steady-state heat conduction problem solved by it will be commented further in the chapter.

Briefly, after the seminal articles of Caputo and Fabrizio [1, 3], where the new fractional derivative $^c_{cf}D^\alpha_t$ (10.1) was postulated, the consequences as the Riemann-Liouville-type version $^R_{cf}D^\alpha_t$ constructed by analogy, as well as $^Y_{cf}D^\alpha_x$ *are directly postulated without any proofs and explanations how they emerge in the modelling equations.* It may be straightforwardly checked that $^R_{cf}D^\alpha_t$ can be developed as a stand of portion of differentiation by parts in the convolution integral of (10.1), but this is beyond the scope of the present chapter. Remember, the classical Riemann-Liouville integral is a direct consequence of the multiple Cauchy integral and its physical origin could be traced back to the Abel integral equation [2].

In this context, this chapter summarizes examples related to the classical diffusion problem, in transient and steady-state, where starting from well-defined constitutive equations relating the flux and the gradient, the construction of $^c_{cf}D^\alpha_t$ defined by (10.1) appears directly when the balance equations of energy (mass) is applied.

10.1.3. Some Applications-Briefly

The Caputo-Fabrizio time-fractional operator (derivative) immediately attracted the interest of the researchers and for about a year after the seminal publication [1] numerous applications have been reported, among them: elasticity [3], resistance and numerical modelling of fractional electric circuit [14-18], electromagnetic waves in dielectric media [19-21], mass-spring damped systems [22, 23], system analysis and modelling [24, 25], flow of complex rheological media [26, 27] and fluids [13, 28-31] the Keller-Segel model [32], Fisher reaction-diffusion equation [33], groundwater flow [34, 35], coupled systems of time-fractional differential problems [36], fractional difference operators [37], approximate solutions of some fractional integro-differential equations [38], initiated new computational techniques [39-48], and boosted many applications of new fractional derivatives with non-singular kernels [43, 44, 46, 47, 49-60].

A common feature of all these publications is the extensive use of *the formalistic fractionalization approach* where the time derivatives in well-known integer-order models are *ad hoc* replaced by fractional ones.

10.1.4. Generalized Fractional Derivative with Non-singular Kernel *via* the Mittag-Leffler Function

As a consequence, provoked by the definition (10.1) and taking into account that the exponential function is a particular case of the Mittag-Leffler function [2], the generalized time-fractional derivative with a non-singular kernel was defined as [61]

Caputo-type Definition

$$\,_{a}^{ABC}D_{t}^{\alpha} = \frac{B(\alpha)}{1-\alpha} \int_{0}^{t} \frac{d}{dt} f(s) E_{\alpha}\left[-\frac{\alpha}{1-\alpha}(t-s)^{\alpha}\right] ds \qquad (10.55)$$

for $f(t) \in H^1(a,b), a < b, \alpha \in [0,1]$

Riemann-Liouville-type Definition

$$\,_{a}^{ABR}D_{t}^{\alpha} = \frac{B(\alpha)}{1-\alpha} \frac{d}{dt} \int_{0}^{t} f(s) E_{\alpha}\left[-\frac{\alpha}{1-\alpha}(t-s)^{\alpha}\right] ds \qquad (10.56)$$

for $f(t) \in H^1(a,b), a < b, \alpha \in [0,1]$

The definitions (10.55) and (10.56) are logical derivations starting from the following fractional ordinary equation [62]

$$\frac{d^{\alpha} y}{dt^{\alpha}} = by, 0 < \alpha < 1 \qquad (10.57)$$

and the solution can be expressed by a single-parameter Mittag-Leffler function defined as [2]

$$E_{\alpha}(-t^{\alpha}) = \sum_{k=0}^{\infty} \frac{(-t)^{\alpha k}}{\Gamma(\alpha k+1)} \qquad (10.58)$$

With the Taylor series expansion of $exp[-b(t-y)] = \sum_{k=0}^{\infty} \frac{[-b(t-y)]^k}{k!}$ and defining $b = \alpha/(1-\alpha)$, the replacement in the basic definition (10.1) yields

$$\,_{cf}^{c}D_t^\alpha = \frac{M(\alpha)}{1-\alpha} \sum_{k=0}^\infty \frac{(-b)^k}{k!} \int_0^1 \frac{df(y)}{dy}(t-y)^k dy \qquad (10.59)$$

Looking for a non-local solution of the problem (10.57), let us replace $k!$ in the denominator of (10.59) by $\Gamma(\alpha k + 1)$ as well as in the nominator $(t - y)^k$ by $(t - y)^{\alpha k}$. The results is the definition (10.55) of Caputo-type derivative. By analogy, it is easy to formulate the derivative expressed by (10.56).

The Laplace transforms of the new derivatives are [61]

$$L_T\big[\,_a^{ABC}D_t^\alpha f(t), p\big] = \frac{B(\alpha)}{1-\alpha}\frac{pL_T[f(t)](p) - p^{\alpha-1}f(0)}{p^\alpha + \frac{\alpha}{1-\alpha}} \qquad (10.60)$$

and

$$L_T\big[\,_a^{ABR}D_t^\alpha f(t), p\big] = \frac{B(\alpha)}{1-\alpha}\frac{pL_T[f(t)](p)}{p^\alpha + \frac{\alpha}{1-\alpha}} \qquad (10.61)$$

The relationship between $\,_a^{ABC}D_t^\alpha$ and $\,_a^{ABR}D_t^\alpha$ is simple [61]

$$\,_a^{ABC}D_t^\alpha f(t) = \,_a^{ABR}D_t^\alpha f(t) - \frac{B(\alpha)}{1-\alpha}f(0)E_\alpha\left(-\frac{\alpha}{1-\alpha}t^\alpha\right) \qquad (10.62)$$

Hence, with zero initial condition, *i.e.* with $f(0) = 0$ both derivatives are identical (as it happens with the classical Caputo and Riemann-Liouville derivatives with singular kernels).

The associate fractional integrals can be straightforwardly determined if we consider the simple equation

$$\,_a^{ABC}D_t^\alpha f(t) = u(t) \qquad (10.63)$$

The unique solution of (10.63) is $u(t) = \,_a^{ABC}I_t^\alpha[f(t)]$, where the associate fractional integral $\,_a^{ABC}I_t^\alpha[f(t)]$ is defined as [61]

$$\,_a^{ABC}I_t^\alpha[f(t)] = \frac{1-\alpha}{B(\alpha)}f(t) + \frac{\alpha}{B(\alpha)\Gamma(\alpha)}\int_a^t f(y)(t-y)^{\alpha-1}dy \qquad (10.64)$$

The Atangana-Baleanu derivative attracted the attention of the researchers and several articles on applied problems have been published [21, 31, 53, 63].

10.1.5. Fractional Operator of Bi-Order

As a natural continuation of the fractional operators with kernels expressed by the Mittag-Leffler function, Atangana [64, 65] suggested a bi-order fractional operator which amalgamates the structures of both the classical derivatives with singular kernels and the ones with exponential memory. The Atangana-Riemann fractional operator is defined as [64, 65].

$$
{}_{a}^{AB}D_{t}^{\alpha,\beta} = \frac{B(\beta)}{n-\beta}\frac{1}{\Gamma(n-\beta)}\frac{d^{n}}{dt^{n}}\int_{0}^{t} f(t)(t-s)^{1-\alpha-n}E_{\beta}\left[-\frac{\beta}{n-\beta}(t-s)^{\beta+\alpha}\right] \quad (10.65)
$$

where E_{β} is a Mittag-Leffler function defined as

$$
E_{\beta,\phi[t]} = \sum_{m=0}^{\infty}\frac{t^{m}}{\Gamma(m+\phi)} \quad (10.66)
$$

The Laplace transform of (10.65) is

$$
L_{T}\left\{{}_{a}^{AB}D_{t}^{\alpha,\beta}[f(t),p]\right\} = \frac{B(\beta)}{1-\beta}\frac{p}{\Gamma(1-\alpha)}L_{T}\left\{t^{-\alpha}E_{\beta}\left[-\frac{\beta}{n-\beta}(t-s)^{\beta+\alpha}\right]\right\} \quad (10.67)
$$

Similarly, the Atangana-Caputo derivative of bi-order is defined as [64, 65]

$$
{}_{a}^{AC}D_{t}^{\alpha,\beta}[f(t)] = \frac{B(\beta)}{n-\beta}\frac{1}{\Gamma(n-\alpha)}\frac{d^{n}}{dt^{n}}f(t)(t-s)^{1-\alpha-n}E_{\beta}-\frac{\beta}{n-\beta}(t-s)^{\beta+\alpha}ds
$$

$$(10.68)$$

$$
n-1<\beta<n, \quad n-1, \alpha<n
$$

with a Laplace transform (for $n = 1$)

$$
L_{T}\left\{{}_{a}^{AC}D_{t}^{\alpha,\beta}[f(t),p]\right\} = \frac{B(\beta)}{1-\beta}\Gamma(1-\alpha)\{pf(p)-f(0)\}\times
$$

$$(10.69)$$

$$
\times L_{T}\left\{t^{-\alpha}E_{\beta}\left[-\frac{\beta}{1-\beta}(t-s)^{\beta+\alpha}\right]\right\}
$$

Detailed analysis of these new fractional operators is available elsewhere [64, 65].

10.2. DIFFUSION MODELS WITH TIME-FRACTIONAL CAPUTO-FABRIZIO DERIVATIVE

This section of the chapter stresses the attention on diffusion models of mass (or heat) with properly defined constitutive equations *Flux-Gradient*. This approach is developed in two sub-directions: time-dependent models with time-fractional derivatives and steady-state models with space-fractional derivatives.

10.2.1. Diffusion Models with Constitutive Equations *Flux-Gradient*

Diffusion phenomena of mass, are generally described as a consequence of the mass conservation law by the relationship

$$\frac{\partial c}{\partial t} = -\frac{\partial j}{\partial x} \tag{10.70}$$

The assumption that the mass flux $j(x,t)$ is proportional to the concentration gradient $j(x,t) = -D_0\, \partial C/\partial x$ in fact is a definition of the diffusivity D_0. Then, applying (10.70) we get the ordinary diffusion equation (the Fick law) (10.71)

$$\frac{\partial c}{\partial t} = D_0 \frac{\partial^2 c}{\partial x^2} \tag{10.71}$$

The principle drawback of the model (10.71) is the infinite speed of propagation of the flux which is unphysical.

A relaxation function related to a finite speed of diffusion (heat conduction) in solids was conceived by Cattaneo [66] as a generalization of the Fourier law by a linear superposition of the heat flux and its time derivative related to its history [67, 68]. Hence, the flux j obeys the constitutive equation [66] involving a memory integral.

$$j(x,t) = -\int_{-\infty}^{t} R(x,t)\nabla C(x,t-s)ds \tag{10.72}$$

Setting the lower terminal of the memory integral in (10.72) at zero we get a more convenient, from physical point view, expression of the constitutive equation, namely

$$j(x,t) = -\int_{0}^{t} R(x,t)\nabla C(x,t-s)ds \tag{10.73}$$

If $R(x,t)$ is assumed as the Dirac Delta function δ_D such that $\int_{0}^{0+} \delta_D(s)ds = \int_{0}^{t} \delta_D(s)ds = 1$ this leads immediately to the classical Fick (Fourier) equation (10.71) *since there is no damping effect in the flux propagation.*

The analysis in this section of the chapter will consider four models with constitutive equations relating the flux and the gradient expressed in terms of the Caputo-Fabrizio derivative which are strongly dependent on the definition of the memory kernel, namely:

i)   The model of Caputo-Fabrizio with a special exponential damping function [3]

ii) A model with Jeffrey memory kernel in the constitutive equation [69]

iii) Single-memory model with Cattaneo constitutive equation related to the fractional Dodson diffusion equation [70]

iv) Two-memories model with composite Cattaneo memory kennel as an extension of the fractional Dodson diffusion equation [70]

10.2.1.1. The Diffusion Equation of Caputo-Fabrizio [3]

In the second article of Caputo and Fabrizio [3], the associate fractional integral to the derivative (10.1) was defined as (see section 7 of [3])

$$_0I^\alpha f(t) = \frac{1}{\alpha}\int_0^t f(s)exp\left(-\frac{1-\alpha}{\alpha}(t-s)\right)ds, \alpha \in [0,1] \qquad (10.74)$$

It is noteworthy that the fractional factor (the temporal rate constant depending on α) in the exponential kernel $(1-\alpha)/\alpha$ is reciprocal to the factor used in the kernel of the derivative $\alpha/(1-\alpha)$ (10.1). For $\alpha = 0$ this definition provides directly the function $f(t)$ as well as it follows that

$$\frac{d}{dt}[_0I^\alpha f(t)] = \frac{1}{\alpha}f(t) - \frac{1-\alpha}{\alpha}[_0I^\alpha f(t)) \qquad (10.75)$$

Caputo and Fabrizio suggested the following constitutive equation for the flux-gradient relationship [3]

$$q(t) = -a_0\left(_0I^\alpha f(t)\left[\frac{\partial T(x,t)}{\partial x}\right]\right) = -a_0\frac{1}{\alpha}\int_0^t \frac{\partial T(x,s)}{\partial x}exp\left(-\frac{1-\alpha}{\alpha}(t-s)\right)ds \quad (10.76)$$

Applying the rule (10.75) to the constitutive equation (10.76) we get (eq. (10.33) in [3])

$$\frac{d}{dt}q(t) = -\frac{a_0}{\alpha}\frac{\partial T(x,t)}{\partial x} - \frac{1-\alpha}{\alpha}f(t) \qquad (10.77)$$

Equation (10.77) coincides with the Cattaneo-Maxwell equation (10.78)

$$\frac{\alpha}{1-\alpha}\frac{d}{dt}q(t) = -q(t) - \frac{a_0}{1-\alpha}\frac{\partial T(x,t)}{\partial x} \qquad (10.78)$$

This equation reduces (for $\alpha = 0$) to the Fourier (Fick) law $q(t) = -a_0\frac{\partial T(x,t)}{\partial x}$.

Recall, that if the constitutive equation is defined with a memory kernel (see further in this chapter) $exp\left[-\frac{\alpha(t-s)}{1-\alpha}\right]$ the same result can be derived for $\alpha = 1$, which is in agreement with the definition of the fractional derivative (10.1). In accordance with the definition (10.1) for $\alpha = 1$ there is no time delay.

Now, applying the mass (heat) balance equation (10.70) we obtain a form of the diffusion equation [3] with exponential memory (damped diffusion equation), namely

$$\frac{\partial T(x,t)}{\partial t} = a_0\frac{\partial^2 T(x,t)}{\partial x^2} + \frac{1-\alpha}{\alpha}\left[\frac{1}{\alpha}\int_0^t \frac{\partial T(x,t)}{\partial x}exp\left(-\frac{1-\alpha}{\alpha}(t-s)\right)ds\right] \qquad (10.79)$$

For $\alpha = 1$ we get the diffusion equation without delay.

The specific feature of (10.79) is that *the last term is expressed through the associated fractional integral ((10.76)) instead the Caputo-Fabrizio time-fractional derivative*, as it is done in the diffusion models discussed next.

10.2.1.2. *Diffusion Equation of with Jeffrey's Kernel in the Constitutive Equation*

This model was developed in [69] and we will present its principle steps demonstrating both the similarities and the differences with respect to existing one, such that of Caputo and Fabrizio (10.79) as well with respect to the model with a simple Cattaneo kernel [70] presented in the next point of this section.

For homogeneous rigid heat conductors the damping function $R(x,t)$ is space-independent and can be represented by the Cattaneo kernel [1] $R(t) = exp[-(t - s)/\tau]$ where the relaxation time τ is finite, *i.e.* $\tau = const$. Then, the mass (energy) balance (10.70) yields the Cattaneo equations [66]

$$\frac{\partial T(x,t)}{\partial t} = -\frac{a_2}{\tau}\int_0^t exp\left[-\left(\frac{t-s}{\tau}\right)\right]\frac{\partial T(x,t)}{\partial x} \qquad (10.80)$$

For $\tau \to 0$ the limit of the Cattaneo equation is the Fourier law. Therefore, in the first order approximation, in τ, the modified Fourier law is [71]

$$q(x, t + \tau) \quad = -k_1 \frac{\partial T(x,t)}{\partial x}$$

$$q(x, t + \tau) \quad \approx q(x,t) + \tau \frac{\partial q(x,t)}{\partial x} \tag{10.81}$$

This leads to a first order differential equation [71]

$$\frac{1}{\tau} q(x,t) + \frac{\partial q(x,t)}{\partial x} = -\frac{k_1}{\tau} \frac{\partial T(x,t)}{\partial x} \tag{10.82}$$

The integration of (10.82) results in the Cattaneo equation (10.80), which is the simplest giving rise to finite speed of flux propagation.

Joseph and Preziosi [72] have considered a modified relaxation function replacing the exponential kernel in (10.72) by a Jeffrey's kernel

$$R_{JP} = k_1 \delta_D(s) + \frac{k_2}{\tau} exp\left(-\frac{s}{\tau}\right) \tag{10.83}$$

where δ_D is Dirac delta function, while k_1 and k_2 are the *effective thermal conductivity* and the *elastic conductivity*, respectively. In this case the Fourier law (10.70) leads to a flux defined as [68, 71]

$$q(x,t) = -k_1 \frac{\partial T(x,t)}{\partial x} - \frac{k_2}{\tau} \int_{-\infty}^{t} exp\left(-\frac{t-s}{\tau}\right) \frac{\partial T(x,s)}{\partial x} ds \tag{10.84}$$

In this case, the energy conservation equation (10.70) of the internal energy [73] results in the Jeffrey type integro-differential equation [71]

$$\frac{\partial T(x,t)}{\partial x} = a_1 \frac{\partial^2 T(x,t)}{\partial x^2} + \frac{a_2}{\tau} \int_{-\infty}^{t} exp(-\frac{t-s}{\tau}) \frac{\partial^2 T(x,t)}{\partial x^2} ds \tag{10.85}$$

or equivalently

$$\frac{\partial T(x,t)}{\partial x} = a_1 \frac{\partial^2 T(x,t)}{\partial x^2} + a_2 \beta \int_{-\infty}^{t} e^{-\beta(t-s)} \frac{\partial^2 T(x,s)}{\partial x^2} ds \tag{10.86}$$

Here $a_1 = k_1/\rho C_p$ and $a_2 = k_2/\rho C_p$ are *the effective thermal diffusivity* and *the elastic thermal diffusivity*, respectively; ρ is the density while C_p is the heat capacity of the medium.

For the sake of simplicity, let us consider a virgin material subjected to a thermal load at $x = 0$, that is the following initial and boundary conditions take place

$$T(x,0) = T(0,0) = T(\infty,t) = T_x(x,0) = T_{xx}(x,0) = 0, T(0,t) = T_s \quad \textbf{(10.87)}$$

Denoting $F(x,t) = \frac{\partial^2 T(x,t)}{\partial x^2}$ and integrating by parts in the last term of eq. (10.86) we get [69]

$$\beta \int_{-\infty}^{t} e^{-\beta(t-s)} F(x,s) = e^{-\beta(t-s)}[F(x,s) - F(x,t)]\Big|_{-\infty}^{t} +$$
$$+\beta \int_{-\infty}^{t} e^{-\beta(t-s)}[F(x,t) - F(x,s)]ds \quad \textbf{(10.88)}$$

The first term in the RHS of eq. (10.88) is zero, while the second one mimics the definition of the Caputo-Fabrizio fractional derivative [3], precisely as

$$_{cf}^{c}D_t^{\alpha} = \frac{\alpha}{(1-\alpha)^2} \int_{-\infty}^{t} [F(t) - F(s)]exp\left[-\frac{\alpha}{1-\alpha}(t-s)\right]ds, t > 0 \quad \textbf{(10.89)}$$

Now, *a pro-Caputo* (non-normalized) derivative denoted as $_{PC}D_t^{\beta}$, expressed in two equivalent forms (in accordance with the notations used in [1], can be defined [69]

$$_{PC}D_t^{\beta} F(x,t) = \beta \int_{-\infty}^{t} e^{-\beta(t-s)}[F(x,t) - F(x,s)]ds = \beta \int_{-\infty}^{t} e^{-\beta(t-s)} \frac{dF(x,s)}{dt} ds \textbf{(10.90)}$$

The rate constant β in (10.90) controls the kernel and $\beta \in (0,\infty)$. Refining the integral operator (10.90) in a form controlled by a single parameter α (in a subdiffusive manner) we have to satisfy the conditions: for $\alpha \in [0,1] \Rightarrow 1/\beta \in [0,\infty]$. With $\beta(\alpha) = \alpha/(1-\alpha)$ [1, 3] the desired properties are obtained, namely

$$\frac{1}{\beta} = \frac{1-\alpha}{\alpha} \in [0,\infty], \alpha = \frac{1}{1+\frac{1}{\beta}} = \frac{\beta}{1+\beta} \in [0,1], \frac{\alpha}{(1-\beta)^2} = \frac{\beta}{(1-\alpha)} \quad \textbf{(10.91)}$$

Hence, following the definition of the Caputo-Fabrizio derivative [1, 3] and considering the lower limit of integral at $t = 0$, *i.e.* $a = 0$, we have [69]

$$_{cf}^{c}D_t^{\alpha} = \frac{N(\sigma)}{\sigma} {}_{PC}D_t^{\beta}T(x,t) = \frac{N(\sigma)}{\sigma} \int_0^t e^{-\frac{\alpha}{1-\alpha}(t-s)} \frac{dF(x,s)}{dt} ds =$$

$$= \frac{M(\alpha)}{(1-\alpha)} \int_0^t e^{-\frac{\alpha}{1-\alpha}(t-s)} \frac{dF(x,s)}{dt} ds \quad \textbf{(10.92)}$$

In the terms used here (10.90), $\sigma = 1/\beta$, while $N(\sigma)$ and $M(\alpha)$ are normalization functions [1, 3]. With $M(\alpha) = 1$, as mentioned at the beginning, the last form of (10.92) reduces to Caputo-Fabrizio time-fractional derivative as it is defined by eq. (10.1).

Now, turning on eq. (10.86) in terms of $T(x,t)$ an taking into account (10.89) and (10.91) the last version of (10.92) is the Caputo-Fabrizio time-fractional derivative of $\partial^2 T(x,t)/\partial x^2$. Hence, the balance equation (10.70) results in the following heat diffusion equation

$$\frac{\partial T(x,t)}{\partial t} = a_1 \frac{\partial^2 T(x,t)}{\partial x^2} + a_2(1-\alpha)\left[{}^{c}_{cf}D^{\alpha}_t \frac{\partial^2 T(x,t)}{\partial x^2} \right], t > 0 \qquad (10.93)$$

Equation (10.93) models transient heat conduction with a damping term expressed through the Caputo-Fabrizio fractional operator (derivative). Clearly, for $\alpha = 1$ we get the Fourier equation. With decreasing in α, that physically means increase in the damping effect, the weight of the last term of (10.93) increases.

At this point, it is noteworthy to stress the attention on the fact that *the relaxation effect expressed by the Caputo-Fabrizio derivative appears as an additional time-fractional diffusion term in the right-hand side of the diffusion equation.* The same result exists, but with relaxing term expressed through the time-fractional integral (10.74), in the diffusion model of Caputo-Fabrizio [3] presented by (10.79). We especially, stress to attention on these, practically equal structures despite the differences in the memory kernels in the constitutive equations, since in this case the integer-order time derivative (in the left side) coming from the mass (heat) balance law (10.70) *is not affected by the fractionalization of the diffusion model.* This fact is of primary importance in creation models with derivatives with non-singular kernels when the derivation starts with well-defined constitutive equations in contrast to *formalistic fractionalization* of existing integer-order models [8, 29-32, 35, 40-42, 44, 45, 49, 54, 56, 57, 59] where the time derivatives are directly replaced by fractional ones (see section 4).

If the Cattaneo kernel is only taken into account (that is, the modified relaxation function of Joseph and Preciozi is omitted), then the equivalent form of eq. (10.93) is eq. (10.80) [71], but now *it accounts only the elastic heat diffusion* and can be expressed in terms of $_{pc}D^{\alpha}_t$ as

$$\frac{\partial T(x,t)}{\partial t} = -a_2\beta \int_0^t exp[-\beta(t-s)]\frac{\partial T(x,t)}{\partial x} \qquad (10.94)$$

$$\frac{\partial T(x,t)}{\partial t} = -a_2\left(\frac{\alpha}{1-\alpha}\right)pcD^{\alpha}_t\left[\frac{\partial^2 T(x,t)}{\partial x^2}\right] \qquad (10.95)$$

Koca and Atangana [47] solved numerically (10.96), expressed as

$$\frac{\partial T(x,t)}{\partial t} = -a_2(1-\alpha)_{cf}D^{\alpha}_t\left[\frac{\partial^2 T(x,t)}{\partial x^2}\right] \qquad (10.96)$$

named as *elastic part of the heat conduction model* (10.93).

The exact solution of (10.96) was performed by separation of variables representing the solution as $T(x,t) = T_1(x)T_2(t)$. Then, equation (10.96) takes the form

$$\frac{\partial}{\partial t}(T_1(x)T_2(t)) = a_2(1-\alpha)\left({}_{cf}D_t^\alpha \frac{\partial^2}{\partial x^2}[T_1(x)T_2(x)]\right) \qquad \textbf{(10.97)}$$

This leads to a set of ordinary equations

$$\begin{aligned}
\frac{dT_1(x)}{dx} + \lambda^2(1-\alpha)\frac{d^2 T_1(x)}{dx^2} &= 0 \\
\frac{dT_2(t)}{dt} + \lambda^2{}_{cf}D_t^\alpha T_2(t) &= 0
\end{aligned} \qquad \textbf{(10.98)}$$

The second equation in (10.98) performed by the Laplace transform is $T_2(t) = T_2(0)$ and therefore, the exact solution of (10.96) is

$$T(x,t) = \sum_{n=0}^{\infty} T_2(0)\left(T_1(0)exp\left[-\frac{\lambda_n x}{\sqrt{a_2}(1-\alpha)}\right] + T_1(0)exp\left[\frac{\lambda_n x}{\sqrt{a_2}(1-\alpha)}\right]\right) \textbf{(10.99)}$$

Further, Koca and Atangana [47] used the definition of the pro-Caputo derivative defined in [69] (see (10.90)) and a kernel expressed through one-parameter Mittag-Leffler function $E_\gamma = \sum_0^\infty \frac{z^k}{\Gamma(\gamma k+1)}$ as a generalization of a family of relaxation (damping) functions. In this context, the exponential function in the Jeffrey kernel used to define (10.96) in [69] is a special case of $E_{\gamma=1}$. As commented above, this concept leads to the Atangana-Baleanu derivative. In the solution of (10.96), Koca and Atangana used the Riemann-Liouville type derivative (10.56) with a corresponding fractional integral represented by (10.64).

In terms of Atangana-Baleanu derivative the elastic part (10.96) of the model (10.93) can be presented as

$$\frac{\partial T(x,t)}{\partial t} = a_2(1-\alpha)\left({}_{PAB}^{ABC}D_t^\alpha\left[\frac{\partial^2 T(x,t)}{\partial x^2}\right]\right)$$

$$\textbf{(10.100)}$$

$${}_{pAB}^{ABC}D_t^\alpha\left[\frac{\partial^2 T(x,t)}{\partial x^2}\right] = B(\alpha)\int_0^t \frac{\partial^2 T(x,t)}{\partial x^2}E_\alpha\left[-\frac{\alpha}{1-\alpha}(t-s)^\alpha\right]ds$$

The version (10.100) of (10.96) was solved by the Laplace transform, namely

$$pT(x,p) - T(x,0) = \frac{p^\alpha \frac{\partial T(x,p)}{\partial x} - p^{\alpha-1} \frac{\partial T(x,0)}{\partial x}}{p^\alpha + \frac{\alpha}{1-\alpha}} \tag{10.101}$$

The inverse Laplace transform of (10.101) results in

$$T(x,t) = \sum_{k=0}^{\infty} \sum_{j=0}^{\infty} \frac{x^{j+k}}{j!k!} \frac{t^{\alpha j - j - k - 1}}{\Gamma(\alpha j - j - k)} \tag{10.102}$$

Koca and Atangana [47] performed a numerical solution of the elastic model (10.96) with the Attangana-Baleanu derivative by the Crank-Nicholson approximation of the time component, namely

$$\frac{T_i^{n+1}}{\Delta t} = \frac{T_i^n}{\Delta t} - a_2 \frac{B(\alpha)}{1-\alpha} \sum_{k=1}^{n} \frac{\partial_x T_i^{n+1} - \partial_x T_i^n}{\Delta t} \int_{t_k}^{t_{k+1}} E_\alpha \left[-\frac{\alpha}{1-\alpha} (t-s)^\alpha \right] \tag{10.103}$$

More details about the approximation of the integral in (10.103) and the numerical solutions are available elsewhere [47]. Recently Algahani and Atangna [74] reported a numerical solution of the complete model (10.93).

At this end, the transient model (10.93), is used in the steady-state version when a Caputo-type spatial derivative with a non-singular Jeffrey kernel formulated in [75] is applied (see section 10.3 of this chapter).

10.2.1.3. *Diffusion Equation with Cattaneo Memory Kernel (Single-Memory Model)*

When the damping function in the flux-gradient relationship is only time-dependent and represented by the exponential kernel $R(t) = exp(-(t-s)/\tau)$ with a finite relaxation time $\tau = const.$, then the mass balance (10.70) results in the Cattaneo equation

$$\frac{\partial C(x,t)}{\partial t} = -\frac{\partial}{\partial x} \left(-\frac{D_0}{\tau} \int_0^t exp \left[-\frac{t-s}{\tau} \right] \frac{\partial C(x,t)}{\partial x} ds \right) \tag{10.104}$$

which is equivalent to the heat diffusion equation (10.80)

Equation (10.104) can be presented in two equivalent forms

$$\frac{\partial C(x,t)}{\partial t} = \frac{D_0}{\tau} \int_0^t exp \left[-\frac{t-s}{\tau} \right] \frac{\partial^2 C(x,s)}{\partial x^2} ds \tag{10.105}$$

$$\frac{\partial C(x,t)}{\partial t} = \beta D_0 \int_0^t exp[-\beta(t-s)] \frac{\partial^2 C(x,s)}{\partial x^2} ds, \beta = \frac{1}{\tau} \tag{10.106}$$

Now, we define the Cattaneo equation as a diffusional flux constitutive equation as it represented by (10.104), (10.105) and (10.106). For the sake of simplicity, let us consider a virgin medium subjected to a mass load at $x = 0$, that is the following initial and boundary conditions take place

$$C(x,0) = C(0,0) = C(\infty,t) = C_x(x,0) = C_{xx}(x,0) = 0, C(0,t) = C_s \quad \textbf{(10.107)}$$

Now we focus the attention on eq. (10.106) denoting $F(x,t) = \partial^2 C(x,t)/\partial x^2$. Then, from (10.107) we have $F(x,0) = \partial^2 C(x,0)/\partial x^2 = 0$.

Integrating by parts of the diffusion term of equation (10.106) we get

$$
\begin{aligned}
\beta \int_0^t e^{-\beta(t-s)} F(x,s)ds &= \left[e^{-\beta(t-s)} F(x,t)\right]_{s=0}^{s=t} + \\
&+\beta \int_0^t e^{-\beta(t-s)} [F(x,t) - F(x,s)]ds
\end{aligned}
\quad \textbf{(10.108)}
$$

and finally

$$
\begin{aligned}
\beta \int_0^t e^{-\beta(t-s)} F(x,s)ds &= \left(1 - e^{-\beta t}\right) F(x,t) + \\
&+\beta \int_0^t e^{-\beta(t-s)} [F(x,t) - F(x,s)]ds
\end{aligned}
\quad \textbf{(10.109)}
$$

It noteworthy that if the lower terminal of the memory integral is $-\infty$, as in the original Cattaneo concept (see eq. (10.72)), then the first term in (10.108) and the first exponential term of (10.109) will be lost. Thus, the more realistic way is the use of the second form of the Cattaneo constitutive equation presented by eq. (10.105) or eq. (10.106).

In terms of the original physical variable $C(x,t)$ (10.109) can be presented as

$$
\begin{aligned}
\beta \int_0^t e^{-\beta(t-s)} \frac{\partial^2 C(x,s)}{\partial x^2} ds &= \left(1 - e^{-\beta(t)}\right) \frac{\partial^2 C(x,t)}{\partial x^2} + \\
&+\beta \int_0^t e^{-\beta(t-s)} \left(\frac{\partial^2 C(x,t)}{\partial x^2} - \frac{\partial^2 C(x,s)}{\partial x^2}\right) ds
\end{aligned}
\quad \textbf{(10.110)}
$$

The second term in the right-hand side of (10.110) mimics the definition of the Caputo-Fabrizio fractional derivative. As it was demonstrated earlier this term can be considered as a *pro-Caputo* (non-normalized) derivative (10.90) $_{PC}D_t^\beta$ with a lower terminal at 0. In terms of $C(x,t)$ we may express two equivalent forms of $_{PC}D_t^\beta$, namely

$$_{PC}D_t^\beta \left(\frac{\partial^2 C(x,t)}{\partial x^2} \right) = \beta \int_0^t e^{-\beta(t-s)} \left(\frac{\partial^2 C(x,t)}{\partial x^2} - \frac{\partial^2 C(x,s)}{\partial x^2} \right) ds \qquad (10.111)$$

$$_{PC}D_t^\beta \left(\frac{\partial^2 C(x,t)}{\partial x^2} \right) = \beta \int_0^t e^{-\beta(t-s)} \frac{d}{dt} \left(\frac{\partial^2 C(x,s)}{\partial x^2} \right) ds \qquad (10.112)$$

Since the rate constant $\beta \in (0, \infty)$ controls the exponential kernel, then $_{pc}D_t^\beta$ can be arranged in the form defined by (10.1) with a fractional order α. From this concept, it follows that for $\alpha \in [0,1] \Rightarrow 1/\beta \in [0, \infty]$. Consequently, the following relationships are valid (see the analogous relationships (10.91))

$$\frac{1}{\beta} = \frac{1-\alpha}{\alpha} \in [0, \infty], \alpha = \frac{1}{1+1/\beta} \in [0,1], \frac{\alpha}{(1-\alpha)^2} = \frac{\beta}{(1-\alpha)} \qquad (10.113)$$

From the definition (10.1) [1, 3] we get

$$_{CF}D_t^\alpha \left(\frac{\partial^2 C(c,t)}{\partial x^2} \right) = \frac{N(\sigma)}{\sigma} \, _{pc}D_t \left(\frac{\partial^2 C(x,t)}{\partial x^2} \right) =$$

$$= \beta \frac{M(\alpha)}{(1-\alpha)} \int_0^t e^{-\beta(t-s)} \frac{d}{dt} \left(\frac{\partial^2 C(x,s)}{\partial x^2} \right) ds \qquad (10.114)$$

or equivalently

$$_{CF}D_t^\alpha \left(\frac{\partial^2 C(c,t)}{\partial x^2} \right) = \frac{N(\sigma)}{\sigma} \, _{pc}D_t \left(\frac{\partial^2 C(x,t)}{\partial x^2} \right) =$$

$$= \frac{\alpha}{1-\alpha} \frac{M(\alpha)}{(1-\alpha)} \int_0^t e^{-\beta(t-s)} \frac{d}{dt} \left(\frac{\partial^2 C(x,s)}{\partial x^2} \right) ds \qquad (10.115)$$

with $((1-\alpha)/\alpha)N(\sigma) = M(\alpha)/(1-\alpha)$ and $\sigma = 1/\beta$; here $N(\sigma)$ and $M(\alpha)$ are normalizing functions [1, 3]. Consequently, we get

$$_{CF}D_t^\alpha \left(\frac{\partial^2 C(x,t)}{\partial x^2} \right) = \frac{M(\alpha)}{(1-\alpha)} \int_0^t e^{-\beta(t-s)} \frac{d}{dt} \left(\frac{\partial^2 C(x,s)}{\partial x^2} \right) ds \qquad (10.116)$$

Finally, the new form of equation (10.106) is [70]

$$\frac{\partial C(x,t)}{\partial t} = D_0 \left(1 - e^{-\beta t} \right) \frac{\partial^2 C(x,t)}{\partial x^2} + D_0 (1-\alpha) _{CF}D_t^\alpha \left(\frac{\partial^2 C(x,t)}{\partial x^2} \right) \qquad (10.117)$$

This is the complete version of the diffusion equation expressed through the Caputo-Fabrizio time-fractional derivative when the constitutive relationship *flux-gradient* is defined with the simple exponential kernel (see (10.80)). It was developed in [70] as a single-memory kernel model with a relaxation time τ_0 (we will use this notation hereafter when this model is at issue).

For $t = 0$, when practically no relaxation exists $(\tau \approx 0 \Rightarrow \beta \rightarrow \infty)$, we get $D_0(1 - e^{-\beta t}) \approx D_0$ and the diffusion coefficient has a maximal value D_0.

In the special case when $\beta t << 1$ we may present $D_0(1 - e^{-\beta t})$ as

$$D_0\left(1 - e^{-\beta t}\right) = D_0 e^{-\beta t}\left(\frac{1 - e^{-\beta t}}{e^{-\beta t}}\right) \qquad (10.118)$$

Approximating the exponential term in the second version of (10.118) as a series and taking into account only the first two terms, *i.e.* $e^{-\beta t} \approx 1 - \beta t$ we may approximate the term in the brackets of (10.118) as

$$\frac{1 - e^{-\beta t}}{e^{-\beta t}} \approx \frac{\beta t}{1 + \beta t} \approx O(1) \qquad (10.119)$$

Therefore, with the assumption $\beta t = t/\tau << 1$ we get the fractional version of the Dodson equation [70]

$$\frac{\partial C(x,t)}{\partial t} \approx D_0 e^{-\beta t}\frac{\partial^2 C(x,t)}{\partial x^2} + D_0(1 - \alpha)_{CF}D_t^\alpha\left(\frac{\partial^2 C(x,t)}{\partial x^2}\right) \qquad (10.120)$$

The first term in the right-hand side of (10.120) matches the diffusion term of the Dodson equation [76].

$$\frac{\partial C(x,t)}{\partial t} = D(0)e^{\frac{-t}{\tau}}\frac{\partial^2 c(x,t)}{\partial x^2} \qquad (10.121)$$

or equivalently

$$\frac{\partial C(x,t)}{\partial t} = D(0)e^{-\beta t}\frac{\partial^2 c(x,t)}{\partial x^2} \qquad (10.122)$$

For $\alpha = 1$, from (10.120) formally we get the Dodson equation (10.122) [76].

Hence, the single-memory model can be expressed completely in terms of the fractional order α as [70]

$$\frac{\partial C(x,t)}{\partial t} = \left(1 - e^{-\frac{\alpha_0}{1-\alpha_0}Fo}\right)\left(\frac{\partial^2 C(x,t)}{\partial x^2}\right) + D_0(1 - \alpha_0)_{CF}D_t^{\alpha_0}\left(\frac{\partial^2 C(x,t)}{\partial x^2}\right) \quad \textbf{(10.123)}$$

It is noteworthy that when a proper constitutive equation relating the flux and the gradient is used, as it was done with the Caputo-Fabrizio model (10.74), the model of Hristov (10.93) with Jeffrey's kernel (10.83) as well as the present one with the simple Cattaneo memory (10.104), then the resulting diffusion equations contain relaxing terms expressed through the Caputo-Fabrizio time-fractional derivative (10.1). *This fractionalization does not affect the time-dependent term* $\partial C/\partial t$ defined by the balance (continuity) equation (10.70).

Now, we stress the attention on the definition of the fractional order α and its relation to the physical parameters of the system. The definition of the fractional order α is of primary importance since the models developed should be practically implemented or at least to be used correctly in numerical simulations. Hence, the reasonable question is : *How to calculate the fractional order α if the process parameters such as the relaxation time τ and characteristic time-scale of the diffusion process are known* ?

The definition of the stretched exponential $exp[-\beta(t-s)]$ shows directly that the dimensions of β is $[1/s]$. However, while the fractional order α is dimensionless, the rate constant $\beta(\alpha) = \alpha/(1-\alpha)$ should have dimension of $[1/s]$, or more precisely the ratio $(1-\alpha)/\alpha$ should have dimension of time. Now, the question is: *how this conflict could be avoided?* To overcome the problem we use a time scale that can be defined by the initial conditions of the diffusion process. With D_0 and the length scale L of the area where the diffusion takes place, the *characteristic diffusion time* is $t_D = L^2/D_0$. Therefore, the time t can be scaled as $\bar{t} = t/t_D = D_0 t/L^2$. In fact the dimensionless time t/t_D is the Fourier number $Fo = D_0 t/L^2$ defined through the initial diffusivity D_0. Now, we turn on the stretched exponential which can be rescaled as [70]

$$exp[-\beta_0(t-s)] = exp\left[-\frac{(t-s)}{\tau_0}\right] = exp\left[-\left(\frac{t_D}{\tau_0}\right)(\bar{t} - \bar{s})\right] \quad \textbf{(10.124)}$$

Hence, from the definition of the fractional order α (see (10.91) we have

$$\frac{\alpha_0}{1-\alpha_0} = \frac{t_D}{\tau_0} = \frac{1}{_0De} = \beta_0 t_D \implies \alpha_0 = \frac{1}{1+t_D/\tau_0} \quad \textbf{(10.125)}$$

or equivalently

$$\alpha_0 = \frac{1}{1+(\tau_0 D_0/L^2)} = \frac{1}{1+(D_0/\beta_0 L^2)} = \frac{1}{1+{}_0De} \leq 1 \qquad (10.126)$$

The ratio $\tau_0/t_D = \tau_0 D_0/L^2 = {}_0De$ is the Deborah number for the macroscopic diffusion relaxation process defined by analogy with the non-Fickian diffusion in complex systems [77, 78]. Hence, with known values of L, τ_0, and D_0 we will be able to define ${}_0De$, and then to calculate α_0. When relaxation does not exist, that is for $\tau_0 = 0$ we have ${}_0De = 0$ and $\alpha_0 = 1$.

The plots in Fig. (**10.1**) demonstrate the functional relationship $\alpha = \alpha(De)$. It is obvious that the value of $\alpha = 0.5$, commonly used in numerical simulations, corresponds to $De = 1$, that is when the relaxation time equals the characteristic diffusional time of the system, *i.e.* when $\tau = L^2/D_0$. The relationships (10.124) and (10.125) are quite informative from physical point of view and may be also constructive in interpretations of the phenomena behind the model.

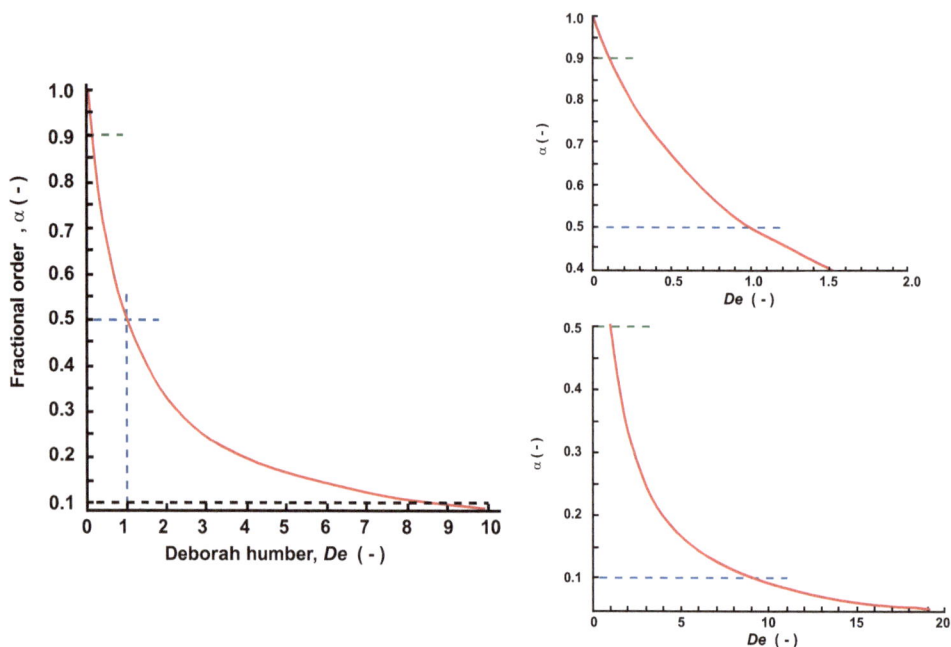

Fig. (10.1). Functional relationship $\alpha = \alpha(De)$ defining the fraction order as a function of the Deborah number. Adapted from [70]

Equations (10.125) and (10.126) are principle relations allowing to define the fractional order on the basis of physical parameters. It is noteworthy that this is a specific consequence of the form of the exponential function where the time is scaled by the relaxation time τ, that is we have $exp(-t/\tau) = exp(-\beta t)$. This form allows the nondimensalization by the characteristic diffusion time t_D. In contrast, when the memory function is defined as $t^{-\alpha}$ as in the classical Caputo and Riemann-Liouville derivatives [2] such nondimensalization is impossible.

10.2.1.4. Diffusion Model with Composite Memory Kernel (Two-Memories Model)

This model, conceived in [70], is based on the assumption that when there are a large time relaxation process (with a relaxation time τ_0) and a short-time relaxation process (with a relaxation time $\tau_s \ll \tau_0$), the transients in the gross diffusion process can be expressed through time-fractional Caputo-Fabrizio derivatives (10.1) with different fractional orders.

This physical conjecture is modelled by a composite memory function relating the flux and gradient by the constitutive equation in a sense of the Cattaneo concept, namely

$$j_a = D_0 \int_0^t R_a(t, \tau_0, \tau_s) \nabla C(x.s) ds \qquad (10.127)$$

The composite kernel $R_a(t, \tau_0, \tau_s)$ is defined by the following constitutive relationship [70]

$$R_a(t) = e^{-\beta_0(t-s)}\left(1 - e^{-\beta_s(t-s)}\right) \quad = e^{-\beta_0(t-s)} - e^{-(\beta_0+\beta_s)(t-s)}$$
$$\beta = \beta_0 + \beta_s \qquad (10.128)$$

The composite memory kernel $R_a(t, \tau_0, \tau_s)$ is a product of a *large-time exponential kernel* $e^{-\beta_0(t-s)}$ and a *short-time fading function* $R_s(t, \tau_s)$ defined as.

$$R_s(t, \tau_s) = \left(1 - e^{-\beta_s(t-s)}\right) \qquad (10.129)$$

In accordance with the conjecture we have $\tau_0 \gg \tau_s$ and therefore, from the constitutive equation (10.127) it follows that $\beta_0 \ll \beta_s$. Thus β_0 corresponds to large-time relaxation processes, while β_s accounts the short-time relaxation mechanism.

Further, the constitutive relation $\beta = \beta_0 + \beta_s$ means that the large-time and short-time relaxations occur simultaneously and overlap. Hence, $\beta = 1/\tau = 1/\tau_0 + 1/\tau_s$ and therefore $\tau = \tau_0\tau_s/(\tau_0 + \tau_s)$.

When the time-scale of the diffusion process is order of magnitude as τ_s, taking into account that $\tau_s \ll \tau_0$, the approximation is $lim_{t \to \tau_s}\tau \approx \tau_s$ and $\beta \approx \beta_s$, that is $\beta = \beta_0 + \beta_s \approx \beta_s$ when only short-time relaxation has to be accounted for. Alternatively, when the time-scale of the process is comparable to τ_0 then $lim_{t \to \tau_0} \tau \approx \tau_0$ and $\beta \approx \beta_0$.

In other words, for short times the term $\left(1 - e^{-\beta_s(t-s)}\right)$ dominates since $e^{-\beta_0(t-s)}$ has little effect due to the fact that $\beta_0 << \beta_s$. For large times we have $\left(1 - e^{-\beta_s(t-s)}\right) \to 1$ and only $e^{-\beta_0(t-s)}$ remains as a memory function.

The short time relaxation is modelled by the diminishing function (10.128) which appears in the single-memory model (10.128) as a factor of the integer-order diffusion term (see more detailed analysis in [70]). It rapidly grows from zero to unity (see Fig. **10.2**).

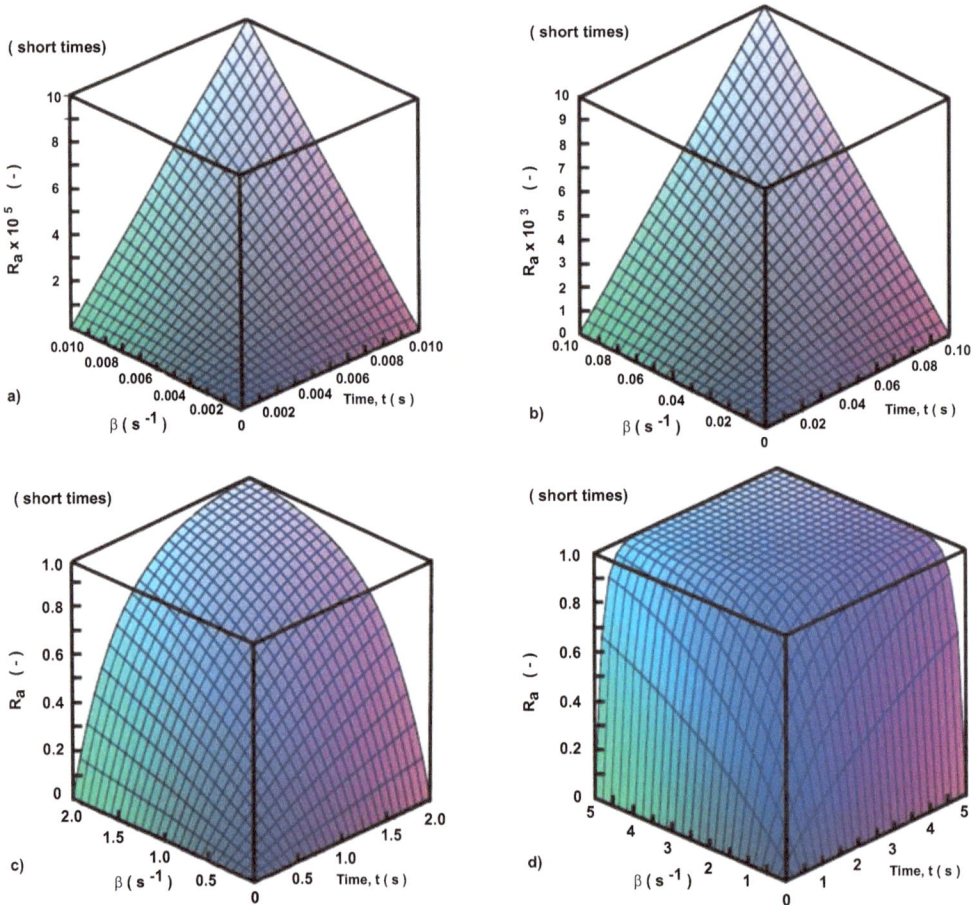

Fig. (10.2). Short time relaxation kernel $1 - e^{-\beta t}$ as a function of the rate constant β and the time t. Adapted from [70]

Now, following the conjecture, the approximation of the memory integral with the composite damping function is

$$(\beta_0 + \beta_s) \int_0^t \left(e^{-\beta_0(t-s)} - e^{-(\beta_0+\beta_s)(t-s)} \right) F(x,s) ds \approx$$

$$\approx \beta_0 \int_0^t e^{-\beta_0(t-s)} F(x,s) ds - \beta_s \int_0^t e^{-\beta_s(t-s)} F(x,s) ds \tag{10.130}$$

where $F(x,t) = \partial^2 C(x,t)/\partial x^2$

Therefore, we have two distinguished memory integrals. Further, we will repeat the technique of integration by parts in the right-hand side of (10.130) as it was done to the single-memory model, but now, to each memory integral separately. Now, recall that from the conjecture $\beta_s \gg \beta_0$ and consequently the factor of the first term in the right-hand side of (10.130) can be approximated as $e^{-\beta_0 t} \approx e^{-\beta_0 t} - e^{-\beta_s t}$ because $e^{-\beta_0 t} \gg e^{-\beta_s t}$. After these adjustments and approximations the new time-fractional equation with two memories is [70]

$$\frac{\partial C}{\partial t} = D_0 [P(x,t,\alpha) - Q(x,t,\alpha)] \tag{10.131}$$

Where the diffusion term in (10.131) has two components

$$P(x,t,\alpha) = \left(1 - e^{-\beta t}\right) \frac{\partial^2 C(x,t)}{\partial x^2} + (1 - \alpha_0)_{cf} D_t^{\alpha_0} \left[\frac{\partial^2 C(x,t)}{\partial x^2} \right] \tag{10.132}$$

$$Q(x,t,\alpha) = \left(1 - e^{-\beta t}\right) \frac{\partial^2 C(x,t)}{\partial x^2} + (1 - \alpha_s)_{cf} D_t^{\alpha_s} \left[\frac{\partial^2 C(x,t)}{\partial x^2} \right] \tag{10.133}$$

Hence, the complete diffusion equation with two memories can be presented as

$$\frac{\partial C(x,t)}{\partial t} = D_0 \left[\left(1 - e^{-\beta_0 t}\right) + (1 - \alpha_0)_{cf} D_t^{\alpha_0} \right] \frac{\partial^2 C(x,t)}{\partial x^2} -$$

$$- D_0 \left[\left(1 - e^{-\beta_s t}\right) + (1 - \alpha_s)_{cf} D_t^{\alpha_s} \right] \frac{\partial^2 C(x,t)}{\partial x^2} \tag{10.134}$$

For $\beta_0 t \ll \beta_s t$ and certainly, when $\beta_0 t \ll 1$ the first term in (10.134) reduces to $D_0 e^{-\beta t}$ as it was demonstrated earlier; this condition leads to the fractional version of the Dodson equation (10.120).

When the short time relaxation (damping effect) is neglected, that is when $\alpha_s = 1$, we get the model with a single memory (10.117). Besides, when the relaxations in the mass flux are generally neglected, that is when $\alpha_0 = \alpha_s = 1$, we obtain the classical integer-order Dodson equation (10.122) with fading diffusion coefficient.

The negative sign of the short-time memory term in (10.134) simply means that *short-time relaxations effects, if they exist, accelerate the total diffusion process*. To be exact, let see the construction of the two-memory relaxation kernel and its logical origin. The expanded expression of $R_a(t, \tau_0, \tau_s)$ can be approximated as (see eq. (10.128))

$$Ra(t) = e^{-\beta_0(t-s)} - e^{-(\beta_0+\beta_s)(t-s)} \approx e^{-\beta_0(t-s)} - e^{-\beta_s(t-s)} \qquad \textbf{(10.135)}$$

Hence, we have a counter-current action of the memory integrals, which means that the short-time memory integral reduces the damping effect of the effect of the one with large-time memory kernel, thus accelerating the diffusion process; the same as it was commented about eq. (10.128). The short-time memory kernel as a function of the fractional order α_s and the Fourier number (*i.e.* for $\alpha_s \to 1$ and short times, *i.e.* low Fourier numbers) is $R_{a(shorttimes)} = 1 - e^{-\beta_s(t-s)} = 1 - e^{[-\alpha/(1-\alpha)]Fo}$ and it rapidly grows to 1 (see Fig. **10.3a**). The decrease in the value of α hinders the increase of $R_{a(shorttimes)}$ at very short times. With increase in Fo the damping effect of α is stronger when $\alpha < 0.5$ (see Figs. **10.3b** and **10.3c**).

10.2.1.5. Brief Concluding Remarks

Therefore, we analyzed in details four models where irrespective of the formulation of the constitutive equation relating the flux and the gradient the final diffusion equations can be expressed in terms of the time-fractional Caputo-Fabrizio derivative (10.1) (and integral in the case of (10.179)).

The developed functional relationship (10.125) (and (10.126), too) allows calculating the fractional order α, a fact that is essentially missing in the existing publications involving time-fractional Caputo-Fabrizio derivatives, as well as in the models with singular fading memories [2].

Moreover, the derivation the fractional Dodson equation, with diffusion coefficient expressed through the fractional order α and the Fourier number (see eq. (10.123)) developed in [70] is a step ahead in modelling with this model; which demonstrates a little progress since the time of its invention.

The two-memory model constitutes a two-kernel composite memory function physically based on the assumption that local disturbances causing short-time transients affect the gross relaxation process. This model reduces simply to the single-kernel memory model when the short-time memory is neglected and further to the original Dodson equation expressed through a fractionalized diffusion coefficient (when the fractional order α is too low, that is when $\beta t \ll$ 1.

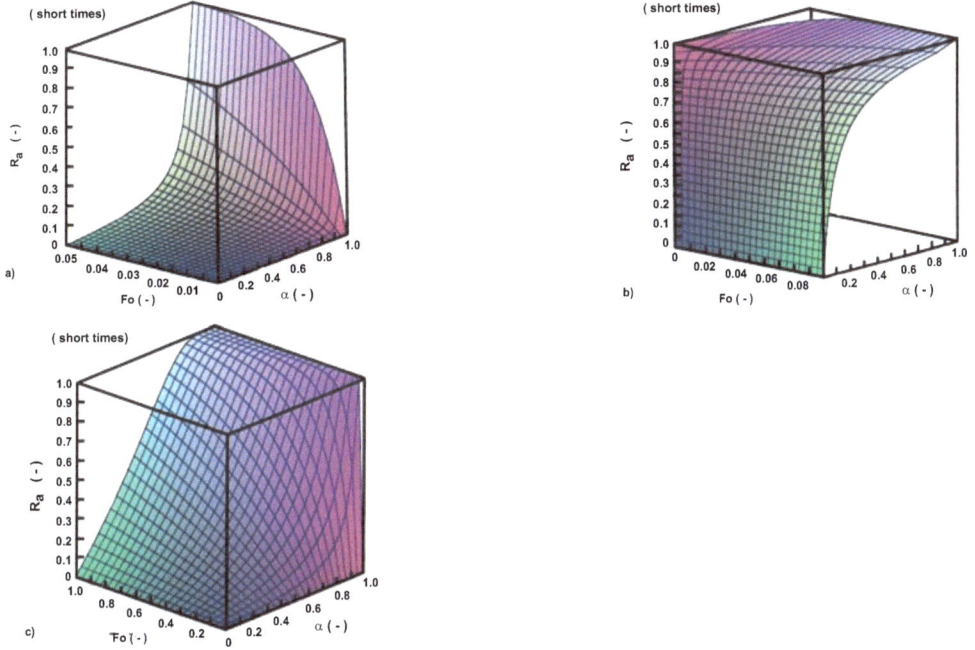

Fig. (10.3). Short time relaxation kernel $1 - exp\left(-\dfrac{\alpha}{1-\alpha}Fo\right)$ as a function of the fractional order α and the Fourier number Fo. Adapted from [70]

10.3. DIFFUSION EQUATION WITH A SPATIAL MEMORY ONLY

10.3.1. Fraction Gradient Operators of Caputo and Fabrizio with Non-Singular Kernels

Parallel to the definition of the time-fractional derivative (10.1) Caputo and Fabrizio conceived spatial fractional derivative, named *fractional gradient operator* [1], namely

$$\nabla^{(\alpha)} u(x) = \frac{\alpha}{(1-\alpha)\sqrt{\pi^\alpha}} \int_\Omega \nabla u(y) exp\left[-\frac{\alpha^2(x-y)^2}{(1-\alpha)^2}\right] dy \qquad (10.136)$$

where $u(\bullet): \Omega \to \mathbb{R}$ is a scalar function with $x, y \in \Omega$ and a set $\Omega \in \mathbb{R}^3$; $\alpha \in [0,1]$.

From the definition (10.136) it follows that

$$lim_{\alpha \to 1} \frac{\alpha}{(1-\alpha)\sqrt{\pi^\alpha}} \int_\Omega \nabla u(y) exp\left[-\frac{\alpha^2(x-y)^2}{(1-\alpha)^2}\right] dy = \delta(x-y) \qquad (10.137)$$

Therefore,

$$\nabla^{(1)} u(x) = \nabla u(x), \nabla^{(0)} u(x) = \int_\Omega u(y) dy \qquad (10.138)$$

For $\alpha = 1$ the operator losses the non-local properties, while for $\alpha = 0$ the operator $\nabla^{(0)}$ relates to the mean value of $\nabla u(y)$ on Ω [1].

The definition of this spatial non-local derivative with non-singular kernel allows transport in materials with non-local properties to be described by fractional constitutive equations. In the same sense, Caputo and Fabrizio [1] defined the fractional Laplacian as

$$(10.\nabla^2)^\alpha f(x) = \frac{\alpha}{(1-\alpha)\sqrt{\pi^\alpha}} \int_\Omega \nabla \cdot \nabla f(x) exp\left[-\frac{\alpha^2(x-y)^2}{(1-\alpha)^2}\right] dy, 0 < \alpha < 1 \quad (10.139)$$

and

$$(10.\nabla^2)^\alpha f(x) = \nabla \cdot \nabla f(x) = \nabla^\alpha \nabla f(x) \qquad (10.140)$$

Now, we stop the explanations related to the fractional in space derivatives of Caputo and Fabrizio and will turn on a definition of a spatial derivative with simple exponential kernel like that used in the time-fractional derivative (10.1). An example of such derivative was commented briefly in section 1.2. in the context of the Yang-Srivastava-Machado space derivative (10.54).

10.3.2. From the Cattaneo Concept of Time-Memory Effects to a Spatial Jeffrey Kernel

Starting from the conservation law (10.70), the transient heat conduction, in accordance with the Fourier law *without a damping function with respect to the heat flux*, is generally described as

$$\rho C_p \frac{\partial T}{\partial t} = -\frac{\partial q}{\partial x}, q(x,t) = -k \frac{\partial T(x,t)}{\partial x} \tag{10.141}$$

$$\rho C_p \frac{\partial T}{\partial t} = k \frac{\partial^2 T(x,t)}{\partial x^2} \tag{10.142}$$

However, as mentioned earlier, the Cattaneo concept [66] is a generalization of the Fourier law through a linear superposition of the heat flux and its time derivative related to its history [67, 68, 71] related to the time-delay by eq. (10.72). In this context, we may express the flux in (10.72) with memory delays with respect to both the time and space coordinates [75] as

$$q(x + \lambda, t + \tau) = -k_1 \frac{\partial T}{\partial x} \tag{10.143}$$

The *memory distance* λ is finite (*i.e.* $\lambda = const.$) and it is the *length scale* of spatial effects on the heat flux correlation to the temperature gradient. The first order approximation, in τ and λ, which can be simply developed by a conventional Taylor series expansion, results in a modified Fourier law, namely

$$q(x + \lambda, t + \tau) \approx q(x,t) + \tau \frac{\partial q(x,t)}{\partial t} + \lambda \frac{\partial T(x,t)}{\partial x} \tag{10.144}$$

For the steady-state heat conduction in a medium with space memory effects only we may define by analogy that the heat flux is related to the temperature gradient [75] as

$$q(x) = -\int_{-\infty}^{t} R(x) \nabla T(x - u) du \tag{10.145}$$

which is a spatial analogue of the Cattaneo constitutive equation (10.72).

If the kernel is defined as $R(x) = exp[-(x - u)/\lambda]$ we get from (10.145) the following constitutive relationship [75]

$$q(x,t) = -\int_{-\infty}^{x} exp\left[-\left(\frac{x-u}{\lambda}\right)\right] \frac{dT(u)}{dx} du \tag{10.146}$$

As mentioned earlier, λ is the *spatial memory length scale* while its inverse $\gamma = 1/\lambda$ is a *space memory constant*.

The first order approximation of the heat flux with respect to λ is (see (10.144))

$$q(x + \lambda) = -k_1 \frac{dT(x)}{dx}, q(x + \lambda) \approx q(x) + \lambda \frac{dq(x)}{dx} \tag{10.147}$$

Repeating, by analogy of (10.83), the concept of a modified space-related memory function of Jeffrey's type we may define [75]

$$R^*(x) = k_1 \delta_x(u) + \left(\frac{k_{2x}}{\lambda}\right) exp\left(-\frac{u}{\lambda}\right), \int_0^x \delta_x(u)du = 1 \qquad (10.148)$$

Where δ_x is the Dirac Delta function and the definition of the relaxation function (10.148) is of Jeffrey's type.

Then, the flux-gradient relationship can be defined as

$$q(x) = -k_1 \frac{dT(x)}{dx} - \frac{k_{2x}}{\lambda} \int_{-\infty}^x exp\left[-\left(\frac{x-u}{\lambda}\right)\right] \frac{dT(u)}{dx} du \qquad (10.149)$$

Equation (10.149) defines the *effective thermal conductivity* k_1, while the conductivity k_{2x} can be defined as a *structural elastic conductivity* related to the spatial memory effects. If no memory effects exist or they could be neglected (when $\mu \to 1$), then

$$q(x) = -k_0 \frac{dT(x)}{dx}, k_0 = k_1 + k_{2x} \to k_1 \qquad (10.150)$$

After differentiation in (10.149) we get

$$\frac{dq(x)}{dx} = k_1 \frac{d^2T(x)}{dx^2} - \frac{k_{2x}}{\lambda} \int_{-\infty}^x exp\left[-\left(\frac{x-u}{\lambda}\right)\right] \frac{dT(u)}{dx} du \qquad (10.151)$$

Applying the energy balance equation (10.141) we get an integro-differential equation of Jeffrey's type

$$\frac{\partial T(x)}{\partial t} = a_1 \frac{d^2T(x)}{dx^2} + a_{2x}\gamma \int_{-\infty}^x e^{-\gamma(x-u)} \frac{d^2T(x)}{dx^2} du$$
$$a_1 = \frac{k_1}{\rho C_p}, \quad a_{2x} = \frac{k_{2x}}{\rho C_p} \qquad (10.152)$$

10.3.3. Towards the Spatial Fractional Derivative of Caputo-Fabrizio Type

The integration by parts of the last term of eq. (10.152) with the notion $F(x) = dT(x)/dx$ results in (see the calculations related to (10.88))

$$\gamma \int_{-\infty}^x e^{-\gamma(x-u)} F(x)du = e^{-\gamma(x-u)}[F(u) - F(x)]_{-\infty}^x +$$
$$+\gamma \int_{-\infty}^x e^{-\gamma(x-u)}[F(x) - F(u)]du \qquad (10.153)$$

The second term in the RHS of eq. (10.153) resembles the construction of the Caputo-Fabrizio time-fractional derivative (10.1) but now re-written with respect to the space variable x and its alternative form (10.3), namely.

$$^{cf}D_x^\mu = \frac{\mu}{(1-\mu)^2} \int_a^x [F(x) - F_a(u)] exp\left[-\frac{\mu}{1-\mu}(x-u)\right] du, x > 0 \quad \textbf{(10.154)}$$

Here we use for the fractional order the notation μ in order to distinguish this spatial derivative from the time-dependent analogue (10.1).

By analogy of the results from [69] a pro-Caputo (non-normalized) space-derivative denoted as $_{cf}D_x^\gamma$, can be defined as

$$_{cf}D_x^\gamma F(x) = \gamma \int_{-\infty}^x e^{-\gamma(x-u)}[F(x) - F(u)]du = \gamma \int_{-\infty}^x e^{-\gamma(x-u)}\frac{dF(x)}{dx} du \quad \textbf{(10.155)}$$

The space memory constant γ in (10.155) controls the kernel and $\gamma \in (0, \infty)$. If we like to refine $_{PC}D_x^\gamma$ as an integral operator controlled by a single parameter μ we have to satisfy the following conditions: for $\mu \in [0,1] \Rightarrow 1/\gamma \in [0, \infty]$. With $\gamma(\mu) = \mu/(1 - \mu)$ [1, 3] the desired properties are obtained, namely

$$\frac{1}{\gamma} = \frac{1-\mu}{\mu} \in [0, \infty], \mu = \frac{1}{1+1/\gamma} = \frac{\gamma}{1+\gamma} \in [0,1], \frac{\mu}{(1-\mu)^2} = \frac{\gamma}{1-\mu} \quad \textbf{(10.156)}$$

Further, following the basic definition of the Caputo-Fabrizio derivative when the constitutive equations is defined through a Jeffrey kernel (see section 10.2.1.2) and considering the lower limit of integral in eq. (10.156) at zero, we have

$$_{cf}D_c^\mu T(x) = \frac{N(\chi)}{\chi}(_{PC}D_x^2) = \frac{N(\chi)}{\chi} \int_0^x e\left[-_1\frac{\mu}{\mu}(x-u)\right]\frac{dF(x)}{dx} du =$$

$$\textbf{(10.157)}$$

$$= \frac{H(\mu)}{1-\mu} \int_0^x e\left[-_1\frac{\mu}{-\mu}(x-u)\right]\frac{dF(x)}{dx} du$$

In the terms used here $\chi = 1/\gamma$, while $N(\chi)$ and $H(\mu)$ are normalization functions, and in accordance with the general definition (10.1) we should have, by analogy, $H(0) = H(1) = 1$.

As it was commented earlier, Losada and Nieto [4] in the case of time-dependent derivative defined $H(\mu) = 2/(2 - \mu)$. It the present case this definition works correctly for $\mu = 0$ but gives $H(1) = 2$. Further, if we suggest $H(\mu) =$

$1/(2-\mu)$ then we have $H(0) = 1/2$ and $H(1) = 1$. It is obvious that both conditions cannot be satisfied simultaneously by these simple expressions of $H(\mu)$. Then, for convenience, we define $H(\mu) = 1$ as it was suggested by Caputo and Fabrizio in [1]. Consequently, the form (10.156) reduces to a space-fractional derivative, defined by analogy of eq. (10.1), in [75], namely

$$_{cf}^{H}D_x^\mu T(x) = \frac{1}{1-\mu}\int_0^x e^{\left[-\frac{\mu}{1-\mu}(x-u)\right]}\frac{dF(x)}{dx}du \qquad (10.158)$$

The definition of $_{cf}^{H}D_x^\mu T(x)$ is of Caputo-type in contrast to the Yang-Srivastava-Machado space derivative $_{cf}^{Y}D_x^\mu T(x)$ (presented by eq. (10.54) which is of Riemann-Liouville type (see the discussion in section 10.1.2).

Moreover, the definition (10.158) differs from the fractional gradient operator [1] presented by (10.136). For $\mu = 1$, it follows from (10.158) that $_{cf}^{H}D_x^\mu T(x) = dT/dx$, while when $\mu \to 0$ we have $_{cf}^{H}D_x^\mu T(x) \to T(x) - T(0)$ since the lower terminal in the integral is chosen to be at zero.

The Laplace transform of $_{cf}^{H}D_x^\mu T(x)$ is [75]

$$L_T\left[_{cf}^{H}D_{x_\lambda}^\mu T(x_\lambda), p\right] = \frac{pL_T[T(x)-T(0)]}{p+\mu(1-p)} \qquad (10.159)$$

From the general rule of differentiation established for the time-fractional analogue [1] the following rules are valid for $_{cf}^{H}D_x^\mu T(x)$ [75], too

$$D^{(n)}\left[_{cf}^{H}D_x^\mu f(x)\right] =_{cf}^{H} D_x^\mu\left(D^{(n)}f(x)\right) \qquad (10.160)$$

For $n = 1$ we may derive equivalent expressions (10.161) namely:

$$D^{(1)}\left[_{cf}^{H}D_x^\mu f(x)\right] = \frac{1}{1-\mu}\left\{\frac{df(x)}{dx} - \frac{\mu}{1-\mu}\int_0^x \frac{df(x)}{dx}exp\left[-\frac{\mu(x-u)}{1-\mu}\right]du\right\} \qquad (10.161)$$

Hence,

$$D^{(1)}\left[_{cf}^{H}D_x^\mu f(x)\right] =_{cf}^{H} D_x^\mu\left[\frac{df(x)}{dx}\right] \qquad (10.162)$$

It is worthnoting that from the definition of the exponential kernel the physical dimension of γ is *length* $[m]$. However, the fractional order μ is dimensionless and as it will be mentioned further in this work that the ratio $(1-\mu)/\mu$ has a dimension of length. In order to resolve this conflict we may present $\lambda = 1/\gamma = l_s[(1 - \mu)/\mu]$ where the dimension of the factor l_s is length [m]. Without loss of generality we may assume that $l_s = 1$. This conflict was resolved in [75] and the solution related to the definition of the fractional order μ will be commented further in this section.

10.3.4. Steady-State Heat Conduction Equation with a Spatial Memory

From equation (10.151) in terms of $T(x)$ the complete expression about the heat flux, expressed by the space-fractional derivative ${}^H_{cf}D^\mu_x$, is [75]

$$q(x) = a_1 \frac{dT(x)}{dx} + a_{2x}(1 - \mu){}^H_{cf}D^\mu_x\left[\frac{dT(x)}{dx}\right], a_1 = k_1/\rho C_p, a_{2x} = k_{2x}/\rho C_p \quad \textbf{(10.163)}$$

After differentiation, we may present the energy balance equation relationship (10.149) in three equivalent forms [75]

$$\frac{\partial T}{\partial t} = \frac{d}{dx}q(x) = -k_1\frac{d^2T(x)}{dx^2} - \frac{k_{2x}}{\lambda}\int_0^x e^{-\left(\frac{x-u}{\lambda}\right)}\frac{d^2T(x)}{dx^2} \quad \textbf{(10.164)}$$

$$\frac{\partial T}{\partial t} = -a_1\frac{d^2T(x)}{dx^2} - a_{2x}(1 - \mu){}^H_{cf}D^\mu_x\left[\frac{d^2T(x)}{dx^2}\right] \quad \textbf{(10.165)}$$

$$\frac{\partial T}{\partial t} = -a_1\frac{d^2T(x)}{dx^2} - a_{2x}(1 - \mu)\frac{d}{dx}\left({}^H_{cf}D^\mu_x\left[\frac{dT(x)}{dx}\right]\right) \quad \textbf{(10.166)}$$

The *steady-state condition* at $t \to \infty$ yields three equivalent expressions

$$0 = \frac{d}{dx}q(x) = -k_1\frac{d^2T(x)}{dx^2} - \frac{k_{2x}}{\lambda}\int_0^x e^{-\left(\frac{x-u}{\lambda}\right)}\frac{d^2T(x)}{dx^2} \quad \textbf{(10.167)}$$

$$0 = -a_1\frac{d^2T(x)}{dx^2} - a_{2x}(1 - \mu){}^H_{cf}D^\mu_x\left[\frac{d^2T(x)}{dx^2}\right] \quad \textbf{(10.168)}$$

$$0 = -a_1\frac{d^2T(x)}{dx^2} - a_{2x}(1 - \mu)\frac{d}{dx}\left({}^H_{cf}D^\mu_x\left[\frac{dT(x)}{dx}\right]\right) \quad \textbf{(10.169)}$$

10.3.4.1. Short-Range Memory Effects: Space Memory Only

Let us consider a material with a spatial memory arranged as a long bar of length L with thermally insulated surface and subjected to a thermal loads at $x = 0$ and $x = L$, that is

$$T(0) = T_0, T(L) = T_L, {}^H_{cf}D^\mu_x T_0 = {}^H_{cf}D^\mu_x T_L = 0, T_0 > T_L \quad \textbf{(10.170)}$$

Without loss of generality, we may assume that $T_L = 0$ (this is only a shift in the temperature scale), that will simplify the calculations. Moreover, the conditions (10.170) follow from the definition of the Caputo-Fabrizio derivative [1].

If the Cattaneo kernel is only taken into account, that is in the modified relaxation function R^* the Dirac Delta function is omitted, then we may write in terms of ${}_{cf}^{H}D_x^{\mu}$ the flux relation as

$$q_2 = -k_{2x}(1 - \mu){}_{cf}^{H}D_x^{\mu}[T(x)] \tag{10.171}$$

This is the equation used by Yang *et al.* [9] despite the fact that the fractional derivative used by these authors (10.54) is of pseudo-Riemann-Liouville type. However, it is easy to check that with zero initial conditions ${}_{cf}^{H}D_x^{\mu}$ and ${}_{cf}^{Y}D_x^{\mu}$ match (see (10.52) where the Laplace transform of ${}_{cf}^{Y}D_x^{\mu}$ is the same as that of ${}_{cf}^{H}D_x^{\mu}$ (10.159)).

Physically, eq. (10.171) means that the macroscopic temperature field is ignored and the zonal temperature distribution is considered only. Further, eq. (10.171) *accounts only the spatial structural memory affect on the heat diffusion and it is valid only within a zone defined as* $x_0 < x < x_0 + \lambda$ where $0 < x_0 < L$ and $0 < \lambda < L$ and the space-memory related thermal diffusivity a_{2x} accounts only short-distance structural memory effects. For $\mu = 1$ we get the transient zonal model (10.172) and the steady-state version (10.173), namely

$$\frac{\partial T(x,t)}{\partial t} = a_{2x}\frac{d^2T(x)}{dx^2} \tag{10.172}$$

$$0 = a_{2x}\frac{d^2T(x)}{dx^2}, t \to \infty \tag{10.173}$$

assuming $T(x_0) > T(x_\lambda) > T(\lambda)$ where (10.171) is expressed in the local (zonal) coordinate system with origin at $x = x_0$ and the axis extends up to $x = x\lambda = x_0 + \lambda$.

The solution by the Laplace transform [75] yields

$$-T(x_\lambda) = [q_\lambda(1 - \mu) - T_{x0}] + q_\lambda\mu x_\lambda$$
$$\frac{T_{x0} - T(x_\lambda)}{q_\lambda} = (1 - \mu) + \mu x_\lambda \tag{10.174}$$
$$x_0 < x_\lambda \leq x_0 + \lambda, and\ T(x_0) > T(x_\lambda) > T(x_0 + \lambda)$$

The linear temperature profile (10.174) confirms the solution of Yang *et al.* [9] irrespective of the differences in the definition of the space-fractional derivatives used. For $\mu = 1$ the result (10.174) reduces to the classical Fourier solution. Further, for $x = x_0$ and $x_\lambda = x_0 + \lambda$ we have

$$-T_\lambda = [q_\lambda(1-\mu) - T_{x0}] + q_\lambda\mu(x_0 + \lambda) \tag{10.175}$$

$$-T_{x_0} = [q_\lambda(1-\mu) - T_{x0}] + q_\lambda\mu x_0 \tag{10.176}$$

From (10.175) and (10.176) we obtain

$$T_{x0} - T_\lambda = q_\lambda\mu\lambda \Rightarrow q_\lambda = \frac{(T_{x0}-T_\lambda)}{\mu\lambda} \tag{10.177}$$

The relation (10.177) is the linear Fourier law across the memory zone of thickness λ.

Replacing q_λ in (10.174) by (10.177) and with $\lambda = (1-\mu)/\mu$, the dimensionless profile in the zonal coordinate system is

$$\frac{T_{x0}-T(x_\lambda)}{(T_{x0}-T_\lambda)} = 1 + \frac{x_\lambda}{\lambda} = 1 + \frac{\mu}{1-\mu}x_\lambda \tag{10.178}$$

The transition to the global coordinate system defined by $0 < x < L$, by a simple shift $x_\lambda = x - \lambda$ results in

$$\frac{T_{x0}-T(x)}{T_{x0}-T_\lambda} = \frac{x}{\lambda} = \frac{\mu}{1-\mu}z_\lambda, 0 < z_\lambda = \frac{x}{\lambda} < 1 \tag{10.179}$$

Therefore, across the memory zone we have a linear temperature profile controlled by the fractional order μ. Fig. (**10.4**) presents the solutions (10.174) and (10.179) for several values of μ.

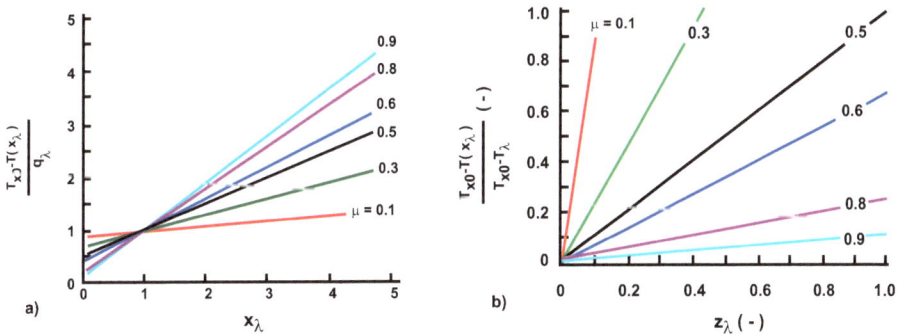

Fig. (10.4). Zonal temperature distribution (space memory only : a) dimensional ; b) dimensionless profiles. Adapted from [75]. By courtesy of *Thermal Science*

10.3.4.2. Space Memory with Extended Relaxation Function (Jeffrey Memory Kernel)

With the complete relaxation function and the rules of differentiation of $_{cf}^{H}D_x^{\mu}$ we have [75]

$$\frac{\partial T(x,t)}{\partial t} = a_1 \frac{d^2T(x)}{dx^2} + a_{2x}\left[\frac{d^2T(x)}{dx^2} + a_{2x}(1-\mu)\frac{d}{dx}\left(_{cf}^{H}D_x^{\mu}T(x)\right)\right] \quad \textbf{(10.180)}$$

Two steady-state equivalent versions of (10.180) can be expressed as

$$0 = (a_1 + a_{2x})\frac{d^2T(x)}{dx^2} + a_{2x}(1-\mu)\frac{d}{dx}\left[_{cf}^{H}D_x^{\mu}T(x)\right] \quad \textbf{(10.181)}$$

$$0 = \frac{d^2T(x)}{dx^2} + m(1-\mu)\frac{d}{dx}\left[_{cf}^{H}D_x^{\mu}T(x)\right], m = \frac{a_{2x}}{a_1+a_{2x}}, 0 < m < 1 \quad \textbf{(10.182)}$$

For $\mu = 1$ we have that the thermal conductivity is $k_0 = k_1 + k_{2x}$ and equivalently $a_0 = a_1 + a_{2x}$, as in the case of a temporal memory (see section 2.1.2). The partition coefficient m accounts for the contribution of the space memory effect (the elastic structural thermal diffusivity a_{2x}) to the total thermal diffusivity a_0. For $\mu = 1$, we get $m = 0$.

The integration of eq. (10.182) yields

$$C_3 = \frac{dT(x)}{dx} = m(1-\mu)_{cf}^{H}D_x^{\mu}T(x) \quad \textbf{(10.183)}$$

Now, applying the Laplace transform solution to (10.183) [75] we have

$$T(x) = T_0 + C_3\left[\frac{a}{b}x + \frac{b-a}{b^2}(1 - e^{-bx})\right] + C_4, a = \frac{\mu}{1-\mu}, b = m + a \quad \textbf{(10.184)}$$

From the boundary conditions (10.170) the integration constants C_3 and C_4 are defined as

$$C_3 = \frac{T_L-T_0}{L}, C_4 = -T_0 \quad \textbf{(10.185)}$$

Hence, in terms in the process variables the solution (10.184) can be presented as [75]

$$\theta_\mu = \frac{T_0-T(x)}{T_0-T_L} = \frac{a}{b}\frac{x}{L} - \frac{1}{L}\frac{1-b}{b^2}[1 - exp(-bx)] + \frac{b-a}{b^2} \quad \textbf{(10.186)}$$

Alternatively, if we assume $T_L = 0$, which means only a shift of the temperature scale, we get

$$\theta_\mu = 1 - \frac{T(x)}{T_0} = \frac{a}{b}\frac{x}{L} - \frac{1}{L}\frac{1-b}{b^2}[1 - exp(-bx)] + \frac{b-a}{b^2} \qquad (10.187)$$

The equivalent forms of (10.186) and (10.187) are

$$\theta_\mu = \frac{T_0 - T(x)}{T_0 - T_L} = \frac{\mu}{\mu + m(1-\mu)}\left(\frac{x}{L}\right) + \frac{1}{L}\frac{m(1-\mu)^2}{[\mu+m(1-\mu)]^2}\left[1 - e^{-\left(m+\frac{\mu}{1-\mu}\right)x}\right] \qquad (10.188)$$

or when $T_L = 0$

$$\theta_\mu = \frac{T(x)}{T_0} = 1 - \frac{\mu}{\mu + m(1-\mu)}\left(\frac{x}{L}\right) + \frac{1}{L}\frac{m(1-\mu)^2}{[\mu+m(1-\mu)]^2}\left[1 - e^{-\left(m+\frac{\mu}{1-\mu}\right)x}\right] \qquad (10.189)$$

Now, bearing in mind that $\gamma = 1/\lambda = \mu/(1-\mu)$ we may express (10.187) in a dimensionless form

$$\theta_\mu = \frac{\mu}{\mu + m(1-\mu)}z + \frac{\lambda}{L}\frac{m\mu(1-\mu)}{[\mu+m(1-\mu)]^2}\left[1 - e^{-(\gamma+m)x}\right] \qquad (10.190)$$

The ratio λ/L and the product $(\gamma + m)x \approx x/\lambda$ are dimensionless. Obviously, the contribution of the non-linear term depends on the ratio $\frac{\lambda}{L} < 1$. The linear approximation of the exponential term in (10.190) as a series within the range where the space memory takes place (that is for $0 < x < \lambda$ and $\lambda << L$, as well as $x << L \Rightarrow x/L << 1$)   is: $exp[-(\gamma + m)x] \approx 1 - (\gamma + m)(x - \lambda) + O[(x - \lambda)^2]$. Then, we may approximate (10.189) as

$$\theta_\mu \approx \frac{\mu}{\mu + m(1-\mu)}\frac{x}{L} + \frac{1-\mu}{\mu}\frac{m\mu(1-\mu)}{[\mu+m(1-\mu)]^2}\left[\frac{(\gamma+m)(x-\lambda)}{L}\right] \qquad (10.191)$$

Now, with $(x - \lambda)/L << 1$ we may see that the first term of (10.190) dominates. However, if we look only at the range $x_0 < x < \lambda$ we may neglect the first term in (10.190), that is assuming $x/L \approx 0$ because $x/L < \lambda/L << 1$, then eq. (10.190) reduces to two equivalent forms:

$$\theta_{\mu(x<\lambda)} \approx \frac{1-\mu}{\mu}\frac{m\mu(1-\mu)}{[\mu+m(1-\mu)]^2}\left[\frac{(\gamma+m)(x-\lambda)}{L}\right] \qquad (10.192)$$

$$\theta_{\mu(x<\lambda)} \approx \frac{\lambda}{L}\frac{m\mu(1-\mu)}{[\mu+m(1-\mu)]^2}[(\gamma + m)(x - \lambda)] \qquad (10.193)$$

The approximations (10.191) and (10.192) represent the zonal solution (within the range defined by space memory effects) but with a temperature field defined by the thermal loads at the macroscopic boundaries. However, λ/L may reach values of order of magnitude of unity as it is shown further in this article and therefore, the results (10.191) and (10.192) are valid for cases with small memory effects.

Alternatively, a solution by direct integration was developed in [75]. This approach leads to a fractional differential equation

$$\substack{H\\cf}D_x^\mu\theta - \frac{1}{m(1-\mu)}\frac{d\theta}{dx} + \frac{1}{L}\frac{1}{m(1-\mu)} = 0, z = x/L, \theta_{(z=1)} = 1, \theta_{(z=0)} = 0 \quad \textbf{(10.194)}$$

The Laplace transform solution of (10.194) yields

$$\theta(z) = \left(\frac{\gamma}{m+\gamma}\right)z - \frac{1}{L}\left\{\frac{1}{m+\gamma}exp[-(m+\gamma)x] + \frac{1}{m+\gamma}\right\} \quad \textbf{(10.195)}$$

In terms of $z = x/L$ the solution is

$$\theta(z) = \left(\frac{\gamma}{m+\gamma}\right)z - \frac{1}{L}\left\{\frac{1}{m+\gamma}exp[-L(m+\gamma)z] + \frac{1}{m+\gamma}\right\} \quad \textbf{(10.196)}$$

Equation (10.195) is an analogue of the solutions (10.186) and (10.188). In addition, the coefficient of the fist term in the RHS of (10.189) in terms of γ is $\mu/(m+\mu) = \gamma/[\gamma + m(1+\gamma)]$. Moreover, the dimension of $\gamma = 1/\lambda$ is $[m^{-1}]$ and therefore the product $(m+\gamma)x$ as well as $L(m+\gamma) \approx L\gamma = L/\lambda = (\lambda/L)^{-1}$ are dimensionless. Physically, this is the ratio of *the macroscopic length scale L to the spatial memory length scale λ* and we have to define its relationship to the factional order μ.

10.3.4.3. The Ratio λ/L and the Fractional Order μ

The ratio λ/L defines the effect of the zone of length λ, where the spatial memory takes place, on the transport in the total area of interest of length L. Obviously, $0 \leq \lambda/L \leq 1$ and we have to relate this ratio to the fractional order because by the definition of the space fractional derivative (10.158) we have $\lambda = (1-\mu)/\mu$. Denoting $\lambda/L = K_L$ we have three equivalent relations [75]:

$$\frac{1-\mu}{\mu}\frac{1}{L} = K_L, \mu = \frac{1}{1+K_L}, K_L = \frac{1-\mu}{\mu}, L = 1, 0 \leq K_L \leq 1 \quad \textbf{(10.197)}$$

The first relation in (10.197) is the scaled memory length λ, *i.e.* λ/L. *Since, the global length L always can be defined as $L = 1$, irrespective of the real physical* dimension and the units used for it, this leads to the definition of μ by the second relation $K_L = K_L(\mu)$ in (10.197). Recall the definition of the fractional order μ in (10.197) has the same form as (10.126) where α is defined by the Deborah number.

Therefore, from known K_L we may define the fractional order μ or at least to estimate the range where it could vary. The relationship $\mu = f(K_L)$ expressed by (10.197) is shown in Fig. (**10.5**). Since $0 < \lambda/L \leq 1$, it follows that $0.5 \leq \mu \leq 1$. The lower boundary $\mu = 0.5$ corresponds to $\lambda/L = 1$ where the entire area is covered by the memory zone, while $\mu = 1$ means $\lambda = 0$ (no memory effects). The decrease in size of λ increases the fractional order μ and reduces the memory effect on the global transport process. The reasonable question is: What happens if $\mu < 0.5$? Simply if $\mu < 0.5$ it follows that $\lambda/L > 1$ which violates the physical meaning of a memory zone embedded in a large homogeneous area of length L.

10.3.4.4. The Redistribution Coefficient m and the Fractional Order μ

The solutions define the redistribution coefficient m and it is natural to ask what the relationship between m and fractional order μ is. From the physics we have that when $\mu \to 1$, consequently $m \to 0$, and *vice versa*. Hence, if we define the relation $m = (1 - \mu)^N$ the limits are obeyed. From the analysis of the heat waves [72], which is used as a template of the present analysis, we have that the slower relaxation effects, the larger the value of k_2 and *vice versa*. Replacement of m by $(1 - \mu)^N$ in the solutions makes the fractional order μ the only parameter controlling the process. The tests with different N of the coefficient of the linear terms $M_\mu = \mu/[\mu + m(1 - \mu)]$ in the solution (10.187) and (10.188) and $m = (1 - \mu)^N$ are shown in Fig. (**10.6**). Since the conditions $K_m = 1$ and $m = 0$ are satisfied simultaneously only for $N = 1$, the numerical simulations demonstrated next use $N = 1$.

10.3.4.5. Temperature Profiles

The temperature profile accounting the memory effects are shown in Figs. (**10.6** and **10.8**). The profiles corresponding to the two terms of the solution (10.187) are especially separated in order to demonstrate how they are affected by the value of the fractional order μ (see Fig. **10.6**). With low $\mu \to 0.5$ the zone with memory effects will cover the entire area of interest and the ratio $\lambda/L \to 1$. The increase in μ reduces the non-linearity of the temperature profiles which is clearly demonstrated by behaviour the exponential term (the lower parts of Figs. **10.8a, b,**

c, d) and the first term (see Fig. **10.6**). With increase in μ the effect of the non-linearity decreases (precisely the effect of the exponential term) and hence, the complete temperature profile approaches the linear Fourier solution.

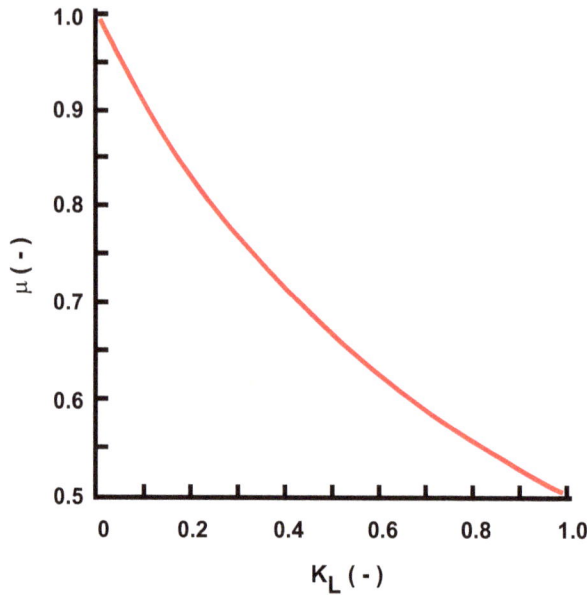

Fig. (10.5). Functional relationship $\mu = K_L$ defining the fraction order as a function of the Deborah number. Adapted from [75].

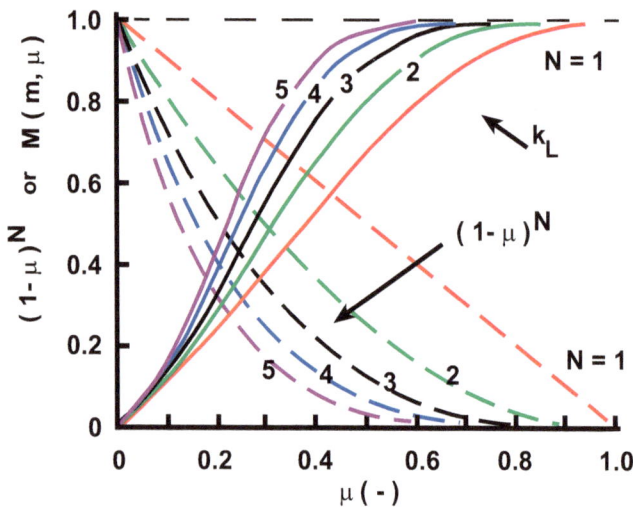

Fig. (10.6). Effect of the fractional order μ and the exponent N on the functional relationship $m = f(\mu) = (1 - \mu)^N$ and the coefficient Mz of the first term of the solution (10.187). Adapted from [75].

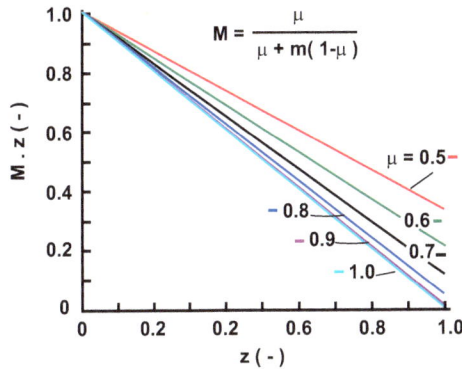

Fig. (10.7). Profiles determined by the first term of the solution (10.187) in the range $0 \leq \mu \leq 1$ corresponding to $0 \leq \lambda/L \leq 1$. Adapted from [75]

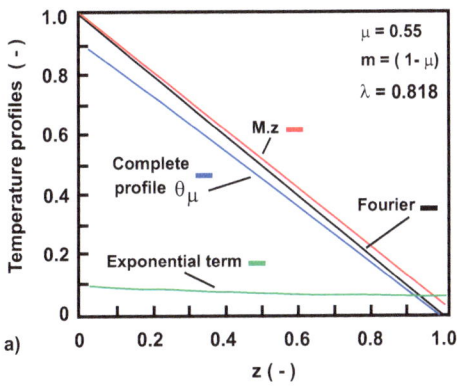

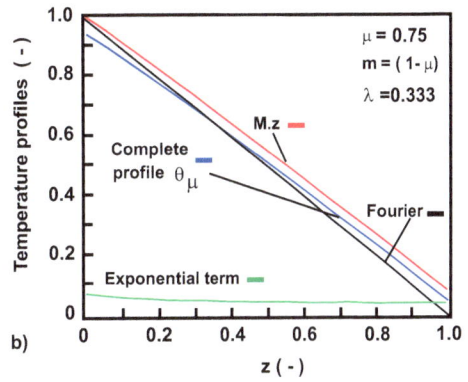

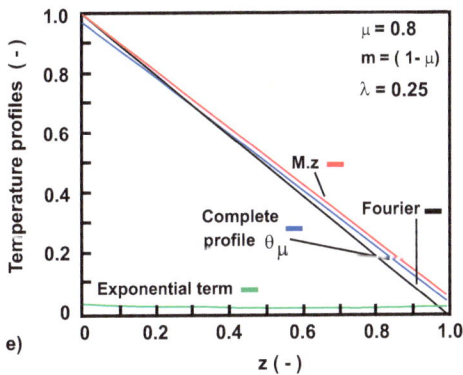

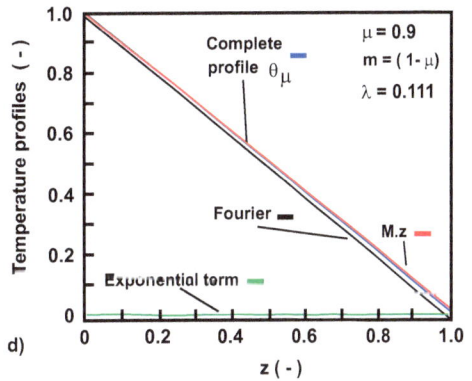

Fig. (10.8). Temperature profiles accounting memory effects for various values of the fractional order μ. Adapted from [75].

10.4. FORMALISTIC FRACTIONALIZATION OF THE DIFFUSION EQUATION: SOME COMMENTS AND SUGGESTIONS BY ANALOGY

Despite the initial intention to analyze diffusion problems with Caputo-Fabrizio derivatives where only initial constitutive equations flux-gradient are initially defined we like to express some comments on the formalistic fractionalization.

10.4.1. Examples Coming from Fractional Models with Singular Kernels and Consequent Formalisms

Let us consider the classical Fick (Fourier) equation. In this case, by a simple replacement of the time derivative we get

$$_{cf}D_t^\alpha u(x,t) = D\frac{\partial^2 u(x,t)}{\partial x^2} \tag{10.198}$$

Simply, this model can be deduced if the mass balance equation (10.70) is re-written *ad-hoc* as a constitutive equation

$$_{cf}D_t^\alpha u(x,t) = -\frac{\partial j}{\partial x} \tag{10.199}$$

and the flux is expressed by (10.72) where the damping function is the Dirac delta, *i.e.* $R(x,t) = R(t) = \delta_D$. However, we have to remind that the diffusion process is caused by the gradient of concentration (temperature) which is the reason the heat flux to diffuse into the medium. Hence, the forms (10.198) and (10.199) contradict the physics because the main idea, since the time of Maxwell, is that *the flux should be damped, not the concentration (temperature)*. Using the formalism expressed by (10.198) and (10.199) the flux can be defined also as advection that immediately leads to damped versions of the advection-diffusion differential equation [40], the Korteweg-de-Vries-Burgers equation (10.200) [8] and Burgers equation (10.201) [79], namely

$$_{cf}^{c}D_t^\alpha = v_f\frac{\partial^2 u(x,t)}{\partial x^2} - 2u(x,t)\frac{\partial u(x,t)}{\partial x} - \mu_f\frac{\partial^3 u(x,t)}{\partial x^3} \tag{10.200}$$

$$_{cf}^{c}D_t^\alpha = -u(x,t)\frac{\partial u(x,t)}{\partial x} + \frac{\partial^2 u(x,t)}{\partial x^2} - \lambda u(x,t) \tag{10.201}$$

Similar approach led to damped Keller-Segel chemotaxis [32] model, the Allen-Cahn models [80] and oxygen diffusion equation [43].

Now, the principle question is: *Are these fractionalization techniques reasonable modelling approaches or they simply allow mimicking already existing models?* To answer this question let us remember the same case but with the fractional-time equation with a singular memory. In this case we have two versions [72] of the subdiffusion equation, namely

$$\frac{\partial u(x,t)}{\partial t} = {}_{RL} D_t^{1-\mu}\left[D_\mu \frac{\partial^2 u(x,t)}{\partial x^2}\right] \tag{10.202}$$

where ${}_{RL}D_t^{1-\mu} = D_t^1 I_t^\mu$ is the Riemann-Liouville (RL) fractional derivative of order $1 - \mu$. Equivalently (at zero initial conditions where the Riemann-Liouville and the Caputo derivatives with a singular kernels match) we have

$$_c D_t^\mu u(x,t) = D_\mu \frac{\partial^2 u(x,t)}{\partial x^2} \tag{10.203}$$

In both cases the diffusion coefficient D_μ has a dimension m^2/s^μ.

These two versions model asymptotic subdiffusion arising from a CTRW with a power-law waiting times and can be considered as models of a subordinated Brownian motions stochastic process [81].

Equation (10.202) can be obtained phenomenologically by combining the constitutive relation (10.70) and the following *ad-hoc* fractional Fick law [81]

$$j_\mu(x,t) = -D_\mu\left[{}_0 D_t^{1-\mu} \frac{\partial u(x,t)}{\partial x}\right] \tag{10.204}$$

The fractional integral in this *ad-hoc* constitutive relation provides a weighted average of the concentration gradient over the prior history [81]. Moreover, we have to remember that using (10.204) or (10.202) we may obtain (10.198) as well as all above-mentioned models using the Riemann-Liouville derivative (or Caputo at zero initial conditions) (see (10.200) and (10.201)) *since the time-fractional Riemann-Liouville derivative is left-inverse to the fractional integral. However, such relation between the Caputo-Fabrizio derivative (10.1) and the associated integral (either (10.19) or (10.74)) does not exist.*

10.4.2. Examples Suggesting Flux-Gradient Relations in Caputo-Fabrizio Sense

Now, inspired by the *ad-hoc* fractional Fick law (10.204) let us assume that the fractional derivative is of Caputo-Fabrizio type and therefore we may write intuitively

$$_{cf} j_\alpha = -D_0 \left[{}^{RL}_{cf} D_t^{1-\alpha} \frac{\partial u(x,t)}{\partial x} \right], \alpha \in [0,1] \tag{10.205}$$

where ${}^{RL}_{cf} D_t^{1-\alpha}$ is defined by (10.44) as a Caputo-Fabrizio derivative of Riemann-Liouville type.

With definition of (10.44) we may express

$$
{}^{RL}_{cf} D_t^{1-\alpha} = \frac{1}{1-\alpha} \int_0^t exp\left[-\frac{1-\alpha}{1-(1-\alpha)} \right] (t-s)f(x,s)ds =
$$

$$
= \frac{1}{1-\alpha} \int_0^t exp\left[-\frac{1-\alpha}{\alpha} \right] (t-s)f(x,s)ds \tag{10.206}
$$

Hence, assuming for the sake of simplicity $M(\alpha) = 1$ we get from (10.206) that

$$
{}^{RL}_{cf} D_t^{1-\alpha} f(x,t) = \left(\frac{\alpha}{1-\alpha} \right)_0 I_t^\alpha f(x,t), \alpha \in [0,1] \tag{10.207}
$$

In this expression $_0 I_t^\alpha f(x,t)$ is defined by (10.74) as it was constituted by Caputo and Fabrizio [3]. Moreover, the scaling to the characteristic diffusion time of the process reveals that the pre-factor in the right-hand side of (10.207) is $\alpha/(1-\alpha) = 1/De = \tau/t_D$, where De is the Deborah number (see (10.125)).

Therefore, from the assumed constitutive relation (10.205) we have

$$
{}^{RL}_{cf} D_t^{1-\alpha} \left[\frac{\partial u(x,t)}{\partial x} \right] = \left(\frac{\alpha}{1-\alpha} \right)_0 I_t^\alpha \left[\frac{\partial u(x,t)}{\partial x} \right] = \tag{10.208}
$$

$$
= \frac{1}{1-\alpha} \int_0^t exp\left[-\frac{1-\alpha}{\alpha}(t-s) \right] \left[\frac{\partial u(x,t)}{\partial x} \right] ds
$$

Then, the *ad-hoc* fractional Fick law in terms of Caputo-Fabrizio time integral and derivative can be expressed in three different forms, namely

$$_{cf}j_\alpha = -D_0 \left(\frac{\alpha}{1-\alpha}\right)_0 I_t^\alpha \left[\frac{\partial u(x,t)}{\partial x}\right] \tag{10.209}$$

$$_{cf}j_\alpha = -D_0 \frac{\alpha}{1-\alpha} \int_0^t exp\left[-\frac{1-\alpha}{\alpha}(t-s)\right]\left[\frac{\partial u(x,s)}{\partial x}\right] ds \tag{10.210}$$

$$_{cf}j_\alpha = -D_0 \frac{1}{De} \int_0^t exp\left[-\frac{1-\alpha}{\alpha}(t-s)\right]\left[\frac{\partial u(x,s)}{\partial x}\right] ds \tag{10.211}$$

Now, applying the procedure presented in 2.1.1 we can obtain a diffusion equation with a structure equivalent to (10.79) developed by Caputo and Fabrizio [3] *which differs from* (10.198).

In this context, we may suggest intuitively an *ad-hoc* relation *flux-gradient* with a Cattaneo kernel, namely

$$j_{Catt}^\alpha = -D_0 \left(_{cf}D_t^{-\alpha}\left[\frac{\partial u(x,s)}{\partial x}\right]\right) \tag{10.212}$$

where $_{cf}D_t^{-\alpha}$ is defined as

$$_{cf}D_t^{-\alpha} = \frac{1-\alpha}{\alpha} \int_0^t exp\left[-\frac{\alpha}{1-\alpha}(t-s)f(x,s)\right] ds \tag{10.213}$$

The definition (10.213) is a memory integral of Cattaneo type used in the example (10.104). Then, following (10.212) the constitutive relationship is

$$j_{Catt}^\alpha = -D_0 \frac{1-\alpha}{\alpha} \int_0^t exp\left(-\frac{\alpha}{1-\alpha}(t-s)\left[\frac{\partial u(x,s)}{\partial x}\right]\right) ds \tag{10.214}$$

bringing to mind that from (10.125) we have $(1-\alpha)/\alpha = \tau/t_D = De$. In addition, if the flux-gradient relationship is defined through a Jeffrey kernel we may write *at hoc* that

$$j_{Jeff}^\alpha(x,t) = -D_0[\delta_D(t) + _{cf}D_t^{-\alpha}]\left[\frac{\partial u(x,s)}{\partial x}\right] \tag{10.215}$$

Then, as it was demonstrated in this chapter, the result is equation (10.93).

10.4.3. Some Brief Comments on the Formalistic Fractionalization

The examples commented above demonstrate how to shift from the known integer order models to their damped versions by using fractional derivatives. However, these simple examples invoke for careful applications of basic physical principles and reasonable assumptions when fractional versions of well-known models are at issues but only the replacement of the integer-order time derivative by a fractional one is the core of the model built-ups. At this point we have to refer to Hilfer [82] that: *the definitions of fractional derivatives are problem-oriented, that is the definitions define different type of relaxation processes and the corresponding fractional operators are defined in different classes of functions.*

Despite the simplicity of demonstrated examples, these are *ad-hoc* created models and many points are still unresolved. In this context, all these constitutive relationships and the following diffusion models should pass through checks for thermodynamic consistency as it was already done for the cases when the time-derivative is of Caputo or Riemann-Liouville type with a singular power-law kernel (see the examples in Chapter 3, 6 and 7 of [83]).

However, focusing on the topic of this chapter, using the idea expressed by (10.205) and the equivalent forms (10.209)-(10.211) it would be possible more complex *flux-gradient* relationships to be developed in terms of the Caputo-Fabrizio derivatives. We hope this will motivate the people interested in this new area to work in this direction.

10.5. WHAT TYPE OF RELAXATIONS ARE MODELLED BY THE NON-SINGULAR MEMORY KERNELS AND THE NEW DERIVATIVES?

At the end of this chapter, we are obliged to comment the relation of the new integral operator (10.1) (*ad hoc* termed new fractional derivative in [1, 3]) to real-word problems that are modelled. In fact in the fist article [1] where (10.1) was postulated there are brief notes that this integral operator with non-singular memory focuses on relaxation processes in new materials. However, to our point of view this was not enough to explain to the rapidly publishing modelling society, for which the physics behind the equations is not of primary interest, why this new integral operator is oriented to models with relaxation functions different from the well-known models with weak (power-law) fading memories. Unfortunately, none of the consequent publications (except [3, 69, 70, 75]) does not focus on the physics of the process where this new derivative is applied. Moreover, the avalanche of articles using the formalistic fractionalization approaches caused negative reactions (to my personal communications) among the people well-skilled in the classical fractional modelling with singular fading memories. Most of the features known from this well-developed area are not encountered in the new integral operator (10.1): first of all, there is not well-defined associate integral and some properties such as the semi-group properties are not available. Second, the derivative is not left inverse to the associated integral as we know from classical fractional calculus. Recall, the derivative of a composite function (the Leibniz) rule, well-known from the classic integer-order derivative does not work when either Riemann-Liouville or Caputo derivatives with power-law kernels are at issue, albeit the fact that the Fa di Bruno formula exists ([84]-chapter 24 and [85]-see the comments at page 80).

All these points invoke the question: what really this new definition tries to model and why we cannot see what we already know from the classical fractional derivatives. Simply, the answer is: this integral operator with exponential memory (termed fractional derivative with a non-singular kernel) is oriented to more complex relaxation processes that the ones commonly modelled by the classical fractional derivatives with power-law (singular) memories. In the next points of this last section of the chapter we will try to explain what this standpoint means.

10.5.1. The Stretched-Exponentially Kohlrausch Relaxation Function

First of all, we focus the attention on the exponential memory kernel in (10.1) which is the well-known Kohlrausch stretched-exponentially relaxation function. It is well-known fact that for years the relaxation processes are at the focus of intensive studies oriented to description of transport properties of various materials such as amorphous and crystalline semiconductors, insulators, polymers, glasses, exhibiting non-exponential relaxation patterns [86-88]. Moreover, irrespective of the type of relaxing medium, *the empirically determined relaxation model reproduces universal behaviour features of the system at issue* [88]. We especially stress the attention on this standpoint since it is directly related to the construction of the Caputo-Fabrizio fractional operator (fractional derivative) as it will be commented further in this section with support from the real physical world.

The most popular function fitting time-domain relaxation patterns is the Kohlrausch stretched-exponential function [86]

$$\Phi_K(t) = exp[-(At)^{\alpha_k}] \tag{10.216}$$

where $0 \leq \alpha_k \leq 1$ is stretching exponent and A is an inverse of the characteristic material time constant. The corresponding response function expressed by the negative time derivative of $\Phi_K(t)$ exhibits *short-time power-law* properties and decays *stretched-exponentially for long time* [86], namely

$$f_K = -\frac{d}{dt}\Phi_K(t) \equiv \begin{pmatrix} t^{\alpha_k} & \text{for } t \to 0 \\ e^{-(At)^{\alpha_k}} & \text{for } t \to \infty \end{pmatrix} \tag{10.217}$$

Following the analysis in [86], we may focus the attention on these asymptotic behaviours of $\Phi_K(t)$ relaxation function.

Historically, at the very beginning of fitting relaxation processes at macroscopic time scale, the most used model is the classical exponential Debye process with different relaxation times [86, 87]. This relaxation pattern may be mathematically expressed as a weighted average of an exponential decay $exp(-t/\tau)$ with respect to distribution $g(\tau)$ of a random effective relaxation time T_r [86] thus mimicking the stochastic properties of the condensed material in general.

$$\Phi(t) = exp(-t/T_r) \tag{10.218}$$

The theory of probability allows equation (10.218) to be presented in an integral form [89]

$$\Phi(t) = \langle e^{-t/\tau} \rangle = \int_0^\infty e^{-\frac{t}{\tau}} g(\tau) d\tau \tag{10.219}$$

where $T_r = \tau \in [0, \infty)$ and the probability $g(\tau)d\tau$ are the effective relaxation time and the *probability density function* (pdf), respectively.

Alternatively, the relaxation function can be presented through the mean relaxation time constant $\beta = 1/\tau$ [86] and the effective relaxation rate $\rho(b)$, namely

$$\Phi(t) = \langle e^{-(\beta t)} \rangle = \int_0^\infty e^{-bt} \rho(b) db \tag{10.220}$$

With the definition of the random relaxation constant $\beta = 1/\tau$ the stretched exponentially relaxation function (10.216) can be presented as [86]

$$\Phi_K(t) = \langle e^{(At)^{\alpha_k}} \rangle = \langle e^{-(\beta_k t)} \rangle = \int_0^\infty e^{-bt} \rho_k(b) db \tag{10.221}$$

The expression (10.221) is the Laplace transform of the relaxation rate $\rho_K(b)$. In general, β_k is α_k-stable, non-negative random variable with completely asymmetric α_k-stable pdf [86].

It was demonstrated in [86] that for $b \to \infty$ the relaxation rate pdf exhibits power-law properties and its asymptotic behaviour is directly related to the asymptotic behaviour of its Laplace transform. Therefore, the behaviour with $b \to \infty$ corresponds to $t \to 0$ in the time domain. From the Tauberian theorem [86, 89] it follows that for short times the response function f_K exhibits a power-law behaviour $t^{\alpha_k - 1}$ (see eq. (10.217)) and

$$\rho_K(b) \equiv b^{-\alpha_k - 1}, 0 < \alpha_k < 1, b \to \infty \qquad (10.222)$$

This scaling behaviour (tail property) indicates that the $\Phi_K(t)$ relaxation pattern is scale-invariant [86, 89].

Further, since $\rho_K(b) = (1/b^2)g(1/b)$ [86] it is possible to express that

$$g_K(\tau) = \frac{1}{\tau^2} \rho_K \left(\frac{1}{\tau} \right) \qquad (10.223)$$

From eq. (10.223) it is straightforwardly seen that for large b we have $g_K(\tau) \propto \tau^{\alpha_k - 1}$ and $\tau \to 0$

The Kohlrausch relaxation function (10.216) is a kernel derived from experiments [90-92] on decaying fluorescence intensity where for $\Phi_K(t)$ is used the symbol $I_f(t)$. In this context, as mentioned by Berberan-Santos *et al.* [91] the exponential decay through its series expansion, with aim to be generalized, is

$$e^{-at} = \sum_{n=0}^{\infty} \frac{(-at)^n}{n!} = \sum_{n=0}^{\infty} \frac{(-at)^n}{\Gamma(n+1)} \qquad (10.224)$$

As we know, the presentation (10.224) is a special case of the Mittag-Leffler function [2]

$$E_\gamma(-at) = \sum_{n=0}^{\infty} \frac{(-at)^n}{\Gamma(\gamma n + 1)} \qquad (10.225)$$

for $\gamma = 1$ and is used also to model the decaying fluorescence intensity [91].

In this context, the Heaviside exponential function $e_\gamma(x)$ [91, 93]

$$e_\gamma(-at) = \sum_{n=0}^{\infty} \frac{(-at)^n}{\Gamma(n+1+\gamma)}, 0 < \gamma \leq 1 \qquad (10.226)$$

or a normalized Heaviside exponential function $\varepsilon_\gamma(x)$

$$\varepsilon_\gamma = \Gamma(1 + \gamma)e_\gamma(x) = \gamma\Gamma(\gamma)e_\gamma(x) \tag{10.227}$$

are also widely applied to fit experimentally derived data of decaying fluorescence intensity [91, 92].

Regarding the function $g(\tau)$ in (10.219), commonly denoted as $H(\tau)$, it is the inverse Laplace transform of $\Phi(t)$ and called the eigenvalue spectrum (of a suitable kinetic matrix) [91], and it is normalized as $\Phi_K(0) = 1$, that is $\int_0^\infty H(\tau)d\tau = 1$; as it was mentioned above about eq. (10.219) it is a probability density function (pdf).

The $H(\tau)$ for these two decay (relaxation) generalized exponential functions are known and are real probability density functions [91] defined as

$$H\gamma(\tau)_{(Mittag-Leffler)} = \gamma^{-1}\left(\tau^{-(1+\frac{1}{\gamma})}\right)L_\gamma(\tau^{-\frac{1}{\gamma}}) \tag{10.228}$$

where $L_\gamma(x)$ is the one-side Levy's pdf [91]. For the Heaviside's $H(\tau)$ we have [91]

$$H_{\gamma\,(Heaviside)} = \begin{cases} y(1-\tau)^{\gamma-1} & \text{if } \tau \le 1 \\ 0 & \text{if } \tau > 1 \end{cases} \tag{10.229}$$

Both generalized exponential functions are decaying asymptotically with t^{-1} [91, 94].

In accordance with [91] there are at least two practical ways to generalize the exponential relaxation function: i) To select $H(\tau)$ pdf family which may reduce to Dirac delta for some values of the parameters, a property observed in the Lorentzian and the Gaussian distributions;this procedure automatically ensures that chosen $H(\tau)$ is pdf. ii) The second approach is to use the exponential decay function, modifying it by varying the parameters: this is the case of the Kohlraush stretched exponential. This approach requires prior computations of $H(\tau)$ and check is it still pdf or not. Further, Berberan *et al.* [91] especially especially focus the case when series expansions of the exponentially decay function (and modifications) are used that the resulting distribution functions deserve attention.

After these comments, it seems natural that after the definition of the construction of the Caputo-Fabrizio integral operator (derivative) with a simple exponential memory (the Kohlrausch stretched exponent is set to $\alpha_k = 1$) in eq. (10.1)) it was repeated (by analogy) by replacing the simple exponential kernel by the Mittag-Leffler function. Two examples in this chapter are directly related to this approach to construct derivatives with non-singular and exponentially decaying memory kernels: the Atangana-Baleanu derivative (10.55) and (10.56), where replacing the single relaxation time by the construct $\alpha/(1 - \alpha)$ they become unified in the general sense with the definition (10.1). Similar logic, combining the well-known singular kernel $t^{-\mu}$ and the Mittag-Leffler function led to the bi-order fractional operator (10.65). It might be expected in the future that other types of similar constructions would appear in modelling of relaxation process since the aforementioned decaying functions are not the only possible [91, 92, 94, 95] to fit experimental data.

As it was already mentioned above, the choice of the memory kernel is object oriented issue and the recovery of the relaxation function from experiential data and fitting with well-defined function is a task strongly dependent on the physics of the process of interest. We especially stress the attention on this standpoint since the definition (10.1) does not come from the vacuum of the universe but from real world problems. In this context we refer to the article of Caputo [27] and the examples collected in the next points of this section (see 10.5.3).

10.5.2. Exponential Kernel of the Caputo-Fabrizio Derivative: What Type of Fading Memory is Modelled?

Now, let us turn on the exponential kernel of the Caputo-Fabrizio derivative where it was straightforwardly demonstrated that $\tau/t_D = (1 - \alpha)/\alpha = De$ (see eq. (10.125)). The condition $\tau \to 0$ corresponds to $\alpha \to 1$ and $De \to 1$ and therefore the relaxation function approaches the Dirac delta. However, recall that the power-law relaxation function constructs the memory kernel of the classical Riemann-Liouville and Caputo derivatives [2]. Hence, with this asymptotic power-law behaviour for $(\tau \to 0)$ the stretched-exponentially Kohlrausch function approaches the memory kernel behaviour of the classical fractional integral [2]. Moreover, since the kernel chosen in the Caputo-Fabrizio derivative definition is with $\alpha_k = 1$ it follows that for large b and short relaxation times $(\tau \to 0)$ we have $g_K(\tau) = 1$ as well as from (10.222) it follows that $\rho_K(b) \equiv 1/b^2$ for $b \to \infty$.

It is worthnoting that the power-law pattern (see below) of the Kohlrausch relaxation function, known also as algebraic decay $\Phi_K(t) \propto (t/\tau)^{-\gamma}$ [96] for short times is observed in the stress relaxation of viscoelastic materials (for now entirely modelled by classical fractional derivatives) where it models the *weak principle of the fading memory* (WFM) [97]. The latter means that recent values of the relaxation function (memory kernel) around 0^+ or asymptotically for $t \to 0$ are more influential on the output than the remote values $(t \to \infty)$ [97].

However, as it was commented by Fabrzio and Morro [97] if there is an influential function $k(s)$ (which decays monotonically for large t, is defined and n-times Frechet-differentiable in a neighborhood of the zero history in the space S^k. Further, if a real-valued function k on $\mathbb{R}^{++}$ is $k(s) > 0$ almost everywhere on $\mathbb{R}^{++}$ and $k \in L^1(\mathbb{R}^+)$, and the functions

$$
\begin{aligned}
K_1(\tau) &:= ess.\sup \frac{k(s+\tau)}{k(s)}, s \in \mathbb{R}^{++} \\
K_2(\tau) &:= ess.\sup \frac{k(s)}{k(s+\tau)}, s \in \mathbb{R}^{++}
\end{aligned}
\tag{10.230}
$$

are defined and have finite values on $\mathbb{R}^+$, then there exist positive numbers a_F, b_F and c_F such that [97] on $\mathbb{R}^{++}$ we have

$$
a_F\left[e^{(-b_F s)}\right] < k(s) < c_F, sk(s) \to 0, for \ s \to \infty
\tag{10.231}
$$

As commented by Berberan *et al.* [91] the time-dependent rate constant f_K (see eq. (10.217) is

$$
f_K = -\frac{d}{dt}\ln[\Phi_K(t)] = \frac{\alpha_k}{\tau_0}\left(\frac{t}{\tau_0}\right)^{\alpha_k-1}
\tag{10.232}
$$

where $0 < \alpha_k \leq 1$ and $\tau_0 = 1/A$ (see eq. (10.216)) is the characteristic relaxation time, in the common sense used in this chapter. A very interesting feature of this function, based on the properties of f_K is that there are two regimes (relaxation patterns) [91]: 1) an initial decay faster-than-exponential (FTE) with respect to the exponential of the relaxation time τ_0 where f_K is infinite for $t = 0$) and 2) slower-than-exponential (STE) decay with respect to the exponential of the relaxation time τ_0 for times longer than τ_0. These two regimes are very distinct for small values of α_k *but become indistinct for* $\alpha_k \to 1$ as it is in the case of the exponential kernel used in the definition of (10.1). The stretched exponential decay function has undesirable short-time pattern due the infinite initial rate, *i.e.* in the FTE regime [91], but the same behaviour is observed with the power-law decay function, too.

In the analysis of dielectric relaxation process [98], as an illustrative example demonstrating the behaviour of the relaxation function with common notation $\phi(t)$, there exists universal pattern irrespective of the nature of the material. This universal law is modelled by the fractional power-law for large and small times [98], namely

$$f_k = -\frac{d\phi}{dt} = \begin{pmatrix} (t/\tau_0)^{-\sigma_1} & \text{for } t << \tau_0 \\ (t/\tau_0)^{-\sigma_2 - 1} & \text{for } t >> \tau_0 \end{pmatrix} \qquad (10.233)$$

where $\sigma_1 < 1$ and $0 < \sigma_2$

In addition, Lim and Li [98] analyzed the behaviour of truncated series expansion of the Mittag-Leffler decay function, but we only mention this as a fact relevant to the fractional derivatives using a such type of memory kernel, since the topic is out of the scope of the present analysis. In this study, the Kohlrausch stretched exponential function was compared to a decay function expressed by a modified generalized exponential Mittag-Leffler series expansion denoted as ϕ_{GC}. The comparative analysis revealed that for sufficiently small times $\Phi_K(t)$ and ϕ_{GC} are indistinguishable and exhibit locally self-similar properties (see Fig. **10.1** in [98]). However, for large times $\Phi_K(t)$ is decay faster than ϕ_{GC} (see Fig. **10.2** in [98]). We especially mention this investigation, since with the freedom to use different relaxation (memory) functions in the integral operator in the Caputo-Fabrizio sense (10.1), it would be possible to construct adequate fractional derivatives (in the context of the present analysis) relevant to specific problems of interest. In this context, the versions of derivatives summarized in section 1 of this chapter, should be carefully tested in the context of the above-mentioned scaling patterns for small and large times, as well as it is desirable to see their real behaviours in modelling of real-world processes with relaxations.

Hence, without further details, we may suggest that the exponential memory kernel in the Caputo-Fabrizio derivative (10.1) with $\alpha_k = 1$ and time-independent $f_K = 1/\tau_0$ is responsible, precisely has properties, allowing to account, to some extent, the strong principle of the fading memory in contrast to the principle of weak fading memory (WFM) modeled by power-law kernels. Certainly, all these suggestions are results of thorough "mining" in the works of Mauro Fabrizio trying to recover what was not explained in the articles where the definition (10.1) and its properties were directly declared [1, 3]; never physically explained in all published so far articles encompassed in the present analysis (see the entire reference list). The consequent to [1, 3] publication of Caputo [27] is physically clear but hard to be understood by people not involved in the physics of the process (it will be commented further in this section of the chapter).

In addition, we may decide, on the basis of the existing examples using the formalistic fractionalization, that: if the strong fading memory (SFM) is modeled by integral operator (10.1) it is reasonable to expect that properties exhibited by diffusion equations where the time derivatives represents WFM principle cannot be expected to be adequately modelled by the integral operator relevant to SFM. Therefore, a serious analysis of the physics of the process that should be modelled is highly required prior applying either WMF or SMF principles with the adequate integral operator with fading memory. The simple copy-paste operation changing the type of the fractional derivative used in formalistic fractionalization approach is not the right way.

10.5.3. Real-World Examples with Exponentially Decaying Relaxation Functions

Here we will present some examples collected from the literature where the exponential relaxation function is taken from experiments and fit better the experimental data. These examples and the related empirical models may serve as starting points for creation models with the Caputo-Fabrizio fractional derivative (10.1) as well as with versions of it encompassed in this chapter.

10.5.3.1. Example 1: The Caputo Set of Constitutive Equations for Plastic Media

We start with this example since it was provided by one of the creator of the new fractional operator. In the note on new set of constitutive equation describing plastic behaviour of hard materials Caputo [27] correctly refers that in contrast to the long time use of Hook's law describing properties of elastic media, in case of inelastic properties such as plasticity the mathematical models use memory integrals for the strain $\varepsilon(t)$ and the stress $\tau(t)$, namely

$$\varepsilon(t) = \int_0^t r(t - u)\tau(t)du \qquad (10.234)$$

$$\tau(t) = \int_0^t \bar{r}(t - u)\varepsilon(t)du \qquad (10.235)$$

where $r(t)$ and $\bar{r}(t)$ are causal functions relevant to the response of the medium to unit pulse.

As mentioned by Caputo, the range of u in (10.234) and (10.235) *depends on matter of elegance and formality for similarity with the memory operators presently used* (sic!). Assuming zero initial conditions, that is the medium is in rest prior to the impulse application,

Caputo conceives two generic functional relationships, namely

$$h(t), k(t) = exp(-ut) \qquad (10.236)$$

or equivalently

$$h(t), k(t) = \frac{1}{log(e+ut)} \qquad (10.237)$$

For $u = 0$ it follows that $h(0,t) = u(0,t) = 1$. This simply means that operator of order zero simply reproduces the function which is monotonically decreasing in time such that $h(\infty) = k(\infty) = 0$.

The experimental tests on the plastic behaviour of hard materials (the Polycrystalline Halite (PH) [27], for a cylindrical sample to a lateral confining pressure $\sigma_2 = \tau_{22} = \tau_{33}$ and to pressure $\sigma_1 = \tau_{11}$ parallel to he axis of the cylinder (and based on experimental data reported in [99]), revealed that the creep curves *are accurately fitted by the following exponential relationship* (sic!)

$$\varepsilon_{11} = A + Bt + Cexp(-Dt) \qquad (10.238)$$

Moreover, in accordance with this study the integral operator (10.1) gives an acceptable approximation of the memory of the tested material (PH). This is a good example demonstrating that each medium may need to be studied *without any preconception regarding the use of any derivatives* (sic!) since the laboratory experiments *will suggest the correct form of the memory kernel* [99]. This remark of Caputo completely agrees with the previous comments on the Kohlrausch exponential function and related to them relationships which are experimentally proved approximations.

Using the formalism with simple memory kernels $h(t) \equiv exp(-ut)$ and $k(t) \equiv exp(-vt)$ of orders $0 < u < v < 1$, the classical Caputo derivative (10.2) transforms it to the definition (10.1) and using (10.238) reveals that [99]

$$C = \frac{u}{1-u}, D = \frac{v}{1-v} \qquad (10.239)$$

This allows to determine the fractional orders by the simple relationships

$$u = \frac{C}{1+C}, v = \frac{D}{1+D} \qquad (10.240)$$

It is worth marking that result (10.240) matches exactly the definition of the fractional order by the Deborah number (10.125) which follows directly after nondimensalization of the exponential factor of the memory kernel. Then fitting experiential data and determining the value of the Deborah number De the fractional order we can determine the fractional order straightforwardly. We have to mention especially that the results (10.125) and (10.240), as well as (10.197) (space-fractional problem (10.180)) follow independently from models of different diffusion phenomena with the only common feature: the exponential memory kernel of Cattaneo type, as it was developed in the example with the Dodson equation (see point 2.1.3).

10.5.3.2. Example 2: Anomalous Diffusion of Vapors Through Solid Polymers

Anomalous diffusion is commonly observed in transport of vapors through solid polymers near the glass transition temperature and are related to the relaxations in the polymer structures [100-102]. The problem how best this phenomenon could be modelled mathematically, through analogy with rheological behaviour of viscoelastic materials is given by Fredrickson [103] by a memory integral

$$j_s = -\int_0^t \eta(t - t') \frac{\partial Cs(x, t\prime)}{\partial x} dt' \qquad (10.241)$$

The flux of vapors j_s defined by eq. (10.241) was used as basic formulation of Neogi [104] who (through a rigorous thermodynamic analysis) derived that the relaxation function $\eta(t - t')$ should have the form (in terms of the original variables in the works of Neogi [104])

$$\eta(t) = D^{in}\delta_D(t) + \frac{D^0 - D^{in}}{t^*} exp(-\beta t) \qquad (10.242)$$

or in a equivalent form [105] as

$$\eta(t) = D^{in}\delta_D(t) + \beta(D^0 - D^{in})e^{-\beta t} \qquad (10.243)$$

where t^* is a time constant, equivalent to the relaxation time τ_0 in previously analyzed examples, and $\beta = 1/t^* = 1/\tau_0$. The initial value of the diffusion coefficient is D^{in}, while the diffusivity at infinite time is denoted as D^0. $\delta_D(t)$ is Dirac delta function.

In the context of the problems analyzed in this chapter the expression (10.243) defines a Jeffrey type relaxation kernel used in the example developed in point 2.1.2 (see also [69]). The equation solved by Neogi [105] for geometry of a slab of length l is

$$\frac{\partial C}{\partial t} = \frac{\partial C}{\partial x}\left[\int_0^t \eta(t-t')\frac{C_s(x,t')}{\partial x}dt'\right] \tag{10.244}$$

with boundary conditions of symmetry

$$\frac{\partial C(0,t)}{\partial x} = 0, C(x = \pm l/2) = C(t) \tag{10.245}$$

It would be easy for the reader to check that the form of the diffusion equation developed in terms of the Caputo-Fabrizio derivative (10.93) is simpler that the form developed by Neogi [104, 105] (see eq. 25 in [105]). The new solution in terms of Caputo-Fabrizio derivative would be a challenging task.

Moreover, in the computer simulations of the diffusion modelled by the above equations, Neogi [106] observed decaying oscillations discussed also in [107, 108]. The solution of Neogi, in a simplified form [106], given by the ratio of the mass of vapors $M(t)$ absorbed by the polymer at time t to the mass absorbed at the equilibrium M_∞ $(t \to \infty)$ is

$$\frac{M(t)}{M_\infty} = 1 - \theta e^{-\frac{T}{De}} - \sum_{m=1}^{2}\sum_{k=0}^{\infty}\frac{Y_1}{Y_2} \tag{10.246}$$

where

$$Y_1 = \left(1 - \frac{y_k}{y_k+\frac{1}{De}}\theta\right)exp\left(-\frac{y_k T}{\frac{1}{De}+\omega\zeta_k^2}\right), Y_2 = \frac{\xi_k}{2} - \frac{De\zeta_k}{2}\frac{(1-\omega)}{(y_k De+1)^2} \tag{10.247}$$

$$y_k = -\frac{1}{2}\left(\frac{1}{De} + \omega\zeta_k^2\right) \pm \left[\left(\frac{1}{De} + \omega\zeta_k^2\right)^2 - \frac{4\zeta_k^2}{De}\right]^{\frac{1}{2}} \tag{10.248}$$

where $m = 1$ and $m = 2$ refer to two values of y_k; $\zeta_k = (2k + 1)/(\pi/2)$, $k = 0,1,2...$ In this solution, $T = 4D^0 t/l^2$, $\omega = D^{in}/D^0$ and $\theta = (Cs_{final} - Cs_{initial})/(Cs_{final} - Cs_i)$, where Cs_i is the concentration of the vapors at time t; The Deborah number is defined as $De = \tau_0/(l^2/4D^0)$.

It is easy to see that the solution (10.246)-(10.248) is more complex than the analytical solution of the elastic part of (10.93) developed by Atangana in [47] and the numerical simulations of the complete equation [74] when the model is expressed by the Caputo Fabrizio derivative.

An interesting information in the solution of Neogi [106] is that for $De = 10^{-3}$ (very short times) the diffusion is classical, while for $De \approx 1$, where the viscoelastic effects are stronger, there are well seen oscillations. Regarding to the model (10.93) developed in terms of the Caputo-Fabrizio derivative and the relation (10.125) we have that for $De = 1$ the fractional order is $\alpha = 0.5$. Moreover, oscillations appear even for $De = 10^3$ that corresponds to $\alpha \approx 9.10^{-3}$. However, for large Deborah numbers such as $De = 10^5$ and $De = 10^7$ (corresponding to $\alpha \approx 0$) no oscillations were observed. In view of the model (10.93), the large Deborah number range ($De >> 1$ meaning $\tau_0 >> t_D = l^2/D^0$) corresponds to over-damped diffusion equation (10.93) and a strong contribution of its last term.

10.5.3.3. Example 3: Diffusion of Solvents Through Swelling Solid Polymers

When the polymer expands (swells) due to the solvent penetration the change of the volume causes "extra pressure". Camera-Roda and Sarti [109] in a model on solvent diffusion through polymers (accounting coupling of stress, concentration and chemical potential fields [104, 105]) suggested a reasonable continuum formulation. This formulation is based on the solvent conservation law (see for example (10.70) and lumping the relaxation properties to the constitutive equation for the diffusion flux. This is the same approach used in the two examples with systematically presented solutions: relaxation with a Jeffrey memory kernel (in 2.1.2) and diffusion with simple Cattaneo kernel (in 2.1.3). The model construction of Camera-Roda and Sarti constitutes that the volumetric diffusive flux through the polymer has two components

$$J = J_f + J_r \tag{10.249}$$

$$J_f = -D_f grad\varphi, J_r = -D_r grad\varphi - \tau\frac{D}{Dt}J_r \tag{10.250}$$

where φ is the penetrant (dissolved solvent) volume fractions φ

The derivative $\frac{D}{Dt}J_r$ is an objective *material time derivative* and this formulation of the problem, mimics a construction similar to that of the Neogi model, commented above. Simply, Camera-Roda and Sarti [109] accept the constructions of the Cattaneo concept with extended Fick (Fourier) law which results in the Jeffrey memory kernel. In this formulation, the relations of the diffusion coefficient and the relaxation time τ to the penetrant volume fractions φ are expressed as

$$D_\infty(\varphi) = D_{eq}exp[g(\varphi - \varphi_{eq})] \tag{10.251}$$

$$\tau_\infty(\varphi) = \tau_{eq}exp[K(\varphi_{eq} - \varphi)] \tag{10.252}$$

where D_{eq} and τ_{eq} correspond to the equilibrium state of the solvent sorption for $\varphi = \varphi_{eq}$. As commented by the authors, the integration in time of the constitutive equation results in the expression of Neogi [104] discussed above. This study does not consider memory effects, but this is a matter of philosophy in the model build-up and we know that (in the context of the present chapter) the aforementioned equation could be easily re-arranged as a model with exponential memory and expressed in terms of Caputo-Fabrizio derivative.

The relationship of the *extra pressure P* and the classical diffusion flux was constituted by Thomas and Windle [110] as

$$J = -D(c)\frac{\partial c}{\partial x} - A(c)\frac{\partial P}{\partial x} \tag{10.253}$$

where $A(c)$ is constant depending on the temperature, diluent molar volume and the *thermodynamic diffusion coefficient* (sic!)

In the context of exponential memory of diffusant flux through polymers we have to mention the model of Durning and Tabor [111] which is a step ahead from the model of Thomas and Windle, namely

$$J = -D(c)\frac{\partial c}{\partial x} - \bar{D}\int_{-\infty}^{t}\Phi(t-s)\frac{\partial c(x,s)}{\partial x}ds, \Phi(s) = e^{-\frac{s}{\tau_0}} \tag{10.254}$$

where the effective coefficient $\bar{D}$ has the same meaning as $A(c)$ in the model of Thomas and Windle [110]

Hence, we have again exponential memory (relaxation function) which could be interpreted in terms of the Caputo-Fabrizio derivative.

10.5.3.4. Example 4: Darcy Flow with Relaxation Effect

The viscous fluid flow through a porous body at low velocity is governed by the Darcy law

$$\frac{\partial(\varepsilon\rho)}{\partial t} + \nabla \cdot (\rho V) = 0 \tag{10.255}$$

where $V(r,t$ is the filter (superficial) velocity, $P(x,t)$ is the pressure and $\rho(r,t)$ is the density of the fluid. $r = (x,y,z)$ is the coordinate vector

The Darcy law is an empirical equation postulating a linear relationship between the fluid velocity (flux per unit area) and the driving force (the pressure gradient).

$$V = -\frac{K}{\mu_f}\nabla P \qquad (10.256)$$

The transport coefficient in (10.256) is a ratio of the porous body permeability K and the fluid dynamic viscosity μ_f. If all quantities in the above equations are constant, then $\nabla^2 P = 0$

However, when the fluid is compressible, the equations of state relating the fluid properties and the pressure are (10.257) and consequently (10.258) [112]

$$\rho = \rho_0 e^{\beta_e(P-P_0)}, P_0 < P < P_{max} \qquad (10.257)$$

where ρ_0 and P_0 are reference values, while β_e is inverse of the elasticity of the liquid.

The permeability $K(p)$ and the porosity $\varepsilon(P)$ are functions of the pressure P expressed with analogy with (10.257)

$$\varepsilon = \varepsilon_0 e^{\alpha_\varepsilon(P-P_0)}, K = K_0 e^{k(P-P_0)} \qquad (10.258)$$

All these relationships are *constitutive equations* [112].

Assuming an extension of the Darcy law by incorporating an acceleration term, Rebinder [112] formulated that

$$\tau_\varepsilon \frac{\partial V}{\partial t} + V = -\frac{K}{\mu_f}\nabla P \qquad (10.259)$$

where τ_ε is a relaxation time which becomes equal to zero in the classical (no-accelerated) Darcy flow.

Moreover, the continuity equation

$$\frac{\partial \varepsilon \rho}{\partial t} + \nabla \cdot (\rho V) = 0 \qquad (10.260)$$

allows to define the velocity (the flow flux per unit area) relationship with the pressure gradient through a memory integral with exponential kernel [112]

$$V = -\frac{1}{\mu_f \varepsilon}\int_{-\infty}^{t} K\nabla P(r,s)e^{-\frac{t-s}{\tau_s}} ds \qquad (10.261)$$

Hence, a flux-gradient relation with exponential memory appears again in the model equation. As it shown by the Rehbinder this results in a damped wave equation (eq. 2.22 in [112]).

Therefore, this model demonstrates that the existence of exponential fading memory in fluid flow allows the formalism of the Caputo-Fabrizio fractional operator (derivative) to be applied. For now, this is an open problem waiting for its solution. In order to be correct, flux-gradient relation with a memory in porous media was studied by Caputo [113-115] but in terms of the classical Caputo derivative with power-law kernel (10.2).

10.5.3.5. Some Briefs on the Real-World Examples

The real-world examples of practically important processes modeled with equations incorporating exponential memories try to stress the attention of the reader on the fact that the target is *the damping (the memory) of the flux propagation, not the retardation of the concentration (temperature or fluid velocity)*. This concept does not affect the time derivative in the left side of (10.70) but only the right side, *i.e.* the term $- \partial j / \partial x$. We repeat again this point, since the real-world examples contradict the published models formulated *ad hoc* by the formalistic fractionalization approach. We avoid citing specific publications in this context, but the reader will understand what is incorrect when reading them.

10.6. CONCLUDING REMARKS

This chapter, in fact, is the first attempt to collate the information about the research on development and applications of time and space fractional operators (derivatives) with non-singular kernels. Despite the short 2 years story after the seminal article of Caputo and Fabrizio in 2015 a serious flux of articles has been published in diverse directions, but mainly on solutions of formalistically fractionalized existing models. Analyzes of such a type of models was generally skipped here (with only short comments and suggestions at the end of this chapter) and the focus was on derivation of diffusion equations, starting from properly defined constitutive equations relating the flux and the gradient. The demonstrated derivations of four models (three models with the time-dependent Caputo-Fabrizio derivative and one with a spatial non-singular derivative of Caputo-type) resulted in well-arranged diffusion equations with relaxation terms expressed by the Caputo-Fabrizio time derivative. Generally, such derivations form a small group in the continuously growing pile of publications in the area, but it is quite important to demonstrate the relation between the physical basis and the model build-up. We hope this chapter did a step in this direction.

In addition to the analysis of the versions of fractional derivatives developed on the basis of the concept of Caputo and Fabrizio simply expressed by (10.1) a special attention was paid to see what really (as a memory operator) this new derivative models. This challenging problem was discussed on the basis of the properties of the Kohlrausch stretched-exponentially relaxation function and its asymptotic short-time and large-time patterns. As an outcome of this analysis it was suggested that the exponential kernel exhibits stronger fading memory than the power-law kernel.

At the end, four real-world examples were briefly presented with the main idea to demonstrate that the memory formalism affects the flux evolution and propagation in the medium. These models exist for long time and were solved in different ways by help of the classical integer-order derivatives. Now, we have a new task applying the Caputo-Fabrizio formalism. These examples, to our personal point of view, would help the people interested in application of the Caputo-Fabrizio derivative and its versions with different non-singular kernels, to understand that the exponential memory (relaxation) is natural, comes from the physics of the transport phenomena and it is not a result of *ad hoc* replacement of one memory function by another one.

We hope this chapter and collated information would help the reader to understand adequately the main ideas (and the areas where they could be applied) of this new trend in fractional calculus.

CONSENT FOR PUBLICATION

Not applicable.

CONFLICT OF INTEREST

The authors declare no conflict of interest, financial or otherwise.

ACKNOWLEDGEMENTS

Declared none

REFERENCES

[1]   M. Caputo and M. Fabrizio, "A new definition of fractional derivative without singular kernel," *Progr. Fract. Differ. Appl*, vol. 1, no. 2, pp. 1-13, 2015.

[2]   I. Podlubny, *Fractional differential equations: an introduction to fractional derivatives, fractional differential equations, to methods of their solution and some of their applications*, vol. 198. Academic press, Oct. 1998.

[3]   M. Caputo and M. Fabrizio, "Applications of new time and spatial fractional derivatives with exponential kernels," *Progr. Fract. Differ. Appl*, vol. 2, no. 2, pp. 1-11, 2016.

[4]   J. Losada and J. J. Nieto, "Properties of a new fractional derivative without singular kernel," *Progr. Fract. Differ. Appl*, vol. 1, no. 2, pp. 87-92, 2015.

[5]   V. Ciancio and B. F. F. Flora, "Technical note on a new definition of fractional derivative," *Progr. Fract. Differ. Appl.*, vol. 3, pp. 233-235, 2017.

[6]   J. Hristov, "The non-linear Dodson diffusion equation: Approximate solutions and beyond with formalistic fractionalization," *Math. Natur. Sci.*, vol. 1, no. 1, 2017.

[7]   E. F. D. Goufo and A. Atangana, "Analytical and numerical schemes for a derivative with filtering property and no singular kernel with applications to diffusion," *Eur. Phys. J. Plus*, vol. 131, p. 269, Aug. 2016.

[8]   E. F. Doungmo Goufo, "Application of the caputo-fabrizio fractional derivative without singular kernel to korteweg-de vries-burgers equation," *Math. Model. Anal.*, vol. 21, no. 2, pp. 188-198, 2016.

[9]   X.-J. Yang, H. M. Srivastava, and J. Machado, "A new fractional derivative without singular kernel: application to the modelling of the steady heat flow," *Therm. Sci.*, vol. 20, no. 2, 2016.

[10]   X. J. Yang, D. Baleanu, and H. M. Srivastava, *Local Fractional Integral Transforms and Their Applications*. Academic Press: San Diego, Oct. 2015.

[11]   A.-M. Yang, Y. Han, J. Li, and W.-X. Liu, "On steady heat flow problem involving yang-srivastava-machado fractional derivative without singular kernel," *Therm. Sci.*, vol. 20, pp. S719-S723, Jan. 2016.

[12]   Y. Xiao-Jun, Z. Zhi-Zhen, and S. H. Mohan, "Some new applications for heat and fluid flows *via* fractional derivatives without singular kernel," *Therm. Sci.*, vol. 20, pp. 833-839, Jan. 2016.

[13]   F. Gao and X.-J. Yang, "Fractional maxwell fluid with fractional derivative without singular kernel," *Therm. Sci.*, vol. 20, pp. s873-s879, Jan. 2016.

[14]   A. Atangana and B. S. T. Alkahtani, "Extension of the resistance, inductance, capacitance electrical circuit to fractional derivative without singular kernel," *Adv. Mech. Eng.*, vol. 7, pp. 1-6, Jun. 2015.

[15]   A. Atangana and J. J. Nieto, "Numerical solution for the model of rlc circuit *via* the fractional derivative without singular kernel," *Adv. Mech. Eng.*, vol. 7, pp. 1-7, Oct. 2015.

[16]   J. Gómez-Aguilar, J. Escalante-Martinez, C. Calderón-Ramón, L. Morales-Mendoza, M. Benavidez-Cruz, and M. Gonzalez-Lee, "Equivalent circuits applied in electrochemical impedance spectroscopy and fractional derivatives with and without singular kernel," *Adv. Math. Phys.*, vol. 2016, May 2016.

[17]   J. Gómez-Aguilar, T. Córdova-Fraga, J. Escalante-Martinez, C. Calderón-Ramón, and R. Escobar-Jiménez, "Electrical circuits described by a fractional derivative with regular kernel," *Revista mexicana de fsica*, vol. 62, pp. 144-154, Apr. 2016.

[18]   J. F. Gómez-Aguilar, V. F. Morales-Delgado, M. A. Taneco-Hernández, D. Baleanu, R. F. Escobar-Jiménez, and M. M. Al Qurashi, "Analytical solutions of the electrical rlc circuit *via* liouville-caputo operators with local and non-local kernels," *Entropy*, vol. 18, p. 402, Agu. 2016.

[19]   J. Gómez-Aguilar, R. Escobar-Jiménez, M. López-López, and V. Alvarado-Martinez, "Atangana-baleanu fractional derivative applied to electromagnetic waves in dielectric media," *J. Electromagnet. Wave. Appl.*, vol. 30, pp 1937-1952, Oct. 2016.

[20]   J. Gomez-Aguilar, R. Escobar-Jimenez, M. Lopez-Lopez, V. Alvarado-Martinez, and T. Cordova-Fraga, "Electromagnetic waves in conducting media described by a fractional derivative with non-singular kernel," *J. Electromagnet. Wave. Appl.*, vol. 30, pp. 1493-1503, Jul. 2016.

[21]   J. Gómez-Aguilar, R. Escobar-Jiménez, M. López-López, and V. Alvarado-Martinez, "Atangana-baleanu fractional derivative applied to electromagnetic waves in dielectric media," *J. Electromagnet. Wave. Appl.*, vol. 30, pp. 1937-1952, Oct. 2016.

[22]   N. Al-Salti, E. Karimov, and K. Sadarangani, "On a differential equation with caputo-fabrizio fractional derivative of order $1 < \beta \leq 2$ and application to mass-spring-damper system," *Progr. Fract. Differ. Appl.*, vol. 2, no. 4, pp. 257-263, 2016.

[23]   J. F. Gómez-Aguilar, H. Yépez-Martnez, C. Calderón-Ramón, I. Cruz-Orduña, R. F. Escobar-Jiménez, and V. H. Olivares-Peregrino, "Modeling of a mass-spring-damper system by fractional derivatives with and without a singular kernel," *Entropy*, vol. 17, pp. 6289-6303, Sep. 2015.

[24]   T. Kaczorek, "Reachability of fractional continuous-time linear systems using the Caputo-Fabrizio derivative," in *Proc.39th Europ. Conf. on Modelling and Simulation*, pp. 53-58, 2016.

[25]   T. Kaczorek and K. Borawski, "Fractional descriptor continuous-time linear systems described by the Caputo-Fabrizio derivative," *Int. J. Appl. Math. Comput. Sci*, vol. 26, pp. 533-541, Sep. 2016.

[26]   M. Fabrizio, "Fractional rheological models for thermomechanical systems. dissipation and free energies," *Fract. Calc. Appl. Anal.*, vol. 17, pp. 206-223, Sep. 2014.

[27]   M. Caputo, "The unknown set of memory constitutive equations of plastic media," *Progr. Fract. Differ. Appl.*, vol. 2, no. 2, pp. 77-83, 2016.

[28]   F. Ali, M. Saqib, I. Khan, and N. A. Sheikh, "Application of Caputo-Fabrizio derivatives to mhd free convection flow of generalized Walters'-B fluid model," *Eur. Phys. J. Plus*, vol. 131, p. 377, Oct. 2016.

[29]   N. A. Shah and I. Khan, "Heat transfer analysis in a second grade fluid over and oscillating vertical plate using fractional Caputo-Fabrizio derivatives," *Eur. Phys. J. C*, vol. 76, pp. 1-11, Jul. 2016.

[30]   A. Zafar and C. Fetecau, "Flow over an infinite plate of a viscous fluid with non-integer order derivative without singular kernel," *Alexandria Engineering Journal*, vol. 55, pp. 2789-2796, Sep. 2016.

[31]   N. A. Sheikh, F. Ali, M. Saqib, I. Khan, and S. A. A. Jan, "A comparative study of Atangana-Baleanu and Caputo-Fabrizio fractional derivatives to the convective flow of a generalized casson fluid," *Eur. Phys. J. Plus*, vol. 132, p. 54, Jan. 2017.

[32]   A. Atangana and B. S. T. Alkahtani, "Analysis of the keller-segel model with a fractional derivative without singular kernel," *Entropy*, vol. 17, pp. 4439-4453, Jun. 2015.

[33]   A. Atangana, "On the new fractional derivative and application to nonlinear Fisher's reaction-diffusion equation," *Appl. Math. Comput.*, vol. 273, pp. 948-956, Jan. 2016.

[34]   A. Atangana and R. T. Alqahtani, "Numerical approximation of the space-time Caputo-Fabrizio fractional derivative and application to groundwater pollution equation," *Adv. Difference Equ.*, vol. 2016, p. 156, Jun. 2016.

[35]   A. Atangana and B. S. T. Alkahtani, "New model of groundwater flowing within a confine aquifer: application of Caputo-Fabrizio derivative," *Arabian Journal of Geosciences*, vol. 9, p. 8, Jan. 2016.

[36]   A. Alsaedi, D. Baleanu, S. Etemad, and S. Rezapour, "On coupled systems of time-fractional differential problems by using a new fractional derivative," *J. Funct. Spaces*, vol. 2016, Jan. 2016.

[37]   T. Abdeljawad and D. Baleanu, "Monotonicity results for fractional difference operators with discrete exponential kernels," *Adv. Difference Equ.*, vol. 2017, p. 78, Mar. 2017.

[38]   D. Baleanu, A. Mousalou, and S. Rezapour, "A new method for investigating approximate solutions of some fractional integro-differential equations involving the Caputo-Fabrizio derivative," *Adv. Difference Equ.*, vol. 2017, p. 51, Dec. 2017.

[39]   V. F. Morales-Delgado, J. F. Gómez-Aguilar, H. Yépez-Martnez, D. Baleanu, R. F. Escobar-Jimenez, and V. H. Olivares-Peregrino, "Laplace homotopy analysis method for solving linear partial differential equations using a fractional derivative with and without kernel singular," *Adv. Difference Equ.*, vol. 2016, p. 164, Jun. 2016.

[40]   D. Baleanu, B. Agheli, and M. M. Al Qurashi, "Fractional advection differential equation within caputo and Caputo-Fabrizio derivatives," *Adv. Mech. Eng.*, vol. 8, pp. 1-8, Dec. 2016.

[41]   N. Al-Salti, E. Karimov, and S. Kerbal, "Boundary-value problem for fractional heat equation involving Caputo-Fabrizio derivative," *NTMSCI*, vol. 4, pp. 1-8, Mar. 2016.

[42]   J. Gómez-Aguilar, M. López-López, V. Alvarado-Martnez, J. Reyes-Reyes, and M. Adam-Medina, "Modeling diffusive transport with a fractional derivative without singular kernel," *Phys. A*, vol. 447, pp. 467-481, Apr. 2016.

[43]   B. S. Alkahtani, O. J. Algahtani, R. S. Dubey, and P. Goswami, "Solution of fractional oxygen diffusion problem having without singular kernel," *J. Nonlinear Sci. Appl.*, vol. 10, Jan. 2017.

[44]   R. T. Alqahtani, "Fixed-point theorem for Caputo-Fabrizio fractional nagumo equation with nonlinear diffusion and convection," *J. Nonlinear Sci. Appl*, vol. 9, pp. 1991-1999, Jan. 2016.

[45] Q. Rubbab, I. A. Mirza, and M. Z. A. Qureshi, "Analytical solutions to the fractional advection-diffusion equation with time-dependent pulses on the boundary," *AIP Advances*, vol. 6, p. 075318, Jul. 2016.

[46] Z. Liu, A. Cheng, and X. Li, "A second-order finite difference scheme for quasilinear time fractional parabolic equation based on new fractional derivative," *Int. J. Computer Math.*, pp. 1-16, Feb. 2017.

[47] I. Koca and A. Atangana, "Solutions of Cattaneo-Hristov model of elastic heat diffusion with Caputo-Fabrizio and Atangana-Baleanu fractional derivatives," *Therm. Sci.*, 2299-2305, Dec 2017.

[48] J. Gómez-Aguilar, H. Yépez-Martnez, J. Torres-Jiménez, T. Córdova-Fraga, R. Escobar-Jiménez, and V. Olivares-Peregrino, "Homotopy perturbation transform method for nonlinear differential equations involving to fractional operator with exponential kernel," *Adv. Difference Equ.*, vol. 2017, p. 68, Feb. 2017.

[49] B. S. T. Alkahtani and A. Atangana, "Analysis of non-homogeneous heat model with new trend of derivative with fractional order," *Chaos Solitons Fractals*, vol. 89, pp. 566-571, Aug. 2016.

[50] B. S. T. Alkahtani and A. Atangana, "Modeling the potential energy field caused by mass density distribution with eton approach," *Open Physics*, vol. 14, pp. 106-113, Jan. 2016.

[51] B. S. T. Alkahtania, A. Atanganab, and I. Kocac, "Huge analysis of hepatitis c model within the scope of fractional calculus," *J. Nonlinear Sci. Appl.*, vol. 9, Jan. 2016.

[52] A. Atangana and I. Koca, "On the new fractional derivative and application to nonlinear baggs and freedman model," *J. Nonlinear Sci. Appl*, vol. 9, pp. 2467-2480, Jan. 2016.

[53] A. Coronel-Escamilla, J. F. Gómez-Aguilar, D. Baleanu, T. Córdova-Fraga, R. F. Escobar-Jiménez, V. H. Olivares-Peregrino, and M. M. A. Qurashi, "Bateman-feshbach tikochinsky and caldirola-kanai oscillators with new fractional differentiation," *Entropy*, vol. 19, p. 55, Jan. 2017.

[54] J. Gómez-Aguilar, L. Torres, H. Yépez-Martnez, D. Baleanu, J. Reyes, and I. Sosa, "Fractional liénard type model of a pipeline within the fractional derivative without singular kernel," *Adv. Difference Equ.*, vol. 2016, pp. 1-13, Dec. 2016.

[55] D. Kumar, J. Singh, M. Al Qurashi, and D. Baleanu, "Analysis of logistic equation pertaining to a new fractional derivative with non-singular kernel," *Adv. Mech. Eng.*, vol. 9, pp. 1-8, Dec. 2017.

[56] D. Kumar, J. Singh, and D. Baleanu, "Modified kawahara equation within a fractional derivative with non-singular kernel," *Therm. Sci.*, 2017.

[57] J. Singh, D. Kumar, M. A. Qurashi, and D. Baleanu, "Analysis of a new fractional model for damped bergersâ€™ equation," *Open Physics*, vol. 15, pp. 35-41, Jan. 2017.

[58] H. Jafari, A. Lia, H. Tejadodi, and D. Baleanu, "Analysis of riccati differential equations within a new fractional derivative without singular kernel," *Fundamenta Informaticae*, vol. 151, pp. 161-171, Jan. 2017.

[59] A. Babei and A. Mohammadpour, "A coupled method to solve reaction-diffusion-convection equation with the time fractional derivative without singular kernel," *Progr. Fract. Differ. Appl.*, vol. 3, no. 3, pp. 199-205, 2017.

[60] W. Chen and Y. Liang, "New methodologies in fractional and fractal derivatives modeling," *Chaos Solitons Fractals*, vol. 102, pp. 72-77, Apr. 2017.

[61] A. Atangana and D. Baleanu, "New fractional derivatives with nonlocal and non-singular kernel: theory and application to heat transfer model," *Thermal Science*, vol. 20, pp. 763-769, Jan. 2016.

[62] A. A. Kilbas, H. M. Srivastava, and J. J. Trujillo, *Theory and applications of fractional differential equations*. Elsevier: Amsterdam, 2006.

[63] J. Gómez-Aguilar, "Irving-mullineux oscillator *via* fractional derivatives with Mittag-Leffler kernel," *Chaos Solitons Fractals*, vol. 95, pp. 179-186, Feb. 2017.

[64] A. Atangana, "Derivative with two fractional orders: A new avenue of investigation toward revolution in fractional calculus," *Eur. Phys. J. Plus*, vol. 131, p. 373, Oct. 2016.

[65] J. Gómez-Aguilar and A. Atangana, "Fractional hunter-saxton equation involving partial operators with bi-order in Riemann-Liouville and Liouville-Caputo sense," *Eur. Phys. J. Plus*, vol. 132, p. 100, Feb. 2017.

[66] C. Cattaneo, "On the conduction of heat (in italian)," *Atti Sem. Mat. Fis. Universita Modena*, vol. 3, pp. 83-101, 1948.

[67] S. Carillo, "Some remarks on materials with memory: heat conduction and viscoelasticity," *J. Nonlinear Math. Phys.*, vol. 12, pp. 163-178, Jan. 2005.

[68] J. Ferreira and P. de Oliveira, "Qualitative analysis of a delayed non-fickian model," *Appl. Anal.*, vol. 87, pp. 873-886, Agu. 2008.

[69] J. Hristov, "Transient heat diffusion with a non-singular fading memory: from the Cattaneo constitutive equation with Jeffrey's kernel to the Caputo-Fabrizio time-fractional derivative," *Therm. Sci.*, vol. 20, no. 2, pp. 757-762, 2016.

[70] J. Hristov, "Derivation of fractional Dodson equation and beyond:transient mass diffusion with a non-singular memory and exponentially fading-out diffusivity," *Progr. Fract. Differ. Appl.*, vol. 3, no. 4, pp. 255-270, 2017.

[71] A. Araújo, J. Ferreira, and P. d. Oliveira, "The effect of memory terms in diffusion phenomena," *J. Comput. Math.*, vol. 24, pp. 91-102, Jan. 2006.

[72] D. D. Joseph and L. Preziosi, "Heat waves," *Reviews of Modern Physics*, vol. 61, p. 41, Jan 1989.

[73] M. E. Gurtin and A. C. Pipkin, "A general theory of heat conduction with finite wave speeds," *Arch. Ration. Mech. Anal.*, vol. 31, pp. 113-126, Jan. 1968.

[74] B. S. T. Alkahtani and A. Atangana, "A note on Cattaneo-Hristov model with non-singular fading memory," *Therm. Sci.*, vol. 21, pp. 1-7, Jan. 2017.

[75] J. Hristov, "Steady-state heat conduction in a medium with spatial non-singular fading memory: derivation of Caputo-Fabrizio space-fractional derivative with Jeffrey's kernel and analytical solutions," *Therm. Sci.*, vol. 21, no. 2, pp. 827-839, 2017.

[76] M. H. Dodson, "Closure temperature in cooling geochronological and petrological systems," *Contrib. Mineral. Petrol.*, vol. 40, pp. 259-274, Sep. 1973.

[77] J. Vrentas, C. Jarzebski, and J. Duda, "A Deborah number for diffusion in polymer-solvent systems," *AIChE Journal*, vol. 21, pp. 894-901, Sep. 1975.

[78] J. Wu and N. A. Peppas, "Modeling of penetrant diffusion in glassy polymers with an integral sorption deborah number," *Journal of Polymer Science Part B: Polymer Physics*, vol. 31, pp. 1503-1518, Oct. 1993.

[79] J. Singh, D. Kumar, M. A. Qurashi, and D. Baleanu, "Analysis of a new fractional model for damped bergers' equation," *Open Physics*, vol. 15, pp. 35-41, Jan. 2017.

[80] O. Algahtani, "Comparing atangana-baleanu and Caputo-Fabrizio derivative with fractional order: Allen cahn model," *Chaos Solitons Fractals*, vol. 89, pp. 552-559, Aug. 2015.

[81] B. I. Henry, T. A. Langlands, and P. Straka, "An introduction to fractional diffusion," in *Complex Physical, Biophysical and Econophysical Systems* (R. Dewar and F. Detering, eds.), vol. 9, (Canberra), pp. 37-89, Australian National University, Dec. 2010.

[82] R. Hilfer, *Applications of Fractional Calculus in Physics*. World Scientific, 2000.

[83] T. M. Atanackovic, S. Pilipovic, B. Stankovic, and D. Zorica, *Fractional Calculus With Applications in Mechanics: Vibrations and Diffusion Processes*. John Wiley & Sons, 2014.

[84] M. Abramowitz and I. A. Stegun, *Handbook of Mathematical Functions: With Formulas, Graphs, and Mathematical Tables*, vol. 55. Courier Corporation, 1964.

[85] K. Oldham and J. Spanier, *The fractional calculus theory and applications of differentiation and integration to arbitrary order*, vol. 111. Elsevier, 1974.

[86] J. Trzmiel, K. Weron, J. Janczura, and E. Placzek-Popko, "Properties of the relaxation time distribution underlying the kohlrausch-williams-watts photoionization of the dx centers in cd1-xmnxte mixed crystals," *Journal of Physics: Condensed Matter*, vol. 21, p. 345801, Aug 2009.

[87] C. J. F. Böttcher, O. C. van Belle, P. Bordewijk, and A. Rip, *Theory of Electric Polarization*, vol. 2. Elsevier: Amsterdam, 1978.

[88] C. J. F. Böttcher, O. C. van Belle, P. Bordewijk, and A. Rip, *Theory of electric polarization*, vol. 2. Elsevier Science Ltd, 1978.

[89] W. Feller, *An Introduction To Probability Theory and Its Applications*, vol. 2. Wiley: New York, 1996.

[90] V. Souchon, I. Leray, M. N. Berberan-Santos, and B. Valeur, "Multichromophoric supramolecular systems. recovery of the distributions of decay times from the fluorescence decays," *Dalton Transactions*, no. 20, pp. 3988-3992, 2009.

[91]   M. N. Berberan-Santos, E. N. Bodunov, and B. Valeur, "Luminescence decays with underlying distributions of rate constants: General properties and selected cases," in *Fluorescence of Supermolecules, Polymers, and Nanosystems*, pp. 67-103, Springer, 2007.

[92]   M. N. Berberan-Santos and B. Valeur, "Luminescence decays with underlying distributions: General properties and analysis with mathematical functions," *Journal of Luminescence*, vol. 126, pp. 263-272, Oct. 2007.

[93]   A. Wintner, "On heaviside's and Mittag-Leffler's generalizations of the exponential function, the symmetric stable distributions of Cauchy-Lévy, and a property of the γ-functions," *J. Math. Pur. Appl. IX*, vol. 38, pp. 165-182, 1959.

[94]   M. Berberan-Santos, E. Bodunov, and B. Valeur, "Mathematical functions for the analysis of luminescence decays with underlying distributions 1. Kohlrausch decay function (stretched exponential)," *Chemical Physics*, vol. 315, pp. 171-182, Aug. 2005.

[95]   M. Berberan-Santos, E. Bodunov, and B. Valeur, "Mathematical functions for the analysis of luminescence decays with underlying distributions: 2. becquerel (compressed hyperbola) and related decay functions," *Chemical Physics*, vol. 317, pp. 57-62, Oct. 2005.

[96]   H. Schiessel and A. Blumen, "Hierarchical analogues to fractional relaxation equations," *J Phys A*, vol. 26, p. 5057, Oct. 1993.

[97]   M. Fabrizio and A. Morro, *Mathematical problems in linear viscoelasticity*. SIAM, Jan. 1992.

[98]   S. Lim and M. Li, "A generalized cauchy process and its application to relaxation phenomena," *J. Phys. A*, vol. 39, p. 2935, Mar. 2006.

[99]   M. Caputo, "Linear and nonlinear inverse rheologies of rocks," *Tectonophysics*, vol. 122, pp. 53-71, Feb. 1986.

[100]   T. Alfrey, E. Gurnee, and W. Lloyd, "Diffusion in glassy polymers," in *Journal of Polymer Science: Polymer Symposia*, vol. 12, pp. 249-261, Wiley Online Library, 1966.

[101]   A. Berens and H. Hopfenberg, "Diffusion and relaxation in glassy polymer powders: 2. separation of diffusion and relaxation parameters," *Polymer*, vol. 19, pp. 489-496, May. 1978.

[102]   N. Panyoyai and S. Kasapis, "A free-volume interpretation of the decoupling parameter in bioactive-compound diffusion from a glassy polymer," *Food Hydrocolloids*, vol. 54, pp. 338-341, Mar. 2016.

[103]   A. G. Fredrickson, *Principles and Applications of Rheology*. Prentice-Hall: Englewood Cliffs, 1964.

[104]   P. Neogi, "Anomalous diffusion of vapors through solid polymers. part i: Irreversible thermodynamics of diffusion and solution processes," *AIChE Journal*, vol. 29, pp. 829-833, Sep.

[105]   P. Neogi, "Anomalous diffusion of vapors through solid polymers. part ii: Anomalous sorption," *AIChE journal*, vol. 29, pp. 833-839, Sep. 1983.

[106]   F. Adib and P. Neogi, "Sorption with oscillations in solid polymers," *AIChE journal*, vol. 33, pp. 164-166, Jan. 1987.

[107]   J. Vrentas, J. Duda, and A.-C. Hou, "Anomalous sorption in poly (ethyl methacrylate)," *J. Appl. Polym. Sci.*, vol. 29, pp. 399-406, Jan. 1984.

[108]   V. Lyubimova, S. Frenkel, "Auto-oscillating regime of sorption of low molecular solvent by ionomers," *Polym. Bull.*, vol. 12, pp. 229-236, March 1984.

[109]   G. Camera-Roda and G. C. Sarti, "Mass transport with relaxation in polymers," *AIChE journal*, vol. 36, pp. 851-860, Jun. 1990.

[110]   N. L. Thomas and A. Windle, "A theory of case ii diffusion," *Polymer*, vol. 23, pp. 529-542, Apr. 1982.

[111]   C. Durning and M. Tabor, "Mutual diffusion in concentrated polymer solutions under a small driving force," *Macromolecules*, vol. 19, pp. 2220-2232, Aug. 1986.

[112]   G. Rehbinder, "Darcyan flow with relaxation effect," *Appl. Sci. Research*, vol. 46, pp. 45-72, Mar. 1989.

[113]   M. Caputo, "Models of flux in porous media with memory," *Water Resources Research*, vol. 36, pp. 693-705, Mar. 2000.

[114]   G. Iaffaldano, M. Caputo, and S. Martino, "Experimental and theoretical memory diffusion of water in sand," *Hydrol. Earth Syst. Sci. Discussions*, vol. 2, pp. 1329-1357, Aug. 2005.

[115]   E. Di Giuseppe, M. Moroni, and M. Caputo, "Flux in porous media with memory: models and experiments," *Transport in Porous Media*, vol. 83, pp. 479-500, Jul. 2010.

Fractional Order Nonlinear Systems: Some Open Problems in Numerical Computations and Chaos Theory

Sachin Bhalekar[*]

Department of Mathematics, Shivaji University, Kolhapur 416004, India

Abstract: In this chapter, we discuss some open problems in the theory of nonlinear fractional order systems. We emphasize on the open issues related to numerical computations and chaos in such systems.

Keywords: Open problems, fractional order, analytical methods, numerical methods, Chaos.

AMS Subject Classification: 26A33, 65P20, 34L30.

11.1. INTRODUCTION

Fractional calculus is one of the important branches of Mathematics having applications in various fields of Science, Engineering and Social Science. Researchers are working on various aspects of this branch ranging from mathematical analysis to biological modelling. The detailed review of this subject is available in the paper [11] and two nice posters [12, 13] designed by Machado, Kiryakova and Mainardi. The international journals "Fractional Calculus and Applied Analysis", "Journal of Fractional Calculus" and "Journal of Fractional Calculus and Applications" are solely devoted to this subject. Some international conferences were also conducted where the researchers discussed recent advances in fractional calculus.

Analysis of nonlinear systems is very important because most of the realistic models are nonlinear. During the last few years, the present author came across some challenges and open issues related to the fractional order nonlinear systems (FONS) which motivated him to write this article.

---

[*]**Corresponding author Sachin Bhalekar:** Department of Mathematics, Shivaji University, Kolhapur 416004, India; Tel/Fax: E-mails: sbb_maths@unishivaji.ac.in, sachin.math@yahoo.co.in

The chapter is organized as follows: Section 11.2 deals with different solution methods for FONS and related open problems. The open issues in chaotic FONS are discussed in Section 11.3. The conclusions are summarized in Section 11.4.

11.2. SOLUTION METHODS

In this section, we discuss some solution methods for FONS and related challenges.

11.2.1. Exact Methods

Usually it is possible to find exact solutions of nonlinear first order equations by using methods such as separation of variables. It is not always possible to find such solutions for FONS.

Open Problem 1: Find some new methods that give exact solutions for (a particular class of) FONS.

11.2.2. Approximate Analytical Methods

If it is not possible to find exact solution of FONS then the methods such as Adomian decomposition method (ADM) [1], Daftardar-Gejji-Jafari method (DJM) [3], variational iteration method [8], homotopy perturbation method (HPM) [9] and homotopy analysis method (HAM) [10] are proved useful in finding approximate analytical solutions. These methods provide solutions in the form of series which is assumed to be convergent. In practice, one has to consider only a few terms of these series and hence the solution remains valid only in a small interval. The estimate for the number of terms of these series is not known. The open problems in this field are as below.

Open Problem 2: To search for new methods which provide analytical approximate solutions to given FONS in a larger interval with sufficiently small number of series terms.

Open Problem 3: Provide an estimate for the number of series solution terms obtained by a particular method so that the error is less than permissible error bound ε in the given interval I.

11.2.3. Numerical Methods

The phenomena such as chaos in FONS require the solutions in quite larger intervals. In this case, one has to consider numerical solutions. The popular numerical method for solving FONS is Fractional Adams Method (FAM) [6] developed by Diethelm and coworkers. The matrix approach for solving these systems is proposed by Podlybny in [15]. Recently, Daftardar-Gejji, Sukale and Bhalekar proposed a novel method [4, 5] based on DJM to solve FONS. This new method is faster than other existing methods and hence can be used efficiently to solve chaotic dynamical systems. Some of the open problems in numerical analysis of FONS are listed below.

Open Problem 4: Since the fractional derivative is nonlocal in nature, one has to provide all the history up to the point of application. Hence the time required for computations increases tremendously as we consider larger intervals. It takes few minutes to compute the numerical solution of given nonlinear system with integer order derivative but may requires some weeks or months for the same task with fractional order derivatives. The open problem here is to find numerical methods with better speed than existing methods.

Open Problem 5: It is observed that the numerical methods described above fail in some cases. For example, consider the Example 7.1 discussed in [4]

$$D^\alpha y(x) + y^4(x) = \frac{\Gamma(2\alpha+1)x^\alpha}{\Gamma(\alpha+1)} - \frac{2x^{2-\alpha}}{\Gamma(3-\alpha)} + (x^{2\alpha} - x^2)^4, y(0) = 0. \quad \textbf{(11.1)}$$

The step-size was taken as 0.01. The exact solution for this example is $y(x) = x^{2\alpha} - x^2$. It can checked from Fig. (**2a**) in [4] that these numerical methods cannot give the correct solution for $x > 2.2$ which is very small value. The error was not reduced even for the smaller step-size $h = 0.001$ also.

The open problem is to search for numerical methods which provide the correct solution to the above example and similar ones.

11.3. CHAOS IN FONS

A deterministic nonlinear system with order three or higher can exhibit chaos. The chaotic signals are extreme sensitive to initial conditions and are aperiodic for all the time. The popular chaotic systems are Lorenz system, Rossler system, unified system and so on. In 1995, Hartley, Lorenzo and Qammer [7] generalized integer order Chua system to a fractional order as

$$D^\alpha x = a\left(y + \frac{x-2x^3}{7}\right)$$

$$D^\alpha y = x - y + z$$

$$D^\alpha z = -\frac{100y}{7},$$ (11.2)

where α is non-integer value. Surprisingly, the system was generating chaotic signals for fractional orders $\alpha = 0.9$ and $\alpha = 1.1$. This is probably the first fractional order chaotic system described in the literature. Since then, so many chaotic fractional order systems are developed by the researchers. The book [14] by Petras is completely devoted to this topic. Usually, there is a threshold value α_* of fractional order such that $\alpha > \alpha_*$ generates chaos and the stable orbits are observed for $\alpha < \alpha_*$. The open issues in the theory of chaos in FONS are described below.

Open Problem 6: The existence of chaotic attractor in integer order Lorenz system is proved analytically by Tucker in [16]. Prove the existence of chaotic attractor for fractional order chaotic systems.

11.3.1. Detection of Chaos

Usually the chaos is shown by numerical solution in some larger interval such as [0,500]. The aperiodic trajectory in such interval which is looking chaotic may settle later on (*i.e.* transient chaos). Further, the numerical solution may have errors (as discussed in Open Problem 5). Hence the chaos in FONS should be detected using some other methods. One of such methods is finding Lyapunov exponents (LE). Some authors have used Adomian decomposition method (ADM) [2] for computing Lyapunov exponents in case of FONS. The ADM is applied to given FONS in distinct subintervals $[t_{m-1}, t_m]$ (cf. Sec. 3.1 in [2]) which is not correct because the fractional derivative is a nonlocal operator and every time we have to consider all the history from initial point. If ADM is used correctly then also the computations will be very large in size and may introduce errors.

The time series analysis proposed by Wolf *et al.* [17] is another approach for finding LEs. Again the time series is found out using numerical solutions and hence we need correct numerical method for this purpose.

Open Problem 7: Find correct analytical (or numerical) method to find LEs of given FONS.

Open Problem 8: Provide some other analytical results to detect chaos in FONS.

11.4. CONCLUSION

In this article, we have discussed some open problems and challenges in fractional order nonlinear systems (FONS). In particular, we have focused on the need for correct numerical method for solving FONS and the analysis of chaos in FONS. We hope that the discussion will be useful to the researchers working in this field and will motivate them to search for new results.

CONSENT FOR PUBLICATION

Not applicable.

CONFLICT OF INTEREST

The authors declare no conflict of interest, financial or otherwise.

ACKNOWLEDGEMENTS

Author acknowledges CSIR, New Delhi for funding through Research Project [25(0245)/15/EMR-II].

REFERENCES

[1]   G. Adomian, *Solving Frontier Problems of Physics: The Decomposition Method*. Academic Press: San Diego, 1999.
[2]   R. Caponetto, and S. Fazzino, "An application of Adomian decomposition for analysis of fractional-order chaotic systems," *Int. J. Bifurcation and Chaos*, vol. 23, no. 03, p. 1350050, Mar. 2013.
[3]   V. Daftardar-Gejji, and H. Jafari, "An iterative method for solving nonlinear functional equations," *J. Math. Anal. Appl.*, vol. 316, no. 2, pp. 753–763, Apr. 2006.
[4]   V. Daftardar-Gejji, Y. Sukale, and S. Bhalekar, "A new predictor–corrector method for fractional differential equations," *Appl. Math. Comput.*, vol. 244, pp. 158–182, Oct. 2014.
[5]   V. Daftardar-Gejji, Y. Sukale, and S. Bhalekar, "Solving fractional delay differential equations: A new approach," *Fract. Calc. Appl. Anal.*, vol. 18, no. 2, pp. 400–418, Apr. 2015.
[6]   K. Diethelm, N.J. Ford, and A.D. Freed, "A predictor-corrector approach for the numerical solution of fractional differential equations," *Nonlinear Dyn.*, vol. 29, pp. 3–22, Jul. 2002.
[7]   T. Hartley, C. Lorenzo, and H. Qammar, "Chaos in a fractional order Chua system," *IEEE Trans. Circuits Systems I Fund. Theory Appl.*, vol. 42, no. 8, pp. 485–490, Aug. 1995.
[8]   J.H. He, "A new approach to nonlinear partial differential equations," *Comm. Nonlinear Sci. Numer. Simul.*, vol. 2, no. 4, pp. 230–235, Dec. 1997.
[9]   J.H. He, "Homotopy perturbation technique," *Comput. Methods Appl. Mech. Eng.*, vol. 178, pp. 257–262, Aug. 1999.
[10]   S.J. Liao, *"The Proposed Homotopy Analysis Technique For The Solution of Nonlinear Problems"*, PhD thesis, Shanghai Jiao Tong University, China, 1992.

[11]   J.T. Machado, V. Kiryakova, and F. Mainardi, "Recent history of fractional calculus," *Commun. Nonlin. Sci. Numer. Simul.*, vol. 16, no. 3, pp. 1140-â€"1153, Mar. 2011.

[12]   J.T. Machado, V. Kiryakova, and F. Mainardi, "History of fractional calculus," Poster available at the link http://www.math.bas.bg/complan/fcaa/Poster_Old_History_FC_A3.pdf, 2010.

[13]   J.T. Machado, V. Kiryakova, and F. Mainardi, "Recent history of fractional calculus," Poster available at the link http://www.math.bas.bgcomplanfcaaPOSTER_Recent_History_FC_A3.pdf, 2010.

[14]   I. Petras, *Fractional-Order Nonlinear Systems: Modeling, Analysis and Simulation*, Springer-Verlag: Berlin, 2011.

[15]   I. Podlubny, "Matrix approach to discrete fractional calculus," *Fract. Calc. Appl. Anal.*, vol. 3, no. 4, pp. 359–386, Apr. 2000.

[16]   W. Tucker, "The Lorenz attractor exists," *C. R. Acad. Sci. Paris*, vol. 328, no. 12, pp. 1197–1202, Jun. 1999.

[17]   A. Wolf, J. Swift, H. Swinney, and J. Vastano, "Determining lyapunov exponents from a time series," *Physica D*, vol. 16, pp. 285–317, Jul. 1985.

SUBJECT INDEX

A

Adaptive algorithms 3, 15, 24, 26
ADM (Adomian decomposition method) 71,
    98, 233, 234, 245, 258, 265, 343, 345
Adomian decomposition method. *See* ADM
Algebraic equations, nonlinear 234, 238, 239,
    258
Algorithm 14, 15, 16, 24, 25, 164, 208, 233,
    258, 261
Analogy 280, 282, 296, 303, 304, 305, 306,
    316, 324, 330, 334
Analysis, present 101, 313, 327
Analytical solutions 2, 17, 19, 102, 331
    approximate 71, 343
Antisynchronization 176, 177
APIs (Application Program Interface) 208
Application Program Interface (APIs) 208
Approximate profiles 70, 75, 83, 84, 85, 87,
    88, 90, 92, 93, 94, 95, 96, 97, 99, 103
Approximate solutions 70, 71, 72, 83, 85, 89,
    92, 98, 100, 101, 102, 106, 107, 113,
    118, 121, 126, 127, 233, 234, 235, 242,
    258, 264, 280
    developed 98, 102
    numerical 106, 116, 121, 126
Approximation 2, 16, 70, 75, 81, 84, 88, 89,
    111, 114, 115, 117, 122, 129, 132, 133,
    155, 165, 291, 297, 299, 312, 329
Approximation error 23, 24, 25
Assumption 40, 41, 80, 81, 146, 148, 151, 186,
    191, 273, 284, 294, 297, 300, 320
Atangana 278, 283, 289, 290, 291, 331
Attention 3, 78, 92, 106, 233, 269, 270, 271,
    282, 284, 289, 292, 295, 321, 322, 324,
    325, 335

B

Bases 6, 8, 20, 26, 27, 65, 207, 270, 296, 328,
    336

Basis functions 4, 8
Bessel function 206
Bessel Wright Function 200, 206, 207, 220,
    221
Bi-CGSTAB 14, 23
Biological population equation 250
Bi-order 283
Bi-order fractional operator 283, 325
Boundaries 3, 12, 14, 33, 36, 37, 38, 43, 72,
    80, 86
Boundary conditions 37, 46, 47, 50, 52, 54,
    55, 58, 59, 61, 64, 75, 84, 155, 254, 287,
    292, 310, 331
Boundary value problem (BVP) 40, 41, 55,
    234, 237, 238, 253

C

Calculations 17, 82, 98, 191, 207, 213, 221,
    222, 233, 234, 304, 308
Caputo 66, 76, 78, 79, 86, 107, 133, 160, 202,
    317, 320, 325, 327, 328, 329, 335
Caputo and Fabrizio 269, 270, 271, 274, 277,
    285, 286, 301, 302, 306, 318, 319, 336
    article of 278, 280, 285, 335
Caputo derivatives 34, 44, 102, 133, 317, 321,
    325
Caputo-Fabrizio 269, 270, 274, 285, 288, 289,
    305, 308, 317, 318, 324, 325, 327, 331,
    332, 333, 336
Caputo-Fabrizio fractional operator 289, 321,
    335
Caputo-Fabrizio Time-fractional Derivative
    269, 270, 286, 288, 289, 294, 295, 305
Caputo fractional 33, 34, 35, 37, 39, 43, 44,
    164, 240, 259
Caputo fractional derivatives 34, 44, 60, 71,
    116
Cattaneo equation 286, 287, 291, 292

Chain 136, 137, 143, 144
  compact 138, 139, 142, 150
Chaos 159, 160, 169, 173, 175, 180, 342, 343,
    344, 345, 346
Chaotic 160, 161, 168, 169, 345
Chaotic attractor 161, 162, 163, 167, 172
Chaotic fractional order systems 160, 345
Chaotic oscillations 160, 169, 172, 173
Chaotic system, new 159, 176, 180
Chaotic time series 171
Characteristic diffusion time 295, 297, 318
Chen system 160, 161, 162
Closed form solutions 71, 248, 249, 252, 253,
    265
Closed-form solutions, approximate 71, 72
Code 209, 210, 211, 222
Collocation points 19, 20
Compact subset, relatively 136, 137
Computation 200, 211, 222, 344, 345
Computational cost 1, 12, 13, 14, 25
Compute Unified Device Architecture
    (CUDA) 211, 222, 223
Concave 87, 92, 101, 103
Concentration 72, 316, 331, 332, 335
Concept 73, 80, 108, 290, 293, 304, 335, 336
Conditions C1 57, 59, 60
Conflict of interest 30, 67, 103, 130
Confluent hypergeometric function 200, 205,
    206, 215, 216
Conjecture 80, 81, 297, 299
Constitutive equations 204, 269, 271, 284,
    285, 286, 289, 292, 295, 297, 300, 302,
    303, 305, 316, 328, 332, 334
Context 73, 85, 86, 97, 98, 101, 204, 280, 290,
    302, 303, 319, 320, 323, 325, 327, 330,
    333, 335
Continuous functions 43, 45, 56, 62, 134, 149,
    196
Continuous time random walk. *See* CTRW
Contraction 138, 139, 147, 148, 150, 153, 154
Control 159, 161, 173, 175
Control parameter 175
Convergence order 17, 20
Convex 89, 92, 101, 103
Convex profiles 70, 94, 95, 101
Convolution 80, 81, 272, 275, 278, 279, 280

Cores 201, 207, 211, 212, 320
Corollary 41, 42, 47
Cost 8, 12, 14, 16
Coupled Burgers equations 248, 249
Coupled FDDEs, numerical solution of 117,
    129
Coupled Fractional Delay Differential
    Equation by Shifted Jacobi Polynomials
    106
Coupled system 67, 119, 121, 122, 280
CTRW (continuous time random walk) 1, 204,
    317
CUDA (Compute Unified Device
    Architecture) 211, 222, 223

D

Damping function 286, 291, 302, 316

Darcy law 333, 334
Dawson's Function 206, 219, 222

Deborah number 296, 313, 314, 318, 329, 330,
    331
Decaying fluorescence intensity 323, 324
Decays 322, 324, 326, 327
Delay terms 262, 263, 265
Deng 2, 4, 6, 8, 10, 12, 14, 16, 18, 20, 22, 24,
    26, 28, 30
Derivatives 43, 159, 183, 197, 204, 269, 271,
    273, 275, 277, 279, 281, 282, 283, 285,
    287, 289, 291, 293, 295, 297, 299, 301,
    303, 305, 307, 309, 311, 313, 315, 317,
    319, 321, 323, 325, 327, 329, 331, 333,
    335
  new 269, 270, 282, 320
  non-integer order 159, 191
Differential equations 2, 34, 35, 44, 67, 75,
    106, 107, 132, 133, 134, 139, 155, 159,
    160, 200, 203, 233, 238, 256, 262, 263,
    264, 265, 287, 312, 316
Differential operator 2, 39, 42, 45, 46, 49, 53,
    58
Differentiation 75, 200, 279, 280, 304, 306,
    307, 310

Diffusion 71, 89, 92, 98, 102, 270, 284, 295,
    316, 331, 332
  anomalous 1, 70, 330
  degenerate 70, 91, 92, 94
Diffusion coefficient 70, 71, 84, 92, 93, 95,
    96, 97, 98, 99, 100, 101, 102, 294, 300,
    317, 330, 332
Diffusion equations 43, 67, 74, 79, 269, 286,
    289, 291, 294, 295, 301, 316, 319, 328,
    331, 335
  conventional 34, 36
  distributed order fractional 54, 55
  fractional Dodson 285
  general 59, 67
  general time-fractional 35, 56, 59, 60, 62, 63
  linear fractional 47
  linear multi-term fractional 51
  multi-term fractional 42, 43, 50, 51, 52, 66,
    238
  multi-term time-fractional 34, 35, 43, 47, 50,
    56, 67
  non-linear 73, 102, 238
  non-linear distributed order fractional 55
  non-linear fractional 46, 47, 67
  original 84, 85
  single-term time-fractional 35, 36, 39
  time-space fractional 63
Diffusion models 269, 284, 286, 289, 297
Diffusion process 132, 295, 297, 300
Diffusion term 73, 292, 294, 299
Diffusivity 92, 97, 98, 99, 284, 310, 330
Digital Library of Mathematical Functions
    (DLMF) 201
DIM (double-integration method) 73, 74, 75,
    77, 80, 81, 82, 83, 89, 102
Dimension 71, 248, 249, 295, 307, 312, 313,
    317
DIM Solution 76, 80, 82, 83
Dirac Delta function 284, 287, 304, 308, 330
DLMF (Digital Library of Mathematical
    Functions) 201
Dodson equation 294, 299, 330
Domain 36, 37, 38, 43
Double-integration method. *See* DIM
Double-integration technique 71, 72, 76, 93,
    94, 95

Dynamics of Fractional Order Current
    Developments 161, 163, 165, 167, 169,
    171, 173, 175, 177, 179
Dynamics of Fractional Order Current
    Developments in Mathematical Sciences
    161, 163, 165, 167, 169, 171, 173, 175,
    177, 179

E

Effect
  damping 284, 289, 299, 300
  weak 92, 93, 95, 103
Eigenfunctions 34, 65
Eigenvalue distribution 22
Eigenvalues 65, 166, 168, 169, 173, 175, 177
Elliptic 33, 39, 42, 58
Entries 14, 26, 120, 125
Equilibrium points 160, 166, 167, 169, 173,
    180
Errors 25, 82, 113, 114, 115, 165, 177, 179,
    262, 343, 344, 345
  absolute 125, 127, 128
Estimate 38, 39, 44, 48, 59, 62, 63, 85, 86,
    122, 125, 129, 175, 189, 196, 313, 343
Estimate for control parameter 175
Estimates function value 186
Estimations 88, 89, 107, 188, 189, 191, 197
Evaluation 78, 80, 85, 209, 211
Exact solution 2, 15, 17, 24, 85, 102, 106, 125,
    234, 237, 238, 240, 241, 243, 246, 250,
    251, 253, 256, 257, 258, 259, 290, 343,
    344

Execute 207, 208, 209
Execution 201, 208, 209, 211, 221, 222
Execution time 199, 211, 221, 222
Existence 35, 48, 106, 132, 133, 134, 148,
    155, 263, 335
Existence and uniqueness 132, 133, 134, 155,
    263
Existence and uniqueness of solutions 106,
    133
Exponent 80, 82, 83, 84, 85, 86, 87, 88, 89, 94,
    183, 314

optimal 70, 84, 87, 89, 90, 98, 102
Exponential function 40, 203, 271, 276, 281, 290, 296
Exponential kernel 270, 285, 287, 291, 293, 294, 302, 306, 324, 325, 326, 334, 336
Exponential relaxation function 324, 328
Exponential term 276, 277, 294, 311, 313, 314
Expressions 7, 8, 27, 80, 81, 82, 83, 85, 86, 89, 92, 96, 99, 276, 284, 306, 318, 322, 330, 333

F

Fabrizio 269, 270, 271, 273, 274, 277, 278, 280, 285, 286, 301, 302, 306, 318, 319, 335, 336
Factor 82, 92, 94, 96, 186, 208, 285, 298, 299, 307
Factorial 202, 204
  rising 204, 206
Fading memory 320, 325, 327, 328, 335, 336
FAM (Fractional Adams Method) 164, 165, 201, 261, 344
Fast wavelet transform. *See* FWT
  nonlinear 133
FFA (Frozen Front Approach) 80, 81, 82
FFT 1, 3, 8, 14, 15, 20
Fick law, ad-hoc fractional 318, 319
Fig 21, 22, 24, 25, 26, 82, 87, 89, 92, 93, 94, 95, 97, 98, 99, 100, 101, 126, 127, 128, 161, 162, 163, 169, 170, 171, 172, 173, 175, 176, 177, 178, 179, 186, 188, 189, 190, 191, 192, 193, 194, 195, 196, 197, 207, 210, 211, 214, 216, 217, 218, 219, 220, 222, 296, 298, 300, 301, 309, 313, 314, 315, 327, 344
First order 1, 183, 184, 187, 191, 194, 196, 197, 287
First order approximation 286, 303
Fixed point 138, 146, 151, 155
Flux 269, 270, 271, 280, 284, 285, 287, 295, 297, 300, 303, 316, 330, 333, 335
  heat 284, 302, 303, 307, 316
Following system of fractional 259, 260
FONS, given 343, 345

FONS (fractional order nonlinear systems) 342, 343, 344, 345, 346
Form, equivalent 272, 288, 289, 291, 292, 307, 311, 320, 330
Formalistic fractionalization approach 269, 270, 281, 328, 335
Formulation 41, 51, 53, 60, 66, 270, 300, 332
Fourier law 284, 286, 287, 302, 303
Fourier number 295, 300, 301
Fractional 2, 26, 34, 35, 36, 39, 44, 45, 46, 48, 49, 53, 63, 67, 72, 81, 82, 85, 106, 107, 132, 146, 151, 155, 159, 160, 184, 200, 201, 203, 233, 238, 246, 259, 260, 265, 270, 274, 276, 277, 278, 279, 280, 281, 282, 285, 286, 289, 302, 308, 312, 317, 318, 320, 321, 328, 344, 345
  general 35, 43, 56, 58, 59, 67
  numerical solutions of 159, 199
  solutions of 200, 203
Fractional Adams Method. *See* FAM
Fractional calculus 1, 106, 107, 160, 183, 184, 199, 200, 201, 202, 222, 269, 279, 336, 342
  special functions of 200, 222
Fractional delay 106
  neutral Hadamard 132, 133, 134
Fractional Delay Differential Equation Current Developments 107, 109, 111, 113, 115, 117, 119, 121, 123, 125, 127, 129
Fractional Delay Differential Equation Current Developments in Mathematical Sciences 107, 109, 111, 113, 115, 117, 119, 121, 123, 125, 127, 129
Fractional delay differential equations (FDDEs) 106, 107, 132, 133, 134, 139, 149, 155, 164
Fractional derivatives 2, 33, 34, 42, 67, 74, 159, 202, 269, 270, 275, 277, 320, 327, 336
  classical 321, 325
  general 56, 57, 59, 66
  new 133, 280, 320
Fractional differential equations (FDEs) 1, 2, 33, 34, 35, 44, 63, 67, 82, 106, 107, 132, 133, 151, 159, 160, 164, 199, 200, 201, 203, 233, 238, 246, 261, 262, 265

Fractional diffusion equations 33, 34, 40, 41, 42, 43, 47, 52, 58, 66, 67, 80, 159, 274

Fractional equations 15, 76, 273

Fractional integration, operational matrix for 106, 113, 115, 119

Fractionalization, formalistic 289, 316, 320, 328

Fractional operators 2, 3, 8, 26, 80, 269, 270, 283

Fractional order 70, 75, 82, 84, 89, 91, 92, 93, 97, 99, 101, 102, 103, 106, 107, 116, 134, 159, 160, 166, 169, 177, 200, 234, 237, 240, 255, 262, 293, 294, 295, 296, 297, 300, 301, 305, 306, 307, 309, 312, 313, 314, 315, 329, 330, 332, 342, 344, 345, 346

   values of 162, 175

Fractional order BG system 162

Fractional Order Current Developments 161, 163, 165, 167, 169, 171, 173, 175, 177, 179

Fractional order derivatives 132, 344

Fractional order integration 113, 200

Fractional order nonlinear systems. *See* FONS

Fractional Order Nonlinear Systems Current Developments 343, 345

Fractional order systems 160, 167

Fractional PDEs 1, 2, 3, 15, 19, 20, 24, 25

Fractional-time Double-balance Integral (FTDBI) 75

Freedom 1, 3, 25, 102, 327

Frozen Front Approach. *See* FFA

FT-DBI 75, 76

FTDBI (Fractional-time Double-balance Integral) 75

Functional differential equations 133
   neutral 133

Functional equations 233, 265
   following general 234

Functional relationship 296, 314

Functions 3, 5, 37, 38, 39, 44, 45, 46, 47, 48, 49, 50, 51, 52, 53, 54, 55, 56, 59, 60, 61, 62, 71, 83, 88, 92, 93, 95, 96, 97, 99, 107, 114, 132, 133, 134, 137, 138, 139, 140, 141, 146, 148, 149, 152, 155, 159,

160, 183, 184, 185, 186, 189, 196, 199, 200, 201, 202, 203, 205, 206, 207, 208, 209, 211, 222, 256, 258, 271, 272, 274, 285, 290, 296, 298, 300, 301, 314, 320, 324, 326, 327, 329, 334

   boundary 4

   boundary scaling 8, 28

   decay 324, 327

   dual scaling 5, 7

   factorial 202

   generalized exponential 324

   increasing 196

   monotone 58

   non-increasing 47, 50, 51, 55

   normalization 271, 273, 288, 305

   residual 85, 86

   smooth 62, 114

   source 40, 41

   transcendental 2, 275

   unknown 39, 45, 66, 67, 259

Function space 40, 47, 50, 51, 54, 55, 62

Function values 183, 186

FWT (fast wavelet transform) 1, 3, 7, 8, 14, 15, 16

G

Gamma function 107, 184, 185, 200, 202, 203, 204

Gauss hypergeometric function 200, 204, 205, 210, 217, 221

Gauss Hypergeometric function (GHF) 200, 204, 205, 210, 217, 221

GB 212

Generalized Lorenz system (GLS) 168, 169

Generalized solution 39, 61

Geometrical interpretations 186, 187

GHF. *See* Gauss Hypergeometric function

GLS (generalized Lorenz system) 168, 169

GMRES 14, 23

Goodman boundary conditions 84, 87

GPU (Graphics Processing Unit) 199, 201, 202, 207, 208, 209, 210, 211, 213, 221, 222, 223

GpuArray 208, 209, 211

GPU Computing 199, 201, 203, 205, 207, 208, 209, 211, 213, 215, 217, 219, 221, 223

GPU execution 209, 211, 222

GPU memory 209

Gradient 269, 270, 271, 280, 285, 295, 297, 300, 316, 335

Graphics Processing Unit. *See* GPU

Grünwald-Letnikov Derivative Current Developments 185, 187, 189, 191, 193, 195, 197

Grünwald-Letnikov Derivative Current Developments in Mathematical Sciences 185, 187, 189, 191, 193, 195, 197

H

Hadamard fractional 132, 133, 134, 139

Hadamard fractional delay 132, 133, 134, 139, 155

Hadamard fractional delay differential equations. *See* HFDDE

Hadamard Fractional Delay Differential Equations 132, 139

Hadamard fractional delay differential equations, uniqueness and approximations of solutions of 132, 133

Handbook 201, 202

HBIM (Heat-balance Integral Method) 71, 73, 74, 78, 79, 80, 81, 82, 83, 89

HBIM Solution 78, 79, 82

Heat 200, 270, 284, 286, 287, 289

Heat-balance Integral Method. *See* HBIM

HFDDE (Hadamard fractional delay differential equations) 132, 133, 134, 139, 140, 141, 142, 146, 147, 148, 149, 155

HHFDDE (hybrid Hadamard fractional delay differential equations) 134, 149, 150, 151

Homotopy perturbation method. *See* HPM

HPM (Homotopy perturbation method) 71, 98, 234, 242, 243, 265, 343

Hybrid fixed point theorems 132, 134, 135, 155

Hybrid Hadamard fractional delay differential equations. *See* HHFDDE

Hypergeometric function 199, 200, 203, 205, 206, 218

Hypotheses 141, 142, 146, 147, 148, 149, 153, 154

I

ILTM (Iterative Laplace Transform Method) 234, 259

IMFOS 167, 172

Implementation 19, 199, 202, 207, 208

Inclusions 37, 40, 42, 56, 59

Increment 183, 184, 185, 186, 188, 191

Increment of function 183, 185

Independent variable 70, 89, 91, 109, 184, 185, 196

Inequality 37, 39, 40, 41, 44, 45, 46, 47, 48, 49, 50, 52, 53, 55, 60, 62, 64, 65, 66, 115 following 59, 115, 167

Inequations 140, 141, 145, 151, 154

Infinite series 2, 102, 183

Initial-boundary-value problems 33, 34, 35, 37, 38, 39, 41, 42, 46, 47, 50, 51, 53, 54, 55, 58, 59, 60, 61, 62, 63, 66, 67

Initial condition, following 243, 246

Initial functions 118, 122

Initial value problem (IVP) 57, 140, 147, 149, 164, 234, 240, 242, 243, 248, 249, 253, 262, 263

Integer, non-negative 204, 205

Integer-order models 79, 80, 289

Integral-balance solutions 70, 72, 84, 88, 98

Integral inequations 146, 151, 154, 155

Integral operator 270, 288, 305, 320, 321, 324, 327, 328, 329 new 320, 321

Integration 70, 74, 76, 79, 115, 272, 273, 278, 287, 299, 304, 310, 333

Integro-differential equations 200, 234, 238, 255, 256, 263, 265, 280, 287, 304

Interval 20, 37, 43, 48, 52, 59, 108, 109, 110, 117, 118, 119, 121, 122, 263, 343, 344, 345

Iterations, successive 138, 139
Iterative Laplace Transform Method (ILTM) 234, 259
Iterative method 12, 165, 234, 238, 258
IVP. *See* initial value problem

J

Jacobian matrix 166, 167, 169, 173

K

Kernel 35, 57, 59, 159, 269, 271, 283, 285, 288, 290, 303, 305, 323, 325
  non-singular 269, 270, 280, 281, 289, 301, 302, 321, 335, 336
Kohlrausch relaxation function 323, 325
Kummer's function 205

L

Laplace transform 57, 59, 77, 78, 79, 259, 275, 276, 279, 283, 290, 306, 308, 322, 323
  inverse 77, 79, 260, 276, 277, 291, 324
Laplace transform solution 79, 310, 312
Lemma 4, 5, 6, 10, 11, 48, 64, 65, 66, 110, 111, 113, 114, 116, 134, 140, 141, 142, 146, 150, 151, 152
Lemmas 9, 11, 48, 65, 66, 111
Length 89, 98, 306, 307, 312, 313, 330
Leopard Cluster 213, 215, 216, 217, 218, 219, 220, 221
Linear systems 12, 14, 23
Log 133, 134, 135, 141, 142, 143, 144, 145, 147, 150, 151, 152, 153, 155
Lorenz system 160, 161, 162, 344
Losada 273, 274, 278, 305
Lu system 161, 162

M

Magnitude, order of 98, 99, 100, 101, 297
Mapping 136, 137, 138, 141, 149
Marginal values 191, 194

Mass 270, 280, 284, 286, 289, 331
Matching 83, 84
Mathematical Functions 200, 201
MATLAB (Matrix Laboratory) 199, 201, 207, 208, 209, 210, 211, 222, 234, 265
MATLAB codes 208, 209, 222, 223
Matrix 7, 8, 11, 13, 14, 26, 125, 168, 169
  operational 106, 112, 113, 115, 119
Matrix Laboratory. *See* MATLAB
Maximum principles 33, 34, 35, 37, 38, 39, 40, 41, 42, 43, 44, 46, 48, 49, 50, 53, 56, 58, 59, 63, 64, 66, 67
Mean-Value Theorem 183, 197
Memory 80, 81, 132, 270, 284, 292, 299, 319, 321, 327, 329, 330, 334, 335
  exponential 283, 286, 321, 324, 333, 335, 336
  short-time 300, 301
Memory effects 304, 313, 333
Memory function 297, 298, 336
  composite 297, 300
Memory kernel 285, 286, 289, 325, 326, 327, 329, 330
Memory zone 309, 313
Methods
  integral-balance 71, 75, 77, 91
  new 343, 344
MHz 212
Minimization 85, 88, 89
Mittag-Leffler function, 4-parameter 207, 221
Mittag-Leffler function (MLF) 199, 200, 202, 203, 204, 214, 215, 248, 269, 281, 283, 323, 325
Mittag-Leffler function of 3-parameters 203, 213, 214, 221
MLF. *See* Mittag-Leffler function
Modelling 269, 270, 280, 300, 325, 327
Models 71, 72, 73, 77, 80, 84, 102, 159, 160, 200, 204, 207, 208, 212, 238, 269, 270, 284, 285, 286, 290, 294, 295, 296, 297, 299, 300, 316, 317, 320, 321, 322, 323, 325, 330, 331, 332, 333, 335, 336, 342
  single-memory 285, 291, 294, 298, 299
  well-known 320
Modified BG system 160, 161, 173, 176, 180

Monotone Iteration Principle Current
  Developments 135, 137, 139, 141, 143,
  145, 149, 151, 153, 155
Monotone Iteration Principle Current
  Developments in Mathematical Sciences
  135, 137, 139, 141, 143, 145, 149, 151,
  153, 155

N

Natural systems 159, 160
Neogi, solution of 331
New iterative method. *See* NIM
New Iterative Method 233, 234, 235, 237, 239,
  241, 243, 245, 247, 249, 251, 253, 255,
  257, 259, 261, 263, 265
New numerical methods (NNMs) 234, 262
New Predictor-Corrector Method. *See* NPCM
NHFDDEs 155
NHFDE (NEUTRAL HADAMARD
  FRACTIONAL DIFFERENTIAL
  EQUATIONS) 151, 152, 153
Nieto 273, 274, 278, 305
NIM (new iterative method) 233, 234, 235,
  237, 238, 239, 240, 241, 243, 245, 246,
  247, 248, 249, 250, 251, 252, 253, 254,
  255, 257, 258, 259, 261, 262, 263, 265
NIM algorithm 239, 241, 244, 250, 252, 256,
  257, 260, 265
NIM solution 241, 251, 256, 257, 258
  3-term 243, 244, 245
NNMs (new numerical methods) 234, 262
Nondecreasing 136, 138, 142, 147, 153
Non-linear fractional equations modeling 238
Nonlinear functions 133, 148, 155
Nonlinearity 70, 79, 83, 84, 89, 92, 93, 95, 96,
  97, 103, 180, 314
Non-linearity 70, 71, 79, 89, 186, 188, 313
Nonlinear Subdiffusion Equation Current
  Developments 71, 73, 75, 77, 79, 81, 83,
  85, 87, 89, 91, 93, 95, 97, 99, 101, 103
Nonlinear Subdiffusion Equation Current
  Developments in Mathematical Sciences
  71, 73, 75, 77, 79, 81, 83, 85, 87, 89, 91,

93, 95, 97, 99, 101, 103
Nonlinear systems 160, 342, 344
Non-Singular Kernels Current Developments
  271, 273, 275, 277, 279, 281, 283, 285,
  287, 289, 291, 293, 295, 297, 299, 301,
  303, 305, 307, 309, 311, 313, 315, 317,
  319, 321, 323, 325, 327, 329, 331, 333,
  335
Norm 3, 17, 136, 137, 138, 139, 140
  uniform 59, 60, 62
Normed linear space 136, 137
  ordered complete 138, 139
Norm equivalences 6, 7, 13
NPCM (New Predictor-Corrector Method)
  162, 165, 234, 261, 262
Nth-order 256
Number 15, 16, 207, 213, 222, 223, 240, 343
Numerical methods 1, 2, 3, 7, 25, 164, 201,
  202, 237, 342, 343, 344
  new 234, 262
Numerical methods for special functions 200,
  202
Numerical solutions 71, 129, 159, 199, 279,
  291, 343, 344, 345
  approximate 101, 265
NVIDIA 213, 215, 216, 217, 218, 219, 220,
  221, 222, 223
NVIDIA Geforce 212, 213

O

ODE. *See* ordinary differential equation
OME (Out of Memory Error) 213, 215, 216,
  217, 218, 219, 220, 221, 222
Open Problem 335, 343, 344, 345, 346
Operations 79, 184, 208, 211, 222, 223
Operator 26, 27, 42, 43, 45, 46, 49, 50, 54, 55,
  60, 64, 65, 133, 136, 137, 138, 142, 146,
  150, 151, 153, 155, 302, 329
  fractional Laplace 63
Operator equation 138, 139, 151, 153, 155
Order 1, 2, 3, 4, 5, 9, 13, 15, 20, 33, 34, 36, 65,
  67, 76, 84, 87, 89, 98, 107, 109, 110,
  113, 115, 116, 118, 120, 126, 129, 134,

135, 159, 160, 164, 183, 184, 185, 186, 189, 196, 240, 274, 279, 305, 307, 313, 317, 329, 335, 344
distributed 33, 34, 35, 42, 43, 52, 53, 54, 56, 57, 66
fraction 296, 314
integer 71, 88, 106, 107, 159, 344
Order Derivative 194, 196
Order KdV equation 242, 243
Order relation 136, 137, 138, 139, 140, 142
Ordinary differential equation (ODE) 73, 75, 161, 205, 234, 262, 265
Ordinary differential equations 73, 75, 234, 262, 265
Oscillations 160, 331, 332

P

Parabolic profile 70, 83, 87, 88, 96
Parabolic type 33, 40, 45, 49, 51, 53, 66
Parallel computing 200, 201
Parallel Computing Toolbox. *See* PCT
Parameters 4, 24, 79, 83, 84, 89, 91, 92, 93, 97, 101, 102, 103, 107, 117, 120, 161, 169, 221, 222, 313, 324
Partial compactness 132, 133, 134
Partial differential equations 33, 34, 40, 51, 63, 66, 67, 233, 234, 237, 238, 251, 259, 260, 265
non-linear 241
Partial Lipschitz conditions 132, 133, 134
Partially ordered normed linear spaces. *See* PONLS
Partially ordered vector space (POVS) 135, 136
PCT (Parallel Computing Toolbox) 199, 201, 208
Penetration depth equations 76, 78
Penetration depths 73, 76, 77, 78, 79, 80, 82, 83, 92, 93, 98, 99
PH (Polycrystalline Halite) 329
Physics 106, 205, 271, 313, 316, 320, 325, 327, 328, 336
Polycrystalline Halite (PH) 329
Polymers 70, 321, 331, 332, 333

PONLS (partially ordered normed linear spaces) 132, 134, 135, 136, 137, 138
POVS (partially ordered vector space) 135, 136
Power-law function 86, 276
Powers 86, 98, 201
Preconditioned system 12, 13, 22
Preconditioning 2, 14, 21, 23
Predictor-corrector method, new 162, 165, 234, 261
Predictors 165, 166, 258, 262, 264
Pressure 329, 333, 334
Principle 73, 74, 75, 327
monotone iteration 132, 133, 134, 147
Principle equation 76, 77
Probability 322, 324
Pro-Caputo 288, 290, 292, 305
Process 102, 132, 159, 207, 298, 313, 316, 318, 320, 325, 327, 328
Profile expressions 91, 92, 96
Profiling of Gauss Hypergeometric function 210
Proof of lemma 64, 65, 111
Proof of theorem 64, 148, 153

R

Ratio 80, 82, 83, 89, 211, 295, 296, 306, 311, 312, 313, 331, 334
Recurrence relation 109, 235, 241, 242, 243, 248, 249, 256
Relation 44, 56, 58, 109, 135, 147, 183, 185, 194, 202, 234, 274, 295, 309, 313, 317, 320, 332, 336
constitutive 274, 297, 317, 318
Relationships 19, 74, 83, 188, 191, 194, 277, 282, 284, 293, 296, 312, 313, 329, 333, 334
Relaxation function 270, 284, 304, 320, 321, 322, 325, 326, 330, 333, 336
modified 287, 289, 308
Relaxations 290, 294, 296, 300, 320, 324, 327, 330, 332, 336
short-time 297

Relaxation time 286, 294, 295, 296, 297, 322, 326, 330, 332, 334

Result, following 46, 47, 50, 51

Review Current Developments in Mathematical Sciences 235, 237, 239, 241, 243, 245, 247, 249, 251, 253, 255, 257, 259, 261, 263, 265

Riemann-Liouville 44, 48, 75, 77, 78, 102, 107, 133, 202, 240, 270, 273, 317, 321

Riemann-Liouville fractional 12, 28, 33, 34, 43, 44, 46, 48, 52, 81, 133

Riemann-Liouville fractional derivatives 34, 35, 43, 44, 46, 52, 57, 66

Riemann-Liouville Time-fractional Derivative 34, 63, 75

Riemann-Liouville type 56, 269, 271, 277, 290, 306, 318, 320

RL (Riemann-Lioville) 160, 317, 318

Rossler system 160, 344

S

Seconds 207, 211

Sequence 5, 61, 62, 136, 137, 138, 139, 141, 143, 147, 148, 150, 151, 152, 155

Sets 25, 41, 42, 86, 87, 137, 221, 222, 290, 302, 324

  order interval B-spline function 4

SFM (strong fading memory) 328

Shifted Jacobi polynomial degrees 125, 126

Shifted Jacobi polynomials 106, 107, 108, 109, 110, 111, 113, 115, 120, 121, 122, 123, 125

Signals 176, 177

Similarity variable 70, 83, 91, 92, 93, 95, 96, 98, 99, 102

  effective 70, 83, 84, 92, 96, 97

Sinh 177, 243, 245, 246, 260, 261

Slow diffusion 71, 72, 83, 84

Solution methods 343

Solution norm 38

Solutions 3, 15, 16, 17, 25, 33, 34, 35, 37, 38, 39, 40, 41, 47, 50, 51, 54, 55, 59, 61, 64, 65, 70, 71, 72, 75, 77, 80, 83, 84, 87, 89, 96, 97, 98, 101, 102, 103, 106, 132, 133,

134, 138, 139, 140, 141, 146, 148, 149, 150, 151, 152, 155, 160, 166, 200, 203, 204, 205, 233, 234, 237, 242, 246, 250, 255, 256, 261, 263, 270, 273, 281, 290, 307, 308, 309, 310, 312, 313, 314, 315, 331, 332, 335, 343

  correct 344

  existence of 134, 141

  integral 121, 123

  lower 67, 140, 141, 149, 150, 152

  unique 9, 58, 147, 148, 282

  upper 134, 141, 146, 148, 149, 151

Solving 1, 2, 14, 15, 16, 20, 23, 25, 72, 73, 81, 106, 107, 233, 234, 237, 238, 258, 262, 264, 265, 344

Solving boundary value problems 253

Solving Fractional Diffusion Equation Current Developments 3, 5, 7, 9, 11, 13, 15, 17, 19, 21, 23, 25, 27, 29

Solving Fractional Diffusion Equation Current Developments in Mathematical Sciences 3, 5, 7, 9, 11, 13, 15, 17, 19, 21, 23, 25, 27, 29

Solving fractional-order equations 71

Space 1, 3, 16, 37, 38, 48, 59, 62, 85, 89, 139, 326

  fractional Sobolev 39

  ordered Banach 138, 140

  ordered normed linear 135, 136

Space-fractional 306, 307

Space memory 307, 309, 310, 311

Space variable 100, 101, 305

Spatial 39, 42, 269, 302, 303, 305, 308, 335

Spatial memory 269, 301, 307, 312

Special case 205, 206, 207, 290, 323

Special functions 199, 200, 201, 202, 203, 205, 211, 213

  properties of 201, 202

Special mathematical functions 199, 201, 202, 222

Stable solution 169, 172

Stable time series 170

Strong fading memory (SFM) 328

Strong maximum principles 35, 41, 42, 45, 46, 49, 53, 66

Strong solution 37, 38, 39, 40, 41, 59, 60, 61, 62
Subdiffusion 70, 72, 73, 84, 91, 93, 94, 95
Subject 40, 184, 251, 252, 342
Subsection 4, 5, 6, 126
Subset, non-empty 136, 137
Successive approximations 134, 141, 147, 150, 152, 155
Summation 202, 221, 222
Support, compact 4, 6, 11
Synchronization 159, 161
  complete 176, 177
System 4, 14, 106, 107, 109, 117, 118, 121, 122, 123, 125, 129, 132, 159, 160, 161, 162, 166, 167, 168, 169, 172, 173, 175, 177, 180, 200, 211, 213, 223, 238, 239, 264, 295, 296, 321, 342, 344, 345
  error 177

T

Table 13, 17, 18, 19, 20, 21, 23, 26, 27, 87, 89, 90, 126, 128, 129, 208, 209, 211, 212, 213, 215, 216, 217, 218, 219, 220, 221, 222, 243, 244, 245, 247
Temperature 72, 316, 333, 335
Temperature profiles 313
Term NIM solution 255, 264
Terms 80, 84, 86, 87, 92, 98, 159, 167, 168, 200, 203, 205, 211, 233, 234, 237, 270, 273, 285, 286, 288, 289, 290, 292, 294, 298, 300, 304, 305, 307, 308, 310, 312, 313, 319, 320, 330, 331, 332, 333, 335, 343
  first 288, 292, 294, 299, 311, 314, 315
Theorem 6, 37, 38, 39, 41, 42, 43, 44, 45, 46, 47, 48, 49, 50, 51, 52, 53, 54, 55, 59, 60, 61, 63, 64, 66, 113, 115, 138, 139, 141, 142, 146, 147, 148, 149, 150, 151, 152, 153, 155, 166, 167, 237, 241
Theorem
  conditions of 45, 142, 147, 148, 155
  following 44, 45, 49, 50, 53, 55, 59, 115, 237

Theory 33, 34, 35, 43, 44, 63, 129, 133, 135, 322, 342, 345
Three-term NIM solution 247
Time 39, 72, 73, 86, 89, 92, 98, 99, 114, 115, 116, 160, 186, 200, 209, 211, 221, 261, 284, 295, 296, 298, 300, 303, 316, 326, 329, 331, 333, 335, 344, 345
  first-order 34, 36
  fixed 91, 99, 100, 101
  large 270, 298, 327
  short 298, 300, 323, 325, 331
  small 327
Time evolution 98, 99
Time fractional 1, 204
Time-fractional 33, 72, 80, 280, 289, 301, 302, 306
  following 248, 249, 250
Time-fractional Caputo-Fabrizio 284, 297, 300
Time-fractional derivatives 72, 102, 284
Time-fractional diffusion equations 33, 34, 35, 37, 38, 39, 43, 44, 56, 66, 67
Time-Fractional Diffusion Equations Current Developments 35, 37, 39, 41, 43, 45, 47, 49, 51, 53, 55, 57, 59, 61, 63, 65, 67
Time-Fractional Diffusion Equations Current Developments in Mathematical Sciences 35, 37, 39, 41, 43, 45, 47, 49, 51, 53, 55, 57, 59, 61, 63, 65, 67
Time-fractional Riemann-Liouville 70, 81, 317
Time vector 222
  size of 221, 222

U

Uniqueness 33, 34, 35, 38, 39, 55, 59, 61, 132, 133, 263
Uniqueness of solutions 33, 34, 35, 38, 106
Uniqueness result 46, 47, 50, 146, 148

V

Value problem, initial-boundary 40, 41, 55
Values 10, 12, 80, 83, 84, 85, 92, 93, 94, 98,

99, 100, 102, 125, 126, 129, 160, 175,
176, 186, 188, 189, 191, 194, 196, 197,
202, 213, 221, 222, 296, 300, 302, 309,
312, 313, 315, 324, 325, 330, 331
following 126, 128
parameter 161, 168, 169
threshold 94, 96, 160, 345
Vapors 330, 331
Variations 92, 93, 94, 97, 99

Vectorization 209, 222

W

Wavelet basis functions 7, 15
Wavelet coefficients 3, 7, 15, 16
Wavelet preconditioning 1, 2, 3, 15
Wavelets 1, 2, 3, 7, 15, 16, 17, 25

www.ingramcontent.com/pod-product-compliance
Lightning Source LLC
Chambersburg PA
CBHW050803220326
41598CB00006B/102